T0245248

CAMBRIDGE LIBRARY COLLECTION

Books of enduring scholarly value

Earth Sciences

In the nineteenth century, geology emerged as a distinct academic discipline. It pointed the way towards the theory of evolution, as scientists including Gideon Mantell, Adam Sedgwick, Charles Lyell and Roderick Murchison began to use the evidence of minerals, rock formations and fossils to demonstrate that the earth was older by millions of years than the conventional, Bible-based wisdom had supposed. They argued convincingly that the climate, flora and fauna of the distant past could be deduced from geological evidence. Volcanic activity, the formation of mountains, and the action of glaciers and rivers, tides and ocean currents also became better understood. This series includes landmark publications by pioneers of the modern earth sciences, who advanced the scientific understanding of our planet and the processes by which it is constantly re-shaped.

Life, Letters and Journals of Sir Charles Lyell, Bart

Sir Charles Lyell (1797–1875) was one of the most renowned geologists of the nineteenth century. He was awarded the Copley Medal by the Royal Society in 1858 and the Wollaston Medal by the Geological Society of London in 1866 for his contributions to geology. Lyell's most important contribution to modern geology was his refining and popularising of the geological concept of uniformitarianism, the idea that the earth has been formed through slow-acting geological forces. This biography, first published in 1881 and edited by his sister-in-law K.M. Lyell, provides an intimate view of Lyell's personal and professional life through the inclusion of his correspondence with family, friends and academic peers. His changing ideas concerning the validity of the theory of natural selection and other geological ideas are also examined through the inclusion of extracts from his private journal. Volume 2 contains Lyell's later career from 1837 to 1875.

Cambridge University Press has long been a pioneer in the reissuing of out-of-print titles from its own backlist, producing digital reprints of books that are still sought after by scholars and students but could not be reprinted economically using traditional technology. The Cambridge Library Collection extends this activity to a wider range of books which are still of importance to researchers and professionals, either for the source material they contain, or as landmarks in the history of their academic discipline.

Drawing from the world-renowned collections in the Cambridge University Library, and guided by the advice of experts in each subject area, Cambridge University Press is using state-of-the-art scanning machines in its own Printing House to capture the content of each book selected for inclusion. The files are processed to give a consistently clear, crisp image, and the books finished to the high quality standard for which the Press is recognised around the world. The latest print-on-demand technology ensures that the books will remain available indefinitely, and that orders for single or multiple copies can quickly be supplied.

The Cambridge Library Collection will bring back to life books of enduring scholarly value (including out-of-copyright works originally issued by other publishers) across a wide range of disciplines in the humanities and social sciences and in science and technology.

Life, Letters
and Journals of
Sir Charles Lyell, Bart

VOLUME 2

EDITED BY K.M. LYELL

CAMBRIDGE
UNIVERSITY PRESS

CAMBRIDGE UNIVERSITY PRESS

Cambridge, New York, Melbourne, Madrid, Cape Town, Singapore,
São Paolo, Delhi, Dubai, Tokyo, Mexico City

Published in the United States of America by Cambridge University Press, New York

www.cambridge.org
Information on this title: www.cambridge.org/9781108017862

© in this compilation Cambridge University Press 2010

This edition first published 1881
This digitally printed version 2010

ISBN 978-1-108-01786-2 Paperback

SIR CHARLES LYELL, BART.

VOL. II.

Cha.s Lyell

Engraved by G. J. Stodart from a Photograph

London Published by John Murray 50 Albemarle Street 1881

LIFE

LETTERS AND JOURNALS

OF

SIR CHARLES LYELL, BART

AUTHOR OF 'PRINCIPLES OF GEOLOGY' &C.

EDITED BY HIS SISTER-IN-LAW, MRS. LYELL

IN TWO VOLUMES—VOL. II.

With Portraits

LONDON

JOHN MURRAY, ALBEMARLE STREET

1881

CONTENTS

THE SECOND VOLUME.

—————

CHAPTER XXI.

MARCH—NOVEMBER 1837.

octrine of Uniformity—Dinner at Miss Rogers'—Babbage's Ninth Bridgwater Treatise—Letter to Herschel on Coral Islands—Copenhagen—Invention of Electric Telegraph—Visit to Sorgen-fri—Norway—Geology—Paris—Von Buch—Mademoiselle Mars—New railroad—Barrière de l'Etoile—Return to London—Parties   .   .   .   .

CHAPTER XXII.

JANUARY 1838—OCTOBER 1841.

' Elements of Geology'—Sedgwick's Lecture—Whewell's ' Inductive Sciences '—Anniversary at the Geological Society—Darwin on Volcanic Phenomena—Visit to Norwich and Yarmouth—Newcastle Meeting—Darwin's Journal—Coral Reefs—Athenæum Club—Dr. Fitton's Review—Hutton's claims—British Association at Birmingham—Visit to Sir Robert Peel—Travelling in the United States—Rapid growth of the country—Niagara geology   35

CHAPTER XXIII.

JUNE 1842—DECEMBER 1843.

Niagara—Nova Scotia—Subterranean Forest—Unfair statements on the United States—Return to England—Contemplates writing a Journal of his impressions of America—Much geological work in prospect—Giant's Causeway—Lord Rosse's Telescope—Prescott's Mexico   60

LIST OF PLATES.

LETTERS AND JOURNALS

OF

SIR CHARLES LYELL, BART.

———◆◆———

CHAPTER XXI.

MARCH–NOVEMBER 1837.

DOCTRINE OF UNIFORMITY—DINNER AT MISS ROGERS'—BABBAGE'S NINTH
BRIDGEWATER TREATISE—LETTERS TO HERSCHEL ON CORAL ISLANDS
—COPENHAGEN—INVENTION OF ELECTRIC TELEGRAPH—VISIT TO SOR-
GENFRI—NORWAY—GEOLOGY—PARIS—VON BUCH—MDLLE. MARS—NEW
RAILROAD—BARRIÈRE DE L'ETOILE—RETURN TO LONDON—PARTIES.

[Mr. and Mrs. Lyell went to Denmark and Norway in 1837, returning home by Brussels and Paris. The 'Elements of Geology'
was published in the following year 1838, and he attended the British
Association at Newcastle, presiding over the Geological Section. In
1839 he visited Norfolk and Suffolk, and went to Birmingham to the
British Association.

In 1840 he made a tour in Normandy and Touraine, and was at
the British Association at Glasgow, presiding over the Geological
Section. In the following year 1841 he was invited to give a course
of twelve lectures at the Lowell Institution, in Boston, Massachusetts,
and he travelled through much of the United States and Canada,
collecting materials for memoirs which were afterwards published in
the Geological Society's Journal.

In August 1843 he went to Bristol to examine the coal mines in
that district, and he and his wife proceeded to Cork, where the British
Association was held.]

CORRESPONDENCE.

To the Rev. W. Whewell.

16 Hart Street, Bloomsbury Square : March 7, 1837.

My dear Whewell,—As we had some conversation the other day touching the extent to which I carried my doctrine of ' Uniformity,' in the ' Principles of Geology,' I wish to refer you to the first edition of that work (with which you, as the historian of the science, are of course principally concerned), in order to show you that certain passages were somewhat unfairly seized upon by the critics, and not duly considered with and interpreted by others, and by the context generally of the first volume.

You know much more of geology now than when you first read my book, or Sedgwick's comments upon it ; judge me, therefore, by your present knowledge, rather than by the first impression.

Any reader of Sedgwick's Anniversary Address to the Geological Society of 1831, would suppose that I had contended for ' an indefinite succession of similar phenomena,' and coupling what is said about my hypothesis of a ' uniform order of physical events ' with what Sedgwick afterwards says of the recent appearance of man, it might naturally be imagined that I had not made due allowance for this ' deviation,' as I myself styled the creation of man. I brought forward this ' innovation ' prominently as a new cause, ' differing in kind and energy from any before in operation,' and mentioned it as an unanswerable objection against any-one who was contending for *absolute uniformity.* I stated the sense in which I believed the system of terrestrial change to be uniform, and that the modifications produced by man had been exaggerated, but in the following page I admitted that they are ' a real departure from the antecedent course of physical events.'

It was impossible, I think, for anyone to read my work, and not to perceive that my notion of uniformity in the existing causes of change always implied that they must for ever produce an endless variety of effects, both in the animate and inanimate world. I expressly contrasted my system with that of 'recurring cycles of similar events.' If certain passages could easily be selected in which the theory of uniformity was, perhaps, stated too broadly as an abstract proposition, and the most striking of these Sedgwick has cited in his address, still the great point of difference between me and most of my predecessors, and between me and Sedgwick, then and now, is not one which implied on my part any unphilosophical solecism. I never drew a parallel between a geological and an astronomical series or cycle of occurrences. I did not lay it down as an axiom that there cannot have been a succession of paroxysms and crises, on which ' *à priori* reasoning ' I was accused of proceeding, but I argued that other geologists have usually proceeded on an arbitrary hypothesis of paroxysms and the intensity of geological forces, without feeling that by this assumption they pledged themselves to the opinion that *ordinary forces and time* could never explain geological phenomena. The reiteration of minor convulsions and changes, is, I contend, a *vera causa*, a force and mode of operation which we know to be true. The former intensity of the same or other terrestrial forces may be true; I never denied its possibility; but it is conjectural. I complained that in attempting to explain geological phenomena, the bias has always been on the wrong side; there has always been a disposition to reason *à priori* on the extraordinary violence and suddenness of changes, both in the inorganic crust of the earth, and in organic types, instead of attempting strenuously to frame theories in accordance with the ordinary operations of nature.

Read also, if you desire to be a fair arbiter between the philosophical spirit and practical results of the two different methods of theorising in geology, what Sedgwick says in his Address for 1831 of De Beaumont's system of parallel elevations, and my chapter on the same subject, or on the relative antiquity of mountain chains. De Beaumont's system was properly selected by him as directly opposed to my fundamental principles. It was well selected, because it not only assumed returning periods of intense activity, or, as Sedgwick termed them, 'feverish spasmodic energy,' which tore asunder the framework of the globe, but also violent and concomitant transitions from one set of species to another. It also assumed the parallel direction of mountain chains contemporaneously elevated, whereas we know that the modern lines of active volcanos and earthquakes are not parallel in our own times. It was a theory invented not only without any respect to the reconciling geological events with the ordinary course of changes now in progress, but it evinced at every step that partial leaning to a belief in the difference of the ancient causes and operations which characterise the system of my opponents. The astonishing eagerness with which Sedgwick caught up and embraced the whole of what he termed De Beaumont's 'noble generalisation,' his declaration that De Beaumont's theory was 'little short of physical demonstration,' and that it had given him (Sedgwick) 'a new geological sense, a new faculty of induction;' his assertion that he was 'using no terms of exaggeration' in saying all this; was all prompted by the same theoretical bias which assumes the discordance between the former and existing course of terrestrial change. Before Sedgwick published his Address, I had fully explained to Lonsdale my objections to De Beaumont's theory, the same which I have since published. I know not how much of De Beaumont's theory Sedgwick now believes, probably but a small part of it. I can only say that I could not find in Switzerland any one of those who had studied its application in reference to the Alps and Jura, as for example, Studer, Necker, and Thurmann, who believed in it. Conybeare could not reconcile it to England. Sedgwick considered that my mode of explaining geological phenomena, or my bias towards a leading doctrine of the Huttonian hypothesis, had served like a false horizon in astronomy to vitiate the results of my observations. But has he not himself been unconsciously warped by his own method of philosophising, which he has truly stated to be directly at variance with mine? I am willing to test the relative value of our modes of dealing with these subjects by all the examples which he has himself chosen in his review of my book.

In the review in the 'British Critic,' in which more justice was done to the real points of originality to which my work could lay claim, than in any subsequent article upon it (although most of them are much more eulogistic), you stated three formidable theses which I had undertaken to defend, in order to bear out my theoretical views; how much less formidable must they now appear to yourself when you and the science have made seven years' progress, than they did in 1830! I am sure that none of the propositions can now seem to you extravagant or visionary, although you may not wholly agree with all my views. I allude to, first, the adequacy of known causes as parts of one continuous progression to produce mechanical effects resembling in kind and magnitude those which we have to account for; secondly, the changes of climate; thirdly, the changes from one set of animal and vegetable species to another.

In regard to this last subject, as well as to the change of climate, you remember what Herschel said in his letter to me. If I had stated as plainly as he has done the possibility of the introduction or origination of fresh species being a natural, in contradistinction to a miraculous process, I should have raised a host of prejudices against me, which are unfortunately opposed at every step to any philosopher who attempts to address the public on these mysterious subjects. Perhaps you would understand the difference which I conceive to exist between me, as a Uniformitarian, and my opponent the Catastrophist, if I put an imaginary case in your own department of the tides. Suppose it was recorded, on authority which you did not question, that at a certain period of the past, say 10,000 years ago, a remarkable inequality, or an unusual rise or fall of the tides took place, or a long suspension of tidal oscillations. You are asked to explain it. After much thought, you declare that every hypothesis has failed, and that you give it up as inexplicable. Some one proposes a theory founded on a supposed periodical increase or diminution in the quantity of matter contained in the sun or moon, or in both of those heavenly bodies. You object that it is unphilosophical to resort to any such guesses, especially until we have laboured for a much longer time in making and recording observations on the tides, and in speculating on all the possible combinations of astronomical and terrestrial circumstances, which in the course of thousands of ages may affect the tides.

Your antagonist replies: 'Why is it unphilosophical? What right have you to assume that the power of the sun and moon to attract the earth is uniform, or why may there not have been two satellites formerly to the earth instead of one? The same power which created man, and gave him the dominion over the earth, may at any time create and then annihilate a heavenly body, or change the volume or density of those which now exist.' You might reply: 'I admit the possibility of these changes in the system, but they are mere arbitrary conjectures. I feel much more confidence in the probable uniformity in the condition of the heavenly bodies for an indefinite period of time, whether past or future, than I feel in my own knowledge. If I was sure that I could see at a glance the effects of every possible combination of existing causes—such as variations in the eccentricity of the earth's orbit, fluctuations in the shape of the bed of the ocean, or in the position of continents, and many others—then perhaps, all others failing, I should venture upon some such theories as you have proposed. But I hesitate now, more especially when I recollect that several problems in the former state of the tides, once referred to violations in the ordinary conditions of our solar system, can now be explained without resorting to such expedients, and are, in fact, found to be the consequences of known and regular causes.'

I find that I cannot draw the above parallel as closely as I could wish, because I cannot fancy in reference to the tides any stumbling block to arise out of prejudices in regard to *time*. The difficulty which men have of conceiving the aggregate effects of causes which have operated throughout millions of years, far exceeds all other sources of prejudice in geology, and is yet the most unphilosophical of all. On this I have said much in my work, and people are now much better prepared to believe Darwin, when he advances proofs of the slow rise of the Andes, than they were in 1830, when I first startled them with that doctrine.

I will only add that Sedgwick was wrong if he imagined that I began with *à priori* reasoning on any assumed uniformity of physical events. I was taught by Buckland the catastrophical or paroxysmal theory, but before I wrote my first volume, I had come round, after considerable observation and reading, to the belief that a bias towards the opposite system was more philosophical.

Believe me, dear Whewell, ever faithfully yours,

CHARLES LYELL.

To HIS SISTER.

London : March 19, 1837.

My dear Sophy,—I shall trust to Mary giving you the regular journal of our proceedings, and shall indulge myself in recalling to my recollection what passed at our last dinner on Friday at Miss Rogers'. Much was said which I should be sorry to forget, and the guests were some of them remarkable and new to me. The party were twelve in number : Miss Rogers and her brother, Lord and Lady Holland, Mr. and Mrs. Milman, Mr. Allen, Mr. Luttrell, Mr. Empson, Sir David Wilkie, and ourselves. The table, so arranged as that everyone could hear and be heard by others, two sitting at the top, Lord Holland and Miss Rogers, and two at the bottom, Mr. Rogers and Mr. Milman. No epergne or other interference in the centre of the table ; lamp from the ceiling ; table just wide enough, and no more. Mr. Allen has been for thirty years physician to the Holland family ; lives with them on a very independent footing ; wrote much in the early 'Edinburgh Review' on Spanish literature and other subjects; an agreeable man of Lord Holland's age. All we know of Luttrell is that he is a diner-out, and much in a literary set. Empson you have heard us speak of before, as a writer and professor of law at Haileybury. Mary was well placed between him and Lord Holland ; the latter has a cheerful, good-humoured expression, talking in a lively way, but never too much, of literary rather than political subjects, and of anecdotes of political men rather than politics. Mr. Allen was saying how strange a contrast Erskine [1] used to be in and out of his lawyer's wig and gown. Out of it he talked in a most *gauche* and foolish way, in it so that you would trust your life and fortune in his hands. Lord Holland, among other

---

[1] Younger son of the tenth Earl of Buchan, an eminent lawyer, and became Lord Chancellor.

stories to confirm this, said that one day when he and Lord
Erskine were in council in the Cabinet, and Lord Erskine's
opinion on a measure was asked, he said in a hasty manner,
'Oh, yes, depend upon it, it must be, for I remember it was
in an old Presbyterian book of prophecies which my mother
had.' When Erskine first came to the bar, he spoke very
broad Scotch; he had never read more than the Bible,
Shakespeare, Milton; and in three years he spoke eloquent
English, and was quite a gentleman in manners.

A discussion arose as to whether the Iliad and Odyssey
were both written by the same poet. Lord Holland said he
felt satisfied they were not. Milman agreed, but contended
for the unity of the Iliad, and said that the arguments
against it founded on the difficulty of transmitting so long a
poem without printing, were answered by a greater wonder
of the same kind being achieved, in the handing down of
much longer Sanskrit poems. He said that Asia Minor was
never once mentioned in the Odyssey, which was probably
the work of an inhabitant of the Peloponnese, whereas the
allusions in the Iliad are so much confined to Asia Minor.
On the Etruscan tombs being talked of, towards the exhibition of which Mr. Rogers had liberally contributed, we
got upon the old Etruscans, and Niebuhr and his Roman
History. Lord Holland said that he never would give up
the real existence of such men as Romulus and Numa, however much fable might be mixed up with them. Milman
thought Niebuhr more successful in pulling down than in
building up. After the ladies were gone, Lord Holland
asked me about Buckland's book, and whether he knew
much of geology. He seemed not to have formed a high
estimate of the said Bridgewater, so I spoke up in favour of
the body of the work, on fossils. This led to a talk on new
species, and that mystery of mysteries, the creation of man.
Lord Holland said that we were no further on that point than
Lucretius, out of whom he could take mottoes which would
have done for each of my volumes. And now, Sophy, if you
think that I am voting you too blue by administering all
this Greek, Sanskrit, Etruscan, and Latin to you, hand
over the dose to Papa, and he may retaliate whatever he
pleases with some lucubrations equally learned from his den.

But I have said nothing of Lady Holland, who took her share in the talk. She asked me about the Danes and Swedes, knew the names, at least, of many of them distinguished in science, said how much energy and love of truth there was in the Northern men of letters, as compared to her favourites the French and Italians, yet the French could be deep and persevering. I spoke of La Place and Cuvier. She said that the latter once wished her to compliment him on his promotion to a higher political place, but she gave him fairly to understand how much she lamented his having abandoned the line in which he was so great, to meddle with politics (in which he played so inferior and, in her opinion, unworthy a part). It is impossible to say in a letter anything which will give an idea of the singularity of Lady Holland's way of questioning people, like a royal personage. It is impossible not to be sometimes amused, and sometimes a little indignant, with her. I cannot say that I formed so high an estimate of her talent and power as to explain to me how she has righted herself to such an extent, and got on in society after all that happened more than thirty years ago. No doubt she has been in the interval prudent, and more strict in the choice of her society than others who had infinitely more right to be so. She had wealth and beauty, of which last there are still some remains yet, with an expression of temper. But then she had a husband who had not only talent, rank, and political station, but an infinite fund both of wit and good humour.

With love to all at Kinnordy, believe me, my dear Sophy,
Your affectionate brother,
CHARLES LYELL.

To CHARLES BABBAGE, ESQ.

May, 1837.

My dear Babbage,—I have read the 'Fragment'[2] with great interest, and think very favourably of it and its originality. As the machine must come in so much unavoidably, I think one apology at the beginning would do, or rather an explanation that the work is founded on it. The argument

---

[2] The *Ninth Bridgewater Treatise*, a Fragment, by Charles Babbage.

of changes of laws comes home to some of my geological
speculations. No doubt some people would not like any
reasoning which made miracles more reconcilable with possibilities in the ordinary course of the universe and its laws ;
but you do not write to please them.

They are shocked at the idea of an eruption of a volcano
being foreknown, which was to destroy Sodom and Gomorrah,
but your break in the sequence is more of a miracle than a
volcanic eruption. I think your estimate of the Creator's
attributes much higher than theirs, and that everyone may
follow so far.

Ever faithfully yours,

CHARLES LYELL.

To HIS SISTER.

London : May 3, 1837.

My dear Caroline,—Many thanks for your letter, which
Mary tells me she has answered. I rather envy her the
dinner to-day, though if Mr. Whitmore had not known that
I was engaged at the Geological Society Club, neither of us
would have been asked, as he wanted one place, and filled by
a lady. His parties are excellently arranged. Jacquemont's
letters have made us both curious to see Lord William Ben-
tinck, and indeed his lady also. The Bishop of London
talked of Sedgwick having narrowly escaped, according to
report, being made Bishop of Norwich. I told him how
popular Whewell had told me that Sedgwick had become at
Norwich, which interested Dr. Blomfield, and he ended by
saying, 'Though he is so popular, with the ladies in particular, I hardly think he will ever marry now.'

I have at last struck out a plan for the future splitting of
the 'Principles' into a 'Principles' and 'Elements,' as two
separate works, which pleases me very much, so now I shall
get on rapidly. The latest news is, that two fossil monkeys
have at last been found, one in India contemporary with extinct quadrupeds, but not very ancient—Pliocene perhaps—
another in the south of France, Miocene and contemporary
with Paleotherium. So that, according to Lamarck's view,
there may have been a great many thousand centuries for

their tails to wear off, and the transformation to men to take place.

Your affectionate brother,

CHARLES LYELL.

To SIR JOHN F. W. HERSCHEL.

16 Hart Street, Bloomsbury, London : May 24, 1837.

My dear Sir John Herschel,—I shall shortly set out with my wife on a tour to Copenhagen, Christiania, and Berlin, and will not quit London without writing to explain to you some steps which I have taken in regard to the publication of extracts from your letter to me of February 1836. You had given Murchison permission to read, when occasion offered, a note of yours on the subject of the Origin of Volcanos, and the outward movement of the internal isothermal lines on the superposition of new deposits, or 'additional clothing.' You seemed to think that I had confounded this notion with Mitscherlich's expansion of stone by heat, attributed by me to Babbage; but in fact Babbage's theory, which I had alluded to, was different, and in substance the same as yours, and I agreed with Murchison, that as Babbage was going to republish it in his 'Ninth Bridgewater,' it would be well to allow him to print, as he desired, both the extract from your first letter to me on the point, and your note to Murchison especially, as in both you had stated that you had not had time to reason out or state with minute accuracy the whole question. Whewell had both letters read at the Geological Society, which produced a most animated discussion in which Whewell took part. But now for another point. Babbage was very desirous that I would allow him to print another short extract from your letter to me. I objected at first, but he showed me that I need not be alarmed, because he introduced it as a counterpart of a passage from Bishop Butler, and that in such company no one could be otherwise than correct and orthodox.

I hope my willingness to be persuaded to have a passage printed, in which incidentally you had paid me a compliment (one which I certainly prized highly), has not led me to

what you would in any way think an indiscretion. Whewell,
in his excellent treatise on the Inductive Sciences, appears
to me to go nearly as far as to contemplate the possibility
at least of the introduction of fresh species being governed
by general laws.

Your volcanic speculations have at least set our wits to
work, and will produce something.

I am very full of Darwin's new theory of Coral Islands,
and have urged Whewell to make him read it at our next
meeting. I must give up my volcanic crater theory for ever,
though it costs me a pang at first, for it accounted for so
much—the annular form, the central lagoon, the sudden
rising of an isolated mountain in a deep sea, all went so well
with the notion of submerged, crateriform, and conical volcanos, of the shape of South Shetland, and with an opening
into which a ship could sail; and then we had volcanos
inside some circular reefs, as in Dampier's island, and then
we knew that it was not the corals which had any inclination of their own to build in a ring, like mushrooms and
funguses in fairy circles on the green, for the very same
species of corals will form a long barrier reef, or grow in any
shape the ground permits: and then the fact that in the
Pacific we had scarcely any rocks in the regions of coral
islands, save two kinds, coral limestone and volcanic! Yet
spite of all this, the whole theory is knocked on the head,
and the annular shape and central lagoon have nothing to
do with volcanos, nor even with a crateriform bottom. Perhaps Darwin told you when at the Cape what he considers
the true cause? Let any mountain be submerged gradually,
and coral grow in the sea in which it is sinking, and there
will be a ring of coral, and finally only a lagoon in the centre.
Why? For the same reason that a barrier reef of coral
grows along certain coasts, Australia, &c. Coral islands are
the last efforts of drowning continents to lift their heads
above water. Regions of elevation and subsidence in the
ocean may be traced by the state of the coral reefs. I hope
a good abstract of this theory will soon be published. In
the meantime, tell all sea-captains and other navigators to
look to the facts which may test this new doctrine. I
suppose that according to the above theory, if a volcanic

crater should sink, there would be two rings of coral, one on the rim of the crater, and one without.

May 26.—Since writing the above, Babbage has given me a copy of his 'Fragment' for you, so you will see how far we have published your letter.

At Babbage's yesterday, there was much talk of Wheatstone's new plan for telegraphing information by five wires conducting electricity, and carried through an india-rubber tube, or rather rope, each wire representing a letter of the alphabet, or several letters, I believe by different intensities of the charge. The wires isolated by the india-rubber. The experiment Mr. J. Taylor said had, according to Wheatstone, conveyed information through a rope five miles long, and of course instantaneously. These ropes it is said can be laid underground at small expense, under a railroad for instance, and news communicated with the speed of light.

Brewster is full of some new proof derived from the optical properties of the diamond in favour of its vegetable origin; some argument from anomaly or disagreement with any arrangement known to arise from mere mineral and inorganic structure.

Hoping that you and your family are quite well, believe me, dear Sir J. Herschel, yours ever faithfully,

CHARLES LYELL.

To HIS SISTER.

Copenhagen: June 13, 1837.

My dear Carry,—It is only a week to-day since we left London, and having already had two whole days with Dr. Beck, I feel as if I had been here an age. We went through Holstein from Hamburgh and its Danish suburb Altona, to Kiel, on a beautiful bay of the Baltic. In the morning we saw the sun rise in the principal Place, the Jungfernstieg, in Hamburgh, where the river Alster is dammed up so as to form a beautiful lake in the middle; in the evening we saw it set in the Baltic when Kiel was long out of sight.

We came to this capital on Sunday before sunset, and saw the beautiful harbour well. I was rather glad to find Prince Christian absent, in Funen, of which he is Governor, and Dr. Beck at leisure all day. We set to work at once at con-

·chology and the Prince's Museum, and the critical examination of those points of recent and fossil conchology in which
Dr. Beck and Deshayes are at variance, and on which some
of my geological conclusions have been based.

June 21.—We were very glad to find on our table the
day before yesterday your excellent journal. I was able to
comply with a request of Mr. Loch's to introduce him to
Prof. Wheatstone, and to get the latter to promise him a
sight of the grand experiment about to be made of the
electric telegraph. By-the-bye, have you heard of this
wonderful invention? It has made a large figure in my
waking dreams ever since I first heard of it at Babbage's,
and I shall be disappointed if it fail to work a mighty
change in the ' march of intellect,' at least of civilisation. I
must tell you what I have seen. Wheatstone has had for a
year at King's College, in the crypts, a copper wire with silk
turned round it, four miles in length, carried backwards and
forwards, and finds that he can transmit instantaneously
through this distance a feeble current of electricity such as
two plates, four inches diameter, of copper and zinc, can produce. Now you probably know that some fifteen years ago
the famous Oersted, of this Copenhagen University, discovered that a current of common voltaic electricity caused
the magnetic needle to move and place itself always at right
angles to the current, so that you may move a needle at the
end of the four miles of wire instantaneously, i.e. the
current goes with the velocity of light, 200,000 miles in a
second. Now you will ask whether it does not grow very
feeble in passing four miles, and it does so, as it is always
escaping ; but then it only takes twice the power to send it
four miles which it required for two, and it has long been
known that if you coil the wire which has the electricity
passing through it, say fifty times, this multiplies the effect
on the needle by fifty. Now with four wires and four
needles you can make sixty distinct signs, and so communicate any news by all the letters of the alphabet, instantaneously, any distance ; and at all the intermediate places,
coiled wires, called electrometers, being fixed to the main
one which takes off but little of the force. It has been
found that by employing small ropes steeped in india-rubber

gum, you may isolate the wires much cheaper than by silk, and thus each wire shall only cost between 3*l*. and 5*l*. per mile, or for the four wires under 20*l*. Now they had been obliged on the Birmingham and Liverpool railroad to use a very much more expensive telegraph, to give notice of trains coming to tunnels or places where they cross, by means of long iron tubes through which a blast of air is sent which blows a whistle at the end. So when this new rope with the four copper wires is substituted, we shall have not only railway news but all others sent out with the speed of lightning, ciphers being used for private confab. A few days after I left, a tarred rope was to be continued from the end of the four miles of wire, and thrown into the Thames, and then carried to the shot tower on the other side, and the rapidity of conveying intelligence through about five miles was to be tried. So perhaps a rope in the sea may carry news from Dublin to Holyhead, but at least under every railroad we can have it. After seeing the experiment for four miles of wire, I shall be surprised if it fails, and then it will be singular that as one revolution in human affairs was produced by observing how the earth's magnetism affected the needle or compass, so a second will follow the detection of the law by which common or voltaic electricity affects this same needle. But I did not mean to write so much London news from Copenhagen.

Your affectionate brother,

CHARLES LYELL.

To HIS SISTER.

Copenhagen: July 2, 1837.

My dear Eleanor,—I shall begin with telling you some of my Danish news which I think Mary is least likely to give you. Perhaps when you hear that I should like to stay here longer than I can afford to do, after our return from Christiania, you will wish to know what geology the streets here and this flat country of gravel and boulders afford me. This I can easily explain, as you have read my book, and know how much I have alluded in it to the comparison of fossil and living species of shells. In the time of Linnæus this city contained finer collections of

shells, and finer works were published here by Chemnitz
(12 vols.), and others, than in any country of Europe. It is
not wonderful, therefore, that even now some of the Danes
should be far ahead, and that as Prince Christian had a
taste for natural history, he should have a splendid private
collection, and that the curator of his museum (containing
above 8,000 species of living shells), Dr. Beck, should be
one of the two or three best conchologists in the world.
But besides this, Copenhagen possesses in its different
museums most of the identical shells which Linnæus described in the editions of his ' Systema Naturæ,' published
during his life; and here therefore alone, we can verify in-
contestably the species which he really described and named.
As Lamarck and others have in very many cases mistaken the shells which Linnæus meant, great confusion
has arisen, and it is here alone that this confusion can most
readily be cleared up.

I am going over with Beck an examination of all the
fossil species identified by Deshayes with living shells, and
it will probably lead to many modifications of my views, at
least of many of my details, also to many new views, which I
shall perhaps have to test at Paris before I return. But my
plans must be governed by circumstances not wholly in my
power.

On Tuesday I went early to Sorgenfri with Beck, a fine
morning, not too hot, and we discussed all the way points
which had occupied us in the museum. He is an excellent
botanist and entomologist. Soon after 8 o'clock we reached
Sorgenfri, and after a walk in the flower garden, the Prince
joined us. A landau and four nice horses was in waiting, and
away we drove over a very fair macadamised road, through
woods as wild and natural as any part of the New Forest,
having at our left a remarkable chain of small lakes far below
us; that is to say, we looked down a precipice covered with
wood upon the lakes. We discussed the probable origin of
the lakes, the Prince not forgetting here and there to stop the
carriage and take me to points slightly elevated, where beautiful views are seen. In one of them we saw the towers of
Copenhagen, about eight miles off to the south, and from another point the ancient Abbey Church of Roskilde. We then

passed between Lake Füre and a smaller lake to the village of
Farum, and some miles beyond this, came to where the soil
is composed to a great depth of innumerable rolled blocks of
chalk with a few of granite intermixed. Fossils were very
numerous in the chalk, and I hammered out many, and Beck
found a finer fossil fish than had ever been found before in
Denmark. Prince Christian set four men to work, while
the horses were baiting, to clear away the talus and make a
section for me, by which I saw that the boulders of chalk
were in fact in beds, with occasional layers of sand between.
A gamekeeper came to show an eagle which he had shot, a
large osprey, I believe. I took two of his large quills, and
mean some day to write a letter with it; of course it must
be, as Count Blücher said, in a very lofty style. On our
return to the opposite side of Lake Füre, which is twenty
miles round, we drove down to the shore, and found a large
sailing boat of the Prince's, and some men waiting for us.
Before embarking, Beck and I collected a bottle full of shells
inhabiting the borders of the lake, and I got some information
which throws much light on the origin of the indusial limestone of Auvergne.

There was much wind, and we scudded about at the
rate of ten miles an hour from one fishing ground to another,
where everybody but I caught carp and pike. At least I
learnt some ichthyology from Beck. We then found, at the
further end of the lake, the carriage in attendance, with a
second one to take such things as might have crowded us, and
so we drove home. The Prince was very agreeable, and I was
glad to find that his lady had been equally so to my princess,
who has given you already an account of an entertainment
on that and the following day, which was exceedingly amusing. I had no opportunity when here before of seeing more
than that the princess was ladylike, handsome, and gracious;
but from her conversation when alone with Mary, and during
the walk with me from the boat in the evening, I am sure
she is a very superior person, and one who has thought
much, and formed her own opinions for herself on things
and people, manners, religion, and politics. It was curious
to see how much the Danes, both men and women, looked
like English, and how much their accent resembles ours,

but the women, though almost fairer than the English, were
not so handsome as in a fashionable party in London.

With love to all, believe me your affectionate brother,

CHARLES LYELL.

To HIS SISTER.

Copenhagen : July 29, 1837.

My dear Eleanor,—Mary has given you a diary of our
proceedings up to our return to this place, and you will see
that we were often separated when at Christiania, making
independent expeditions.   In my tour with Professor Keilhau to Hakkedal, we went about twenty-five miles north of
Christiania, and I had a pretty good specimen of a Norwegian forest of fir-wood, and steep hills of granite, and
other such rocks, with here and there a deserted iron or copper
mine opened in the middle of the wilderness, and some
tractas by which the woodmen draw out the fir trees which
they fell, leaving enough to supply a natural growth in
succession.   The ground under the trees was often thickly
covered with the *Linnœa borealis.*   I was not aware that it
smelt so sweet.   On the marshy ground there was great
abundance of the *Rubus chamœmorus,* which the natives call
moltk-bœr, which was red when I picked it, but which is
yellow when ripe.   We afterwards had some for dessert at
Sörby. The Norwegians seem to prize it more than any of their
wild fruits.   They told me it had been much destroyed this
year by the hard frosts, which continued during the nights
even of the beginning of July this year, for, as in England,
they had an unusually long winter.   Even when we were there,
though the weather was very hot, they said it was not equal
to their usual summers.   I heard the cuckoo in these woods.
There were a great many farm houses scattered through
those valleys where the granite was covered with clay,
which, as a geologist, I must tell you is a marine deposit
containing recent species of shells such as now inhabit the
fiords of Norway, and as this clay rises sometimes more
than 600 feet high, it fertilises a large district.   There has
been a good, brisk trade in wood of late years, which had
thinned some of the forests we went through, but though

they do not sell as much as formerly, since we have favoured
Canada by our duties, obliging ourselves to get dearer wood
from thence, still a new market has opened in the north of
France, where the French peasants are flooring their houses
with wood, a new luxury to them. Upon the whole, all those
we spoke with agree that Norway has increased wonderfully
in prosperity since she was severed from Denmark, although
the same natives are unanimous in extolling the Danes as a
people above the Swedes. The fact is that Norway is extremely independent of Sweden, and pays nothing to it, and
being no longer drained for the sake of Denmark, the Storthing has been able to diminish, and last year even to take
off all the direct taxes, and they have put their money
system on so good a footing, that their paper money is
worth as much as the metallic, instead of being at a great
discount, as in Sweden and Denmark. We were sorry,
however, to hear some of the natural effects of a democratical
system pushed so far as it is at present, so that the representation is virtually in the hands of the peasants or small
yeomen, among whom almost all the land is divided in small
lairdships. For example, the cathedral at Drontheim, said
to be 800 years old, is falling into ruins, unless repaired,
one of the finest monuments of the early days of Christianity
in the North, but the Storthing will not vote a small sum for
its repair. Several persons assured me that Mr. Laing, in
his book on Norway, was unpardonably erroneous in what
he says about the succession of land, for instead of a law
like that of France, all the land goes now, and always did,
to the eldest, and though he has to pay nominally a share
in money to each of his brothers equal to his own, and half
a brother's share to each sister, yet the customary mode
of valuation has always been such as to give in fact a considerable *majorat,* and to keep properties as long in the
same hands as in England. Norway was always too poor in
the feudal times to be worth invading, so it was never conquered, nor shared out by a feudal lord amongst his vassals,
as happened in other parts of Europe. On the other hand,
the Danes supplied ministers, judges, governors, &c., who if
they made fortunes, retired with them to Denmark, instead
of founding great Norwegian families. Many substantial

small proprietors emigrate annually to the United States, chiefly to Illinois, near the Canadian lakes, and they are excellently fitted to settle there, in a better climate, cut- ting down the wood, and building houses like those in the mother country, and, as there, each having great resources in himself when far from neighbours. Upon the whole we liked much what we saw of the Norwegians, but I have no doubt they would be better off, and especially be more likely to remain permanently independent of Swedish influence, intrigue, and domination, if they had more than one single nobleman of fortune and independence like Count Wedel Jarlsberg, who is much liked and a very intelligent man, but with whom dies out the last remnant of Norsk nobility, the Storthing having provided by law that no other noble should be created. It is curious how many English and especially *Scotch* faces and figures one sees in the tall and fair men and women of Christiania. The rate at which they bowled us down the hills, exceedingly steep, without drags to the wheels, was such as I never witnessed elsewhere.

With love to all, believe me your affectionate brother,

CHARLES LYELL.

To CHARLES DARWIN, ESQ.

Wesel-on-the-Rhine : August 29, 1837.

My dear Darwin,—I write this to you, at least I am beginning it, in a steamboat on the Rhine, so make allow- ance for the tremulous motion. We came in a steamer from Copenhagen to Lübeck, then in a hired carriage to Ham- burgh, across the sand and boulder formation of the Baltic, which for the most part we have been on ever since, although we have crossed the Weser, and Ems, and Lippe. The blocks of red syenitic granite, which I hammered away at in Norway, and which I saw there *in situ*, sending its veins into the trilobite and orthoceratite schist, have been carried with small gravel of the same, by ice of course, over the south of Norway, and thence down the south-west of Sweden, and all over Jutland and Holstein down to the Elbe, from whence they come to the Weser, and so to this or near this. But it is curious that about Münster and Osnabruck, the low secondary mountains have stopped them; hills of chalk, Muschelkalk, old coal &c., which rise a few hundred feet in general above the great plain of north and north-west Germany, effectually arrest their passage.

This then was already dry land when Holstein, and all from the Baltic as far as Osnabruck or the Teutobarger Wald hills, was submerged. At Bremen I saw Olbers, aged seventy-two, the astronomer who discovered Pallas and Vesta, and there and at Osnabruck and Münster I met a warm and German reception from men of whom I had never heard, but who had read my paper on Sweden or something else. I mean by German, that kind of frank expression of enthusiasm for science, or of any emotion, which a well-bred Englishman tries to suppress, at least all outward expression of it, from the dread of being thought ridiculous, or of affecting to feel more than he does, or from *mauvaise honte*. If you ever get sick of that fashionable nonchalance which would blush to admire anything, or at least to confess it, I advise you to plunge into Germany, and you will be soon refreshed, and brought back to a right tone again, whether it be literature, science, or any other pursuit you are following.

I hope to write to Horner a full account of my surprise to find that there is no truth in overlying granite in Norway—no exception to the rule, as I will tell you presently—that he may read it at the British Association. Be it known then to you and others who have read what Von Buch wrote on Norway in the days of his youth, and of his Wernerianism and Neptunianism, that his notion of the granite overlying the transition rock arose from this, that whenever he found schist dipping towards granite regularly up to the point of contact, he assumed that it went under. The granite may lean over a little here and there, but this is accidental, and in general it sends veins only into the transition beds, changing the limestone into marble near the junction, and the shale into micaceous schist, and other metamorphosis, the fossils often escaping in the white marble, and some traces in some crystalline schists.

Now what I have seen in several places, I take Keilhau's word to be the universal fact. Had Von Buch believed as he afterwards did in the igneous origin of granite, he would have found the veins, but without this, somehow or other, he came by false reasoning to the true conclusion, that the granite is newer than the Silurian beds. This is lucky for him. What struck me forcibly was this : after seeing proofs innumerable and beautifully clear of the order of age being first gneiss, then secondly, unconformable transition beds, then granite, and after seeing that the granite was the newest, I then found that the granite not only sent veins into both transition and gneiss, but actually sometimes passed by imperceptible gradation into gneiss. This gneiss, so ancient that it had been crystallised and then thrown into vertical and curved stratification even before the triloites flourished, this most ancient rock is so beautifully soldered on to the granite, so nicely threaded by veins large and small, or in other cases so shades into the granite, that had you not known the immense difference of age, you would be half staggered with the suspicion that all was made at one batch !

Paris, September 5.—Last year when Charlesworth spoke at Bristol about Crag, and the numerical percentage of recent tertiary shells, Sedgwick and Buckland gave some useful impromptu replies, stating that I was aware some modifications would be required, &c., but they would not affect this classification in the main. Now if he should again, as I expect, speak on this subject, and if he should again cite Beck, will you state that you happen to know from correspondence which you have had with me this summer, that I have been engaged with Dr. Beck in a careful exami- nation of the species of fossil shells of the Crag and other tertiary formations which have been identified with recent species; that you have learnt from my letters that Dr. Beck by no means denies the absolute identity of a certain number of Crag species, though he thinks a large proportion of those identified by Deshayes to be distinguishable. Also that I consider that Dr. Beck's views of the conchological fauna of the Crag, drawn from the consideration of 260 species, tend to confirm the classification which places the Crag as older Pliocene and on a parallel with the Sub-Apennine beds, and distinct from and more modern than the Touraine, Bordeaux, and other Miocene deposits. Also, that I am convinced that independently of the relative percentage of recent shells, about which naturalists may differ according to their notions of what constitutes a specific difference, there are other characters in the entire assemblage of forms of shells belonging to each great tertiary epoch, which will enable us to classify the deposits according to the approach which they make to the type of organisation now existing in the neighbouring seas; and that this approach will serve as a chronological test of the eras to which tertiary deposits may respectively belong. I mean the degree of approach to or departure from the assemblage of living shells in the neighbouring seas will be a test of the relative newness or antiquity of the several deposits; also that I consider the terms Eocene, Miocene, and Pliocene to be as convenient and (if I the inventor may say so) happy, in reference to this view, as when they were made to refer more exclusively to the proportion of fossil species identified with recent. For the Pliocene and Miocene will always express greater and less degrees of approximation to the existing fauna, and the Eocene that first dawn of resemblance which characterises the oldest tertiary shells when compared with those of the newest secondary formations. As I have not yet finished my examination of disputed points, and not even begun to hear the pleadings on one side, I will not risk any statement as to how far the identification of fossil with recent shells may have been pushed too far, for I certainly suspect that the error has been chiefly on this side; but I wish you to know that my conviction is stronger than ever, that rules may be given for measuring the approximation of different groups of tertiary shells, and that the degree of this approximation may be used as a test of age, and may lead to the same classification as that which I have adopted in the ' Principles.' I am fully prepared to defend all that is essential in my system of tertiary classification as founded on fossil shells.

Believe me ever most truly yours,

CHARLES LYELL.

To LEONARD HORNER, ESQ.

Paris : September 23, 1837.

My dear Horner,—I was glad to receive your report of the meeting at Liverpool, and to hear the Association was so well supported. I must try to be at the next. I saw a very blundering account of my letter in the ' Athenæum.' I am anxious that when printed by the Association,[3] it should be given as I sent it, and this principally because I have carefully avoided bringing prominently forward any collision with Von Buch, whom I like too well not to be sorry that he should fancy, as he has done before now, that I am always seeking to run against him. I only mentioned him as having been the first to announce the posteriority of the granite to the transition beds, which was a grand step at that time, and true ; though perhaps he did not come at it by a strictly logical course. I know how he drew his inferences, which were quite natural twenty-six years ago, before he had adopted the true theory of granite, but had I brought forward his errors, I should also have dwelt on his praises. Chancellor Brougham said of Serjeant Wilde, that it was rather hard to visit on the barrister the sins which he committed in the flesh as an attorney, and I should also think it hard to show up the mistakes which Von Buch the Huttonian fell into, in the youthful days of his Wernerianism.

With love to all, believe me yours most affectionately,

CHARLES LYELL.

To HIS SISTER.

Paris : September 23, 1837.

My dear Sophy,—We were glad to hear such a good account of the party at Kinnordy and your visitors in your last letter, and hope you still feel as if you could ' jump over the moon,' which we nearly do at our (or perhaps I should say my) mature age, at the thoughts of flying across the Channel once more after a longer absence than usual. We mean to steam it all the way, land and water, for the benefit of my fossil shells and our bones, not the fossil ones, and for the novelty of the thing, for both of us are heartily sick of the everlasting *pavé* between this and Calais. We go to St. Germain by steam carriage, by the Chemin-de-fer, which is advertised in a pamphlet beginning ' Paris vient de s'enrichir d'une *gloire nouvelle*,' and then by steamboat to Havre and

---

[3] British Association at Liverpool.

by another from thence to London. I have bought you a dozen small shuttlecocks, and went to the Quai Voltaire to ask for Grangier's Dante.[4] A complete copy had been sold a few weeks before for the very small price of fifteen francs, and I was told I should probably not get one, but Merlin on the Quai des Augustins might have it. Merlin had only one volume, ' Il Paradiso,' which I got, well thumbed, for four francs, and I took at a venture Montonnet's translation of the ' Inferno,' though Merlin said he believed it was not worth much; but it was only five francs, 1776. The same bibliopolist asked if I was going to London, ' because it was there that the French send for old French books, and there, if anywhere, Grangier might be bought entire.' À propos to the ' new glory ' which Paris has gained, I have had a compliment to-day myself in the same style in a note presenting me a fossil shell. The donor, a stranger, tells me he had been requested to give duplicates away to geologists, and it was impossible for him ' d'adresser à une meilleure *illustration* géologique.' But a more delicate piece of flattery has been administered in my new passport. Even two years ago my ' nez petit ' was changed to ' nez moyen,' but now it is ' nez long.'

After several days of heat, we have now so cool an evening that we have ordered a wood fire. The De Beaumonts, whom we mentioned last spring, have been calling, but we were out, as they were when we called. I mean Gustave and his wife (Mdlle. Lafayette) ; they sent a polite note of regret at missing us. They are only just returned two days. We had such a treat on Wednesday at the Théâtre Français ; two best places in the balcon to hear Mdlle. Mars and Mdlle. Noblet in several pieces. The principal one, an old play, ' Le Jeu d'amour et du hasard.' Mdlle. Mars is not in the least gone off at sixty, and her voice very sweet, and looks like a pretty woman of thirty. Mary has had a pleasant letter from Lord Cockburn, who tells us that the new tower at Bonaly by Playfair is picturesque, and pronounced by Sir T. Lauder to be like the ' peel tower of a Border chief.'

[4] His father being desirous of obtaining every edition of Dante, of which he formed a fine collection.

Maison, on the Seine : September 24.—Here we are, after reaching St. Germain by railroad, and thence to this place by a fiacre over a road which we thought would have upset us. Certainly it was worth while paying some five francs extra to see the Parisians enjoying themselves on a *fête* day. It was a very gay scene. We had to wait three-quarters of an hour in a fine newly built *salle*, in which I suppose a thousand persons were waiting as if in a theatre, divided into compartments according to the price of tickets you had taken out. They made us take nine places, seven being for the luggage, which I am told was a great imposition, but I took only one-franc places. Though there was no cause for hurry, when the bell rang, there was such a bustle and rush, and the agent himself lost his head and was quite nervous. Fortunately, every one of the engineers as well as engines are English, and not one accident has happened, though as many as 10,000 a day have travelled for four weeks since it has been opened. To-day 17,000 have gone by it; allowing that a great many have gone both ways, perhaps some 10,000 persons. Our train contained 1,200 persons, each carriage forty, and many being filled with well-dressed ladies and beaux, was a very gay sight. It was twelve miles and a half, performed in about half an hour. Twice we crossed the Seine. The motion we thought pleasanter even than the Greenwich railroad. All nearly are travelling for pleasure ; an exchange for the Montagnes Russes, Tivoli, or a jaunt to Versailles. The shareholders are quite surprised at their success and profits, and as the machines are tried and strained by the overweight, they have sent to England for more locomotives and more men. What a good thing for our own machinery-makers to supply the Continent ! All the steamboats, I am told, in the Mediterranean, have English men and machinery, and so I found it in the Baltic and in Norway.

The Byrnes [5] came to the railroad to see us off, and just before we left our hotel, Clara Tourguenef (formerly Viaris) came to see Mary. She is a beautiful young woman, and very pleasant. The most singular part, perhaps, of the railroad sight was the length of way, several miles I should think out

[5] Uncle and aunt to Mrs. Lyell.

of Paris, of an unbroken line of lookers-on, on each side of
the road to see the carriages pass.

Normandy : September 25.—We are on our way in a swift
steamer which flies down the Seine at an average rate of
fifteen miles an hour. Have passed Château Gaillard, built
by Richard Cœur de Lion, and what interested me most,
from historical recollections, Sully's country house, Rosny.
It is a picturesque château, one side coming very close to
the river, and all kept up in fine habitable order, having
belonged to the Duchess of Berri lately. We fancied Sully
walking on his favourite terrace. The scenery is always
agreeable on this river, and sometimes striking, and our
rapid pace gives it great variety. I have heard a discussion
on Louis Philippe having married a son and daughter to
Protestants, and being about to give two other daughters to
Protestants—one to the Prince of Wurtemburg, the other to
the reigning Duke of Saxe Coburg. This they say is popular
with a large number of the Liberals in France, who, though
Catholics, have been always fearing the return of the Jesuits,
who were put down when Charles X. was banished. As far
as a visitor can judge, Louis Philippe seems as quietly
established on the throne as any king in Europe.

Havre : September 27.—We have to wait till the evening
tide, having arrived last night at this port, now the Liverpool
of France, and very flourishing. The Seine below Rouen, though
not so beautiful and grand as above, was still very picturesque.
The view back towards Rouen, some villages on the way to
Honfleur, and the beautiful château of La Meillerie, where
Mdlle. La Vallière was brought up, was an interesting
object. We passed the old Castle of Robert le Diable,
William the Conqueror's father. How full of antiquities
Normandy is!

London : September 29.—We arrived last night, and found
all well here.

With love. Believe me your affectionate brother,

CHARLES LYELL.

To His Father.

My dear Father,—I send you three of the last medals which have been struck at the Mint of Paris, and which I hope you will add to your collection as memorials of the time of our last visit to that place. We went to see the triumphal arch of the Barrière de l'Etoile which is now finished, and has cost half a million sterling. I had no idea it was so colossal till we came up to it, and certainly it is very beautiful. Although the list of victories is too long to be marked on the medal, the first two which meet your eye on the side facing the Tuileries, are *Gemappe* and *Valmy*. I thought it almost ludicrous on the reverse of some of the medals of the arch to see two heads looking at each other, Napoleon and Louis Philippe, so I chose the one I send, though less characteristic of 1837.

As to the obelisk, you will see, I think, by the medal, that in itself it is a beautiful thing, but if I mistake not, you will wish it back in Egypt again when you see it. The fact is, hardly any monument could be placed where it stands, in the centre of the Place Louis XV., without being in the way. But this is, to my eyes, as out of place as a huge sphinx would have been just before Westminster Hall or Abbey. The first view I got of it was on entering the Tuileries Garden close to the Palace, and looking down the centre aisle, or rather avenue of wood, where the great fountain is at some distance. The end of this avenue, far off, should have been terminated simply by the great Arc de l'Etoile, and would have been so softened by the red light of the setting sun seen through it and round it. But one saw a black line much sharper than could be formed by any round column, standing before and cutting right in two, from top to bottom, this distant arch. Still worse, when you go to the Faubourg St. Honoré or Madeleine, and wish to look on the bridge at the front of the Chambre des Députés, you see the portico cut in two by the Theban intruder, and I imagine the effect must be equally fatal to the beautiful façade of the Madeleine itself when seen from the Chamber of Deputies.

We did not see the Musée de Versailles. Some told me

it was a ' grand collection of rubbish,' others that it was of
great historical interest. I suspect it was one of Louis
Philippe's good hits, of which I heard many. Among the rest
Byrne told me this:—In Napoleon's time all the army
became acquainted with the Emperor in the field; but Louis
Philippe, seeing that he could never have this advantage in
times of peace, changed the system of having certain regiments
of guards always attached to the Palace, and had three at a
time (each I believe is nearly 2,000 strong), to guard the
Tuileries, of regiments of the line. They are changed after
a year and a half or so, and all the officers dine in their
turn as they mount guard, either with the King (those of
higher rank) or with the Princes. He has already got
through nearly two-thirds of the entire regular army, and
they take pains to get personally acquainted with all the
officers. I send the two shabby-looking old volumes of
Dante I spoke of.

Believe me your affectionate son,

CHARLES LYELL.

To HIS FATHER.

London : October 21, 1837.

My dear Father,—As Mary is sending off a frank, I
have been thinking whether our party on Thursday will afford
any literary gossip worthy of being Boswellised, which I believe it would if my memory served me. I was amused with
an anecdote of Calcott's, who said that some one was praising
the waterfall at Bowood as by far the most beautiful artificial
one in the world. Rogers, on being appealed to, asked if they
would except Terni? When the Milmans were at Bowood
the other day, one of Lord Lansdowne's horses ran away with
Milman in sight of Mrs. Milman and his children; carried
him under the trees, and then charged a high invisible fence of
wire, which was thrown down, the poet rolling on the ground.
But he was not stunned, and, being immediately bled, soon recovered. Mrs. Milman had time to be frightened terribly.

The subject of the bill for a general registration of
mortgages, as in Scotland, was discussed, and declared to be
only delayed by the attorneys. Milman cited some one who

said we were governed by three powers : the power of steam, the power of the press, and the power of attorney. Babbage, on our talking of what books might be stereotyped, said that a great part of his ' Economy of Manufactures ' had been, but it would scarcely do for his Bridgewater treatise. Milman suggested that he might safely stereotype the blank chapter —a joke which Babbage took well. The question was started whether it was best for an author not to read the reviews against himself, as Sydney Smith declared was his system. Milman said it never answered, for when Sydney Smith was so cut up lately in the ' Quarterly Review ' for his sermon on the Queen, he could not help trying to get from Milman what they had said of him.

Babbage had bound up in a volume all the violent attacks against him for his ' Bridgewater.' [6] Van de Weyer told us that he collected all the things which the Dutch wrote against him, and sent them to his mother, begging her to see what a son she had. One of these compared him to Robespierre, declaring the only difference to be ' that he (Van de Weyer) was the most bloody.' There was much joking from Van de Weyer and others about Sydney Smith having made the experiment last spring, in full season, of a trip to the Rhine, of which he got soon tired, and hastened back to London dinners and routs. Milman pronounced it to be a pure piece of coquetry, to try how they would miss him, though against his own rule, for he says ' that in London as in law, de non apparentibus et non existentibus eadem est lex.' We had much talk about Sir Walter Scott's Life, our summer tours, and so forth; but Mary has got her bonnet on, so I must conclude.

Believe me ever your affectionate son,
CHARLES LYELL.

P.S. Having a few minutes more, I must add an observation of Van de Weyer, that he knew many merchants whose families had had galleries of pictures (in Holland and Antwerp) for 200 years, and who added to them the best modern works of their artists. Babbage said that English merchants' families never outlasted two centuries. Milman said the

[6] The *Ninth Bridgewater Treatise*, by Charles Babbage.

reason is that here they are never stationary, and before two
centuries go by they are above or below merchants—part of
the aristocracy, or bankrupts, and forgotten.

To His Father.

London : November, 1837.

My dear Father,—Many thanks for your letter of congratu-
lation on my birthday.   I will begin by way of answer to put
down a few recollections of an agreeable party which we had
yesterday at the Milmans.'

A note from Sydney Smith excusing himself, with a joke
even on the envelope.   To Mrs. *Heart* Milman (instead of
Hart) the seal having a ♡ on it with the motto.   'Il tuo
nome non è la ventura,' if I caught the words rightly.
Milman observed it was strange, for surely Sydney Smith
could not have a premeditated joke purposely engraved on a
seal.   Presently Mr. and Mrs. Milman, Mr. and Mrs. Senior,
Mr. Lockhart, Mr. Whewell, Mr. Rogers, Rev. W. Harness,
Mr. Rich, M.P., and ourselves.   When Mr. Rogers heard that
Mary had been to Norway without seeing any waterfall, he
said, ' That comes of having a man with a hammer for a
fellow-traveller.   What would you have thought when a girl,
if a fortune-teller had predicted that you would marry a man
who lived by his hammer?   I was asking Lady Mary Fox
(Rogers had just come from a visit to Colonel and Lady Mary
Fox in Windsor Palace) what she would have thought if,
when playing as a child in Bushey Park, she had been told
that she would one day be a housekeeper ; aye, and after
your marriage, and such will be your situation that your
friends will all congratulate you on getting the place of
housekeeper.'

Rogers mentioned having known three persons who had
known Pope.   One was Lord Lyttelton, another a boatman
who rowed Rogers over the Thames at Twickenham. ' Pray, as
you have been here so long, did you know Mr. Pope ? ' ' Mr.
Alexander Pope, you mean, sir; we called him Mr. Alexander
Pope.   Yes ; when a boy I often rowed him over, I and my
father, and he came to the river side in a sedan chair.'   This
led to much conversation about Pope, the subject having

first been brought on by Milman alluding to a new German book by one Waagen on paintings in Great Britain, in which Rogers' collection was praised, and an original clay bust or model by Roubilliac of Pope, I think. Whewell thought Pope by far the most influential literary man of his day. Lockhart thought that Dean Swift was superior, and, on our exclaiming at this, said that the Dean would have been too much for Pope had you met them together. Whewell rejoined that doubtless he would have been more than a match for Pope had they come to blows, but he (Whewell) was not thinking of a fight, but of mental qualities higher than he thought Swift possessed. Milman had been amused with an opinion of Goethe, showing his estimate of the relative dignity of poets and peers; he said, if Byron had spoken now and then in the House of Lords, his bad feelings would have found a vent, his spleen would have evaporated, to the great purification of his poetry. Whewell said, good-humouredly alluding to an attack on him by Brewster in a late 'Edinburgh Review,' that perhaps that was the meaning of the proposal to give peerages to scientific men, in which case science also might be purified of its grosser qualities.

Milman asked Mr. Rich (groom of the chambers to the Queen) whether it was true that the Queen had complained that they would not give her a diamond necklace. 'Not a word of truth in it, I assure you; she has several splendid diamond necklaces, but chose to wear a pearl one.'

Is it true that King Ernest has pressed for the restoration of the Hanoverian Crown jewels? 'Undoubtedly; they were joined to those of England at the union with Hanover, and ought, he says, to be separated again now that the crowns are disunited, but they are so mixed together they cannot tell which are which, and I suppose that the Queen thinks that the Salic law was never meant to apply to diamonds.' Senior remarked that she played her part remarkably well in the pageant the other day; that nature had intended her for such performances.

On Mr. Harness dwelling on the risk that a young person ran of being corrupted by flattery and power, Whewell said she would at any rate escape Henry VIII.'s trials, for he (Whewell) could see nothing in the spirit of the present times

which made him apprehend that Victoria would be corrupted
by the possession of exaggerated power. Mr. Rich had been
with her to the Theatre the night before, and said that her
reception had been most enthusiastic. I asked him afterwards whether she had read much; he said, not much of
works of imagination, none of Scott's novels, many of Miss
Martineau's politico-economical tales ! ! and she is now reading O'Driscol's ' History of Ireland.'   When we were in the
drawing-room I saw Mr. Whewell and Mary talking together,
first, as I learnt afterwards, about Bonn, where he had been
lately, and then about all the novels they had read, in which
Rogers afterwards joined, showing a perfect recollection even
of the tales written for children ; Miss Edgeworth's ' Simple
Susan' for example, to which Rogers objected that there
never was a man so bad as Attorney Case, and said that in
all her novels there were some characters overdrawn, in
which Whewell would not agree.   On Rogers joining Milman and me, we talked of Sir David Brewster's article in the
' Edinburgh Review ' on Whewell, and Whewell's pamphlet
in reply, and Rogers then remarked that such striking
articles as that of Macaulay on Bacon did no good to the
Review, but made the dulness of the rest more apparent ;
he regretted much in that article, and said Macaulay would
repent having done anything to lower the name of a great
teacher.   I argued against Macaulay's view of the Utilitarianism of Bacon, and his comparing Bacon's inductive
method to a simple piece of machinery like Aristotle's ' Logic.'
Rogers said that even in the historical part, in which Macaulay was much stronger, he frequently sacrificed truth, because
he had written for effect, and Bacon was not so bad as he
had painted him.   Milman complained that Seneca had been
selected as representing the science of the ancients, and that
even his ' Quæstiones Naturales ' had not been mentioned, his
best work in that line.   They then expressed their admiration of the merits of the article, and among other eloquent
passages Rogers repeated the following : ' It is a philosophy
which never rests. Its law is progress ; a point which was
invisible yesterday is its goal to-day, and will be the starting-
post to-morrow.'   I have since thought of having this
passage as a motto for my ' Elements of Geology.'

Talking of the popularity of Wordsworth's Poems, Senior asked Rogers if he thought Wordsworth had made fifteen thousand pounds. Rogers said he never made anything at all till last year, and then he received one thousand pounds for a new edition.

I was not sorry that Sydney Smith happened to be engaged, for though such a party would have drawn out some of his best fun, he would have overpowered Rogers with his boisterous laugh and sonorous voice, and it is a great pleasure to enjoy quietly some rays of Rogers' sunset; everything he says has a remarkably fine finish in it, but he is very mild and indulgent, and no remains of the epigrammatical sarcasm for which he seems to have been famous. I was amused with the abrupt ending of his confab in the drawing-room with Milman and me above mentioned. Milman's eldest boy came up, and, looking Rogers in the face, said, 'Mr. Rogers, do you like sugar?' There was a pause. I was thinking of the acidity formerly imputed to Rogers' conversation. Milman laughed, and Rogers said in an emphatic tone, 'Indeed I do,' and walked away to the tea-table.

I must not fill another sheet with discussions for and against Scott's partnership with Bannatyne (after Lockhart was gone); Mr. Trollope's new novel; Harriet Martineau's 'America,' and whether she was the plainest woman in the world, or almost handsome, especially on her first return from the New World.

Believe me, with love to all at Kinnordy,

Your affectionate son,

CHARLES LYELL.

CHAPTER XXII.

JANUARY 1838–OCTOBER 1841.

To PROFESSOR SEDGWICK.

January 20, 1838.

My dear Sedgwick,—I have been very busy since my
return from Denmark with my ' Elements of Geology,' which
I hope to get out before the summer, and have just finished
abstracting from Murchison's sheets, which I have read with
no small pleasure, a short account of the Silurian rocks and
their fossils. I am now upon your Cambrians, and as I give
a few woodcuts of fossils in each group I should have liked to
give a few of the earliest ones if possible. Phillips, in his new
treatise in Lardner, has given us a few Snowdonian fossils of his
own collecting in 1836, and I would of course select others,
that I might not interfere with his, and might give as much
as possible original information to the public. I should,
however, be content if there were any figures published of
what you consider true Cambrian organic remains of any
order of animals, to copy them in my small reduced woodcuts, just to give novices an idea of the generic forms, if I
can do no more. I believe there are some specimens, from
Snowdon and elsewhere, of Cambrian fossils in the Geological Society's Museum, but if you can put me in the way of
furnishing a page of woodcut figures, it will be doing me a
great service.

Our cousin Miss Winthrop[1] gave us an account of your second lecture, and of your manner of controverting one of my theories, with which I was well satisfied. But when Sir W. Parish put into my hands the report of the same lecture in the 'Norfolk Chronicle' of January 13, I confess I was out of patience with the reporter. He makes you say, 'Various false theories have been adopted by *infidel* naturalists,' &c. He goes on, 'One of these,' &c., one sentence; then 'another was Lamarck's theory,' three sentences ; then 'Mr. Lyell's theory, that the creation of new species is going on at the present day, was also condemned as rash and unphilosophical.' Had I not known more about the lecture and about you, I should have thought that I had been classed with the infidels. Now touching my opinion, I have no right to object, as I really entertain it, to your controverting it ; at the same time you will see, on reading my chapter on the subject, that I have studiously avoided laying the doctrine down dogmatically as capable of proof. I have admitted that we have only data for *extinction*, and I have left it rather to be inferred instead of enunciating it even as my opinion, that the place of lost species is filled up (as it was of old) from time to time by new species. I have only ventured to say, that had new mammalia come in, we could hardly have hoped to verify the fact. Indeed, if I were asked, where is your new mammiferous species, which has filled up the great vacancy caused by so many removals? I might perhaps say, this is just my strongest case, for there is *man*, a mammifer of yesterday, which has been spreading by millions over the globe. But I certainly wish it to be inferred from my book, that in the ocean, beyond the sphere of man's interference, and in the desert, and in the wilderness, and among the infusoria and insects, the extinction has been going on in the last 6,000 years, and that the substitution of species to supply the vacancies, which must always be occurring, has also been going on ; though *how*, is a point we are as ignorant of as of the manner of God's creating the first man. You will, I hope, allow that to assume that there have been no new creations since man appeared, is at least as 'rash and unphilosophical,' as modestly to hint

[1] Now Mrs. C. Baring Young.

the possibility of such occurrences, which is all you will find I have done. Whewell, I think, has put this well in his 'Inductive Sciences.' He has brought up the subject of extinction and new creations to that point which science has reached, or where our *present* ignorance begins, without discouraging those investigations which may lead to our discovering what laws may still govern this mysterious part of nature's operations, the coming in of new beings. To me it appears that the line you are represented to have taken is to hazard a far bolder hypothesis than I should have dared to do, viz. that no new creatures have begun to exist for the last 6,000 years, or for such time as man has existed, although geology has now brought to light the proofs of an indefinite series of antecedent changes, such repeated failures of species of animals and plants, and their replacement by others. The burden of proof rests on him who ventures to affirm that Nature has, at length, stopped short in her operations, and that while the causes of destruction are in full activity, even where man cannot interfere, she has suspended her powers of repair and renovation.

Believe me ever most truly yours,

CHARLES LYELL.

To LEONARD HORNER, ESQ.

London : February 24, 1838.

My dear Horner,—Mary is going to send you a journal, but, as she did not attend the anniversary,[2] I will send you a short account of that meeting. Not so full as usual, and most of the M.P.s absent on the Irish Poor Law question; nevertheless, a grand display of talent. Whewell in the chair, Sedgwick, Buckland, Sir P. Egerton, Darwin, Owen (who is wonderfully pleased at receiving the Wollaston medal), Fitton, Greenough, Hallam, Milman, Murchison, Lord Burlington, Prof. Jones, Lubbock, Bayley, Clift, Hamilton, and on the whole the great horseshoe table tolerably well filled. A short time before the meeting, Whewell told me that I should have nothing to do, unless I would propose the

[2] Of the Geological Society.

President, which as ex-President fell naturally to me. So I began the speaking, and said it would be impossible for them to appreciate Whewell's services in the chair, unless I reminded them in what an eventful year of his career as author and scientific workman he had rendered those services. I therefore enumerated his doings in the year, beginning with three volumes on the Inductive Sciences. I mentioned that on my return from Norway and Denmark, I mentioned to Lodge, fellow of Cambridge, how good a companion Whewell's last work had been to me, but soon found myself at cross purposes, as he had to tell me of *two* other works published since my absence, viz. the ' Mechanical Euclid,' which had already come to a second edition, and secondly, the ' Studies of the University of Cambridge.' On my return home, I found his paper (eighth series), ' Experiments and Observations on the Tides' in 'Philosophical Transactions,' for which the Royal Society soon after gave him the royal medal. On my expressing to my friend Mr. Jones (who sat opposite me) my wonder that Mr. Whewell, entering so much into literary and scientific society, could possibly, with all his Cambridge professional duties, find time even to pass through the press so much matter, supposing he did not write a word, what was my astonishment when Prof. Jones told me that his friend Whewell, then staying with him in town, was actually then passing through the press *four other works*, viz. two new editions in different forms of his ' Bridgewater Treatise,' and thirdly, a long article in the ' Medical Gazette,' on Physiology of Nerves of Sensation and Volition, relatively to the Bell and Mayo controversy and claims, and fourthly those four sermons preached at Cambridge in November last, which I had read with much pleasure, in which he has treated of Butler's and Paley's views of morals, moral philosophy, &c. There may be other writings, and certainly a controversial letter to the ' Edinburgh Review;' but I asked, if these alone were considered, not as the monument of one year, but of a lifetime, would it not be thought sufficient for the zeal and industry of one man, and yet in this year Whewell had made an efficient President of the Geological Society? I would not dwell on that labour of love, the eloquent address they had heard the opening of that morning, but his sacrifices.

in attending our meetings, and the time given to details of management, of getting a new secretary and other internal rearrangements of the Society's official business, for doing this in so busy a year, called for their acknowledgments. This address was well received, and Whewell replied modestly, fearing that most of them would think when they heard of so many works, that he had done too much to do any well. In proposing Owen, he put very well his being a fellow townsman and schoolfellow of his own. In proposing the different Societies, he made a beautiful allusion to Terence, saying, 'I am a man; homo sum, humani nihil a me alienum puto,' and parodied it well, 'We are geologists, and we regard nothing in physics and natural history as foreign to our purpose.' Sedgwick was uncommonly splendid in replying for Cambridge, and pointing out the connection between abstract science cultivated there, and all general reasoning on particular facts, geological and others; such reasoning alone raised us above the dregs of matter, &c. Whewell drank the 'Strangers,' and Prof. Jones last, who made a truly eloquent speech, and very extemporaneous, on the similarity of the prospects of the two new sciences, different as they are in their subjects, geology and political economy. After the anniversary evening, Lord Cole pressed me so hard to go and eat pterodactyl (alias woodcock) pie at his rooms, that I went, with Whewell, Buckland, Owen, Clift, Egerton, Broderip, Hamilton, Major Clerk, Lord Adair; and there we were till two o'clock, fines inflicted of bumpers of cognac on all who talked any 'ology.' Cigar smoke so strong as half to turn one's stomach. I lost the enjoyment of Murchison's dinner next day, and for five days only did half a day's work or less. It is a serious warning to me how careful I must be.

Yours affectionately,

CHARLES LYELL.

To LEONARD HORNER, ESQ.

London: March 12, 1838.

My dear Horner,—At the last meeting of the Geological Society, Darwin read a paper on the Connection of Volcanic Phenomena, and Elevation of Mountain Chains, in support of my heretical doctrines; he opened upon De la Bêche, Phillips, and others (for Greenough was absent) his whole battery of the earthquakes and volcanos of the Andes, and argued that spaces at least a thousand miles long were simultaneously subject to earthquakes and volcanic eruptions, and that the elevation of the Pampas, Patagonia, &c., all depended on a common cause; also that the greater the contortions of strata in a mountain chain, the smaller must have been each separate and individual movement of that long series which was necessary to upheave the chain. Had they been more violent, he contended that the subterraneous fluid matter would have gushed out and overflowed, and the strata would have been blown up and annihilated. He therefore introduces a cooling of one small underground injection, and then the pumping in of other lava, or porphyry, or granite, into the previously consolidated and first-formed mass of igneous rock. When he had done his description of the reiterated strokes of his volcanic pump, De la Bêche gave us a long oration about the impossibility of strata in the Alps, &c., remaining flexible for such a time as they must have done, if they were to be tilted, convoluted, or overturned by gradual or small shoves. He never, however, explained his theory of original flexibility, and therefore I am as unable as ever to comprehend why flexibility is a quality so limited in time. Phillips then got up and pronounced a panegyric upon the 'Principles of Geology,' and although he still differed, thought the actual cause doctrine had been so well put, that it had advanced the science and formed a date or era, and that for centuries the two opposite doctrines would divide geologists, some contending for greater pristine forces, others satisfied like Lyell and Darwin with the same intensity as nature now employs. Fitton quizzed Phillips a little for the warmth of his eulogy, saying that he and others who had Mr. Lyell always with them, were in the habit of admiring and quarrelling with him every day, as one might do with a sister or cousin whom one would only kiss and embrace fervently after a long absence. This seemed to be Mr. Phillips' case, coming up occasionally from the provinces. Fitton then finished this drollery by charging me with not having

done justice to Hutton, who he said was for gradual elevation.

I replied, that most of the critics had attacked me for
overrating Hutton, and that Playfair understood him as I
did. Whewell concluded by considering Hopkins' mathematical calculations, to which Darwin had often referred.
He also said that we ought not to try and make out what
Hutton would have taught and thought, if he had known
the facts which we now know. I was much struck with the
different tone in which my gradual causes was treated by all,
even including De la Bêche, from that which they experienced in the same room four years ago, when Buckland,
De la Bêche (?), Sedgwick, Whewell, and some others treated
them with as much ridicule as was consistent with politeness
in my presence. Yours affectionately,

CHARLES LYELL.

To LEONARD HORNER, ESQ.

Kinnordy : September 1, 1838.

My dear Horner,—This our second day at Kinnordy has
been, they tell us, the most like summer of any they have
had this year. The place looks beautiful, and I have at last
some leisure to give you an account of our British Association Meeting at Newcastle. Mary, however, has just come
into the library, and reminded me that I must begin
further back, and go to the banks of the Yare, from which
she sent her last journal. I ascertained during my tour in
Suffolk and Norfolk two points respecting the Crag, which
I had never made out before, at least to my own satisfaction.
First, that the mammalia, such as mastodon, elephant, &c.,
were really coeval with the true Crag shells. Secondly,
that the Crag of Norfolk was not, as some have supposed,
of a newer date than that of Suffolk. This and several
other minor points, besides the acquisition of specimens of
shells, would have satisfied me, even if I had not met with
certain vertical lines of flint, or paramondra, which supplied
me with the materials of a small notice which I read at
Newcastle, and of which you will see a good abstract by-and-
by in the 'Athenæum.'

We avoided all but collectors at Norwich; but on the last day there (Sunday), we went to the Cathedral, where the Bishop spied us out from his throne, and sent the verger to ask Mary and me to come to the Palace to luncheon, which we did, and saw Mrs. and Miss Stanley, and walked round the garden. We had to hurry away, as we had to set off in the Yarmouth steamer at three, and had a pretty sail down the river. The day after (Monday) I took a long walk with Mr. Pellew along the sea-shore. We went by steam from Yarmouth to Hull, and thence, after sleeping a night, in another steamer to North Shields, where we landed at six o'clock in the morning, and so perfectly refreshed by our night in the steamboat, that we both enjoyed a good long expedition along the shore to Tynemouth, walking, and then in a gig to a point on the coast, Cullercoats, where the ninety-fathom dike is laid open in the cliffs, the magnesian limestone on one side and the coal on the other. I then went to see another fine section of the same dike, or rather fault, cut open by a railway about a mile inland, and the quarries of magnesian limestone and marl slate in which the fossil fish are found. Next day we crossed the Tyne to South Shields and visited the Marston Rocks, where there are lofty perpendicular cliffs of magnesian limestone, and small isolated rocks or needles of the same in the sea. The coast scenery was very grand, and the brecciated form of the magnesian limestone, which is an aggregate of angular masses of itself, as if broken up and reconsolidated *in situ*. The same evening we got to Newcastle. As soon as I saw Murchison, he told me it was arranged that I should be President of the Geological Section, which was immediately confirmed at the meeting of the general committee. The other Presidents were Herschel, Physical; Babbage, Mechanical: Whewell, Chemical; Sykes, Statistical; Sir W. Jardine, Zoological; Dr. Headlam, Medical. Our section was crowded, from 1,000 to 1,500 persons always present; Sedgwick, Buckland, Daubeny, De la Bêche, Greenough, Phillips, Griffiths, Sir P. Egerton, Owen, and many other good men taking part. Portlock and Torrie my secretaries.

All that I saw of the government of the Association gave me a good idea of the spirit, but no wish to consume my time

in taking a part in it, to which I am invited, I hear, by being
put on the council.   Sedgwick was so eloquent; his lecture
to 3,000 people on the sea-shore made a great impression.
                              Ever affectionately yours,
                                        CHARLES LYELL.

                To CHARLES DARWIN, ESQ.

                                        Kinnordy: September 6, 1838.

My dear Darwin,—I must first read your letter again
which I answered in a great hurry at Newcastle.   I should
like to have a talk over Salisbury Craigs with you, especially
on the spot.   I do hope some day that we shall be able to
examine together some of the volcanic rocks on the coast
near the Red Head, within a day's ride of this place.   It is
a splendid exhibition, and I think we should make out several
points of eruption and sections of the feeders of the old
volcanic islands of the Old Red Sandstone period.   The variety
of porphyries and amygdaloids is quite splendid.   I assure
you my father is quite enthusiastic about your journal, which
he is reading, and he agrees with me that it would have had
a great sale if separately published.   The other day he told
me that he wished to get a copy bound the moment it was
out, and send it as a present to Sir William Hooker, who
more than any one would be delighted with yours.   He was disappointed at hearing that it was to be fettered by the other
volumes, for although he should equally buy it, he feared so
many of the public would be checked from doing so.   I hope
you mean to sell a portion of those copies which I think you
told me you were to have separate, as I think it was a large
number.
    When do you think the book on Coral Reefs and Vol-
canos will be out?   In recasting the 'Principles,' I have
thrown the chapter on De Beaumont's contemporaneous
elevation of parallel mountain chains into one of the Preliminary Essays, where I am arguing against the supposition
that nature was formerly parsimonious of time and prodigal
of violence.   You will, I am sure, find the discussion of that
question much more naturally placed by this new arrangement.   I should like to know, when you next write to me,

how far you consider your gradual risings and sinkings of the spaces occupied by coralline and volcanic islands in the Pacific as leaning in favour of the doctrine that many parallel lines of upheaval or depression are formed contemporaneously. If I remember right, some of your lines are by no means parallel to others, although many are so. In one point of view, your grand discovery proves, I think, in the most striking manner, the weight of my principal objection to the argument of De Beaumont. You remember that I denied that he had proved that the Pyrenees were elevated after the cretaceous period, although it is true that the chalk has been carried up to their summits, and lies in inclined beds upon their flanks; for who shall say that the movement was not going on during the cretaceous period? Now in your lines of elevation, there will doubtless be coralline limestone carried upwards, belonging to the same period as the present, so far as the species of corals are concerned. Similar reefs are now growing to those which are upraised, or are rising.

September 8.—Many thanks for the 'Spectator' which came this morning. I really find, when bringing up my Preliminary Essays in 'Principles' to the science of the present day, so far as I know it, that the great outline, and even most of the details, stand so uninjured, and in many cases they are so much strengthened by new discoveries, especially by yours, that we may begin to hope that the great principles there insisted on will stand the test of new discoveries. I am pleased to think of the improved form in which this part of the work will come before the French for the first time. You hope that the 'Elements' may send many to the 'Principles,' but I am not yet so sanguine as to be free from apprehension lest they should stand in the way of the 'Principles.' This, however, cannot be known yet.

I am very glad to hear you like the Athenæum. I used to make one mistake when first I went there. When anxious to push on with my book, after a 'two hours' spell,' I went there by way of a lounge, and instead of that, worked my head very hard, being excited by meeting with clever people, who would often talk to me, very much to my profit, on the very subject on which I was writing, or I fell in with a Re-

view or Magazine relating to geology. Now this was all
very well, but I used to forget that this ought to count for
work although nothing had been written, and that I ought
consequently to give up my second 'two hours' spell.' By
not doing so I was often brought to a dead stop, so that at
last, for fear even of meeting with anybody in the streets who
would also talk geology, I was sometimes driven for a walk
into Gray's Inn Gardens. But then you will say comes the
difficulty, how to avoid theorising, for nothing substantial is
gained by dwelling on the subject when there is no pen, ink,
or paper before one. After lying two hours fallow the mind
is refreshed, and then in five minutes your fancy will frame
speculations which it will take you the two hours to realise
on paper. As your eyes are strong, you can afford to read
the light articles and newspaper gossip, which I could never
indulge in much with impunity.

My father has been more and more taken up and delighted with your journal, and begged me this morning to
invite you to come here any day this or the next month,
when we shall be here, for as long or short a time as you
like. Steamboats every Wednesday to Dundee, passage from
thirty-six to forty hours; railroad four times a day from
Dundee to Glamis, where the carriage meets you, and brings
you in half an hour to Kinnordy—an easy trip for one who
was never sea-sick except in sailing vessels. Do come, if
you want some fresh air, and if you choose to bring MS.
here and write, as I do whenever I choose, four or five hours
quietly every day, I promise you the means of doing so; or
if you prefer a geological excursion, remember that an autumn on this East coast may be almost always reckoned upon
for fine weather.

Will you be so good, after reading the enclosed note to
Dr. Richardson, to send it post-paid to him? I was glad to
see him at Newcastle. Do not let Broderip, or the 'Times,'
or the 'Age,' or 'John Bull,' nor any papers, whether of saints
or sinners, induce you to join in running down the British
Association. I do not mean to insinuate that you ever did
so, but I have myself often seen its faults in a strong light,
and am aware of what may be urged against philosophers
turning public orators, &c. But I am convinced, although

it is not the way I love to spend my own time, that in this country no importance is attached to any body of men who do not make occasional demonstrations of their strength in public meetings. It is a country where, as Tom Moore justly complained, a most exaggerated importance is attached to the faculty of thinking on your legs, and where, as Dan O'Connell well knows, nothing is to be got in the way of homage or influence, or even a fair share of power, without agitation. The local committee at Newcastle were quite amused at the eager press for tickets, after the meeting began, on the part of those who had most sneered against the whole thing down to a few weeks before its commencement. I can also assure you, as the strongest commendation, that the illiberal party cannot conceal their dislike, and in some degree their fear, of the growing strength of the Association, in which circumstance as geologists we are particularly interested. We must take care not to hint this last argument to the Tories, many of whom are helping forward the cause gallantly at present, and, Heaven be praised, we seemed in no danger of splitting on the rock of politics, which I always fear much more than any occasional squabbles amongst ourselves, which can never come to anything like lasting feuds in a body collected from so many different quarters. The moral of all this is, Go next year to Birmingham if you can, although your adviser has been only to two out of eight meetings. Did you really manage to drink nothing but water at old Jones ' ?

Pray write and gossip at full when lounging at the Athenæum to me, and never imagine you can say enough.

As to the Glen Roy case, I saw, in Orust in Sweden, great beds of stratified gravel and sand like those which cover our Scotch hypogene rocks entirely destitute of shells; yet there were beds of shells like those of Uddevalla, and sea beaches at still higher levels. Mind I tell you about the absence of shells in some of my Norwegian Newer Pliocene beds.

Believe me, my dear Darwin, ever most truly yours,

CHARLES LYELL.

To Dr. Fitton.

My dear Dr. Fitton,—Although we are not far north of
Edinburgh, your review only reached my father's house
last night, and I have this morning read it through without
stopping, and with very great pleasure and profit, for it contains much that is new to me, as well as much that is gratifying. Indeed I suspect you must have left out those
passages of an 'excoriating' character which you hinted at
when we last met, or must have thought me one of the most
sensitive of authors. I have certainly reason to be more
than satisfied, for it is a piece of good fortune which happens
to few even once in their lives, to be criticised by a reviewer
who has read and thought over all they have written, as well
as what others have done and said, and to still fewer who
have that luck, to come off with so much praise. I am very
glad that you have dwelt so clearly, in your ample analysis
of my 'Elements,' on those points in which it differs from the
'Principles.' I do not remember a single point in which I
think you have misunderstood me, geologically speaking.
The first note I think of a slight mistake is where you
ascribe to Scrope a section which is my own of the lava of the
Coupe d'Aysac. I made it on the spot, and to the best of my
recollection you will find nothing like it in Scrope, whose
work is not at hand. But I have borrowed so many illustrations from Scrope (though not, I trust, either in his or any
other case, without acknowledgment), that I am the last who
should grudge him the appropriation of one of mine.

I am pleased to see you have been able to draw out
Lonsdale's just claims to original views in regard to the
Devonian and other subjects, and assert for him what he will
never do for himself. Your application of the lines 'The
times have been,' is admirable, and the whole article runs off
fluently and in entertaining style, and varied until you get
into your elaborate disquisition on Hutton, which to all the
general readers, and to many of the initiated, will, I fear, be
somewhat heavy. It has been useful to me, as I found it
difficult to read and remember Hutton, and though I tried,

I doubt whether I ever fairly read more than half his
writings, and skimmed the rest. Considering at how late a
period, as compared to Steno, Hook, Leibnitz, and Mora, he
came into the field, and consequently how much greater
were his opportunities, I think his knowledge and his original
views were confined to too small a range of the vast science
of geology, to entitle him to such marked and almost exclusive pre-eminence as you contend for in his behalf. If you
had not felt some natural indignation at the unpardonable
neglect with which the French and Germans have treated
him, you would not, I think, on reading downwards from the
theories of the older writers, have considered his merits on
the whole as so transcendant. Your citation (immediately following the woodcut) is a more perfect enunciation of the gradual acquisition of the metamorphic character by successive
modification than I had remembered to exist in his writings.

I remember on one occasion Boué, in reference to my
citations of Macculloch, apologised for having pretended to
such complete originality in regard to the change of strata by
Plutonic action, and Von Buch told me on my citing Playfair
for the elevation of Sweden, that neither he nor any German
had had access to Hutton's and Playfair's works during the
war, &c. I feel, however, by no means disposed to defend the
Continental writers from your charge, and you might have
included Macculloch, who so ably filled up Hutton's rough
sketch of the metamorphic theory, but with scanty reference
as usual to the merits of a great predecessor.

In distinctly alleging that my defective appreciation of
Hutton's claims was the ground of the pages which you
write in his defence and eulogy, a point again enforced in
the last sentence, you should have been careful to distinguish
between the total neglect of Cuvier, Von Buch, Humboldt,
Boué, Brongniart, and others, and the inadequate rank in the
relative scale which in your judgment I have assigned, as the
simple reader of your review must confound me with all the
rest. As you do not complain that my historical sketch was
disproportionate in length to the general plan of my 'Principles,' I must assume that you think Hutton occupies too
small a space; yet in my first chapter I gave Hutton credit
for first separating geology from other sciences, and declaring

it to have no concern with the origin of things, and after
rapidly discussing a great number of celebrated writers, I
pause to give, comparatively speaking, full-length portraits of
Werner and Hutton, giving to the latter the decided palm of
theoretical excellence, and alluding to the two grand points
in which he advanced the science. First, the igneous origin
of granite, secondly that the so-called primitive rocks were
altered strata. I dwelt emphatically on the complete revolution brought about by his new views respecting granite, and
entered fully on Playfair's illustrations and defence of
Hutton, and he is again put prominently forward in the
' Elements,' where no other but Lehman and Werner are
mentioned. The mottos of my first two volumes were
especially selected from Playfair's Huttonian Theory, because
although I was brought round slowly, against some of my
early prejudices, to adopt Playfair's doctrines to the full extent,
I was desirous to acknowledge his and Hutton's priority, and
I have a letter of Basil Hall's in which after speaking of
points in which Hutton approached nearer to my doctrines
than his father, Sir James Hall, he comments on the
manner in which my very title-page did homage to the
Huttonians, and complimented me for thus disavowing all
pretensions to be the originator of the theory of the adequacy of modern causes.

Yet, to how many of your readers, who will never see my
work, will your elaborate advocacy of Hutton seem to imply
that I overlooked, or have been unwilling to acknowledge
even in a moderate degree, his just pretensions! It was my
business, in tracing the progress of our science to its present
state, to estimate the importance of each writer, and adjust
the quantity of space due to him in my historical sketch, not
simply according to his originality and genius, but partly at
least in proportion to his influence; and I still think that
Werner's eloquence, popularity, enthusiasm, and position at
Freyberg, placed him in this point of view as much above
Hutton as I have represented him to fall below him in reference to the truth of his theories. Yet as an admirer of
Hutton all I could have wished is, that your panegyric on
Hutton had appeared as aiding and seconding my efforts,

since I trust that no book has made the claims of Hutton better known on the Continent of late years than mine.

As the 'Edinburgh Review' has been so long in noticing any of my writings, it is really a great point to have at last been honoured by so very long an article, though I doubt not I have lost much by Napier's cashiering. But I have no more room to thank you now for having taken so much pains to write what I regard, in spite of the Huttonian episode, so very favourable a critique on my productions.

With our joint remembrances to Mrs. Fitton,

Believe me ever most sincerely yours,

CHARLES LYELL.

To HIS SISTER.

Dudmaston : September 1, 1839.

My dear Sophy,—After the business of the Section (at the British Association at Birmingham) on Tuesday in which I took a part, as you will see by the 'Athenæum,' a party of us set off by railway to Tamworth, to go twenty-seven miles before reaching Sir Robert Peel's, whereas the direct road is only fourteen north of Birmingham. We first went about fifteen miles on the road to Coventry, and then turning off on a new railway lately opened to Derby, we were taken to Tamworth, where three of Sir Robert's carriages were in attendance to convey us and our bags about three miles to his place, which we reached after eight o'clock, having been rather more than two hours on the road. When in motion we went always at the rate of twenty-two miles per hour, but time had been lost in starting, and when we turned off from the great London road, we had to wait sixteen minutes to see whether there were any passengers in the train from London booked for Derby. During this stop we had a fine view of the great luggage train, in which there were no passengers, but forty-eight waggons laden with baggage covered with canvas. It seemed of interminable length, was preceded by two locomotives, and passed us at full speed, having to keep clear of a train only ten minutes behind it. To prevent this train from running into it, the last carriage contained a huge red lamp which looked like a

conflagration. It was a marvellous sight to see it shoot by.
At Tamworth there is a fine viaduct, and Tamworth Castle is
a fine and very ancient building. We found Lady Peel in
the drawing-room, who is tall, handsome, and ladylike, but
rather grave. Sir Robert received us very pleasantly, and we
should have sat down soon after eight if Whewell had been
dressed, but it was nearly nine before the dinner began.
Whewell and Lord Northampton sat on each side of Lady
Peel, Murchison, Mr. Lloyd, son of the Principal of Dublin
University, Comte de Cambour, a French mathematician, who
has published the best work on the theory of steam engines,
Fox Talbot (photogenic), John Taylor, Treasurer of British
Association, Mr. Yates, Honorary Secretary of ditto, Major
Sabine, late Mathematical Secretary to the Royal Society,
and myself. I sat on Sir Robert's right hand, and during a
conversation of three hours we talked of a great variety of
subjects; antiquities of Tamworth, railways, paintings,
sculpture, chartists of Kirriemuir, Birmingham, &c., British
Association, bearing of geology on Scripture, Wordsworth's
Poems, Chantrey's busts. Some of the party said next day
that Peel never gave an opinion for or against any point from
extra caution, but I really thought that he expressed himself
as freely, even on subjects bordering on the political, as a well-
bred man could do when talking with another with whose
opinions he was unacquainted. He was very curious to
know what Vernon Harcourt had said on the connection of
religion and science. I told him of it and my own ideas, and in
the middle of my strictures on the Dean of York's pamphlet
I exclaimed, ' By-the-bye, I have only just remembered that
he is your brother-in-law.' He said, ' Yes, he is a clever
man and a good writer, but if men will not read any one book
written by scientific men on such a subject, they must take the
consequences.' After he had explained to me how railways
were taxed, I pointed out to him Lord Carnegie's proof that
such a method acted as a bonus towards the imposition of
high fares. This he saw, and admitted as an evil. If I had
not known Sir Robert's extensive acquirements, I should
only have thought him an intelligent, well-informed country
gentleman, not slow, but without any quickness, free from
that kind of party feeling which prevents men from fairly

appreciating those who differ from them, taking pleasure in
improvements, without enthusiasm, not capable of joining
in a hearty laugh at a good joke, but cheerful, and not preventing Lord Northampton, Whewell, and others from
making merry. He is without a tincture of science, and interested in it only so far as knowing its importance in the
arts and as a subject with which a large body of persons of
talent are occupied. He told me he was one of the early
members of the British Association, and that he was glad that
we had persevered in holding our meeting at Birmingham
under discouraging circumstances; yet I learnt afterwards
from the Birmingham Committee of Management, that when
some of them, being personal friends of Sir Robert, asked his
opinion only three weeks before, he could not venture any
opinion at all. Being pressed to come in to the grand dinner
on Thursday, he said he was so unpopular in Birmingham that he would only be an element of discord where all
was harmony as far as the British Association was concerned.
His refusal was a disappointment to the leading members of
the British Association and to his political friends in Birmingham, where he would have been well received. There are
many beautiful pictures and statues, of which he was very
happy to give us a full account. He pressed us most politely
not to hurry back to our business next morning, but to stay
and breakfast with the family, which some did; but most of
us returned early, and I was reading a paper at eleven
o'clock, but I contrived besides breakfasting to see the
garden, which, although not a fine one, pleased me from
having each kind of flower in large masses. In the dining-
room is a single picture by Haydon of Napoleon in St.
Helena. It was very striking. Sir Robert told me that
there is an ode on it by Wordsworth which is given in the
' Quarterly Review' in a late article on Waagen's Tour in
England. It represents Napoleon with his back turned
towards you, looking from a height on the boundless ocean.
Among the pieces of sculpture are some remarkable statues
by Gibson of Liverpool ? artist. But the most striking is a
bust which Roubilliac made for Lord Bolingbroke. Chantrey
made Sir Robert a present of his own bust of Sir Walter
Scott as a pendant for this. I might meet Sir Robert

Peel in society and visit him for years without having
as much talk with him as from having thus by accident sat
next him for three hours. He is looking in wonderfully
better health and spirits than when I met him at the Royal
Society with Sir H. Davy ten years or more ago. He is
much taken up with his boys, and had been showing Liverpool to them.

Ever affectionately yours,

CHARLES LYELL.

To CHARLES BABBAGE, ESQ.

London : August 11, 1840.

Dear Babbage,—I send you two guidebooks for the steam
trip on the Seine, and you will not be whirled by the geology, and old Sully's château and Cœur de Lion's Castle,
&c., as we were, at fifteen miles an hour. Pray read the
accompanying geological notes, and inquire about that
strange coating of salt, and write to me about it. Address to
Hart Street from Paris, for I am writing on it, and would
give much to be able to go and hammer with you at those
fine old inland sea-cliffs. I wish you could land and send
me notes enough for a joint paper on the subject. The salt
at Andelys was liberally laid on the surface. As I once saw
a raised beach with recent marine shells thirty feet high
near the base of the cliff under the lighthouse at Cap la Hève,
there has been an elevation there in geologically modern
times, but still perhaps twenty thousand years ago, and that
cannot explain the inland cliffs having their incrustation of
salt. But bear in mind how little would make the valley of
the Seine once more a deep bay bounded by chalk cliffs. As I
only found out the salt when I got home, I unluckily cannot
say whether any part of the ' roches d'Orival ' (the first, I
believe, which you will see of the grand old cliffs) has a
saline incrustation. The surface is decomposing these and
pulverising.

Ever faithfully yours,

CHARLES LYELL.

[In the autumn of 1841, Mr. Lyell paid his first visit to the
United States, and lectured at the Lowell Institution, Boston. He
was thirteen months absent, and made tours in Canada and Nova
Scotia.]

To LEONARD HORNER, ESQ.

Lockport [near Niagara] : August 26, 1841.

My dear Horner,—I am much obliged to you for your
news of the British Association at Devonport, and the Geological Society Section, and glad that Phillips has come out
with his Devonian fossils. Mr. James Hall, State Geologist of
New York, who is now travelling with me, wishes much to
have the book. I have been getting up the details of their
Silurian system with great pleasure, comparing the supposed
ancient boundaries of Lake Ontario, when it was 150 feet
higher, with its present shore, the phenomena of drift and erratics, and other features of the geology here, with which as
well as with their working scientific men I have been much
pleased. The signs everywhere of rapid progress and sudden
conversion of the wilderness into a region covered with
populous cities and communicating with each other by railways, canals, and river steamboats, which go through still
waters at the rate of fifteen miles an hour, is quite exhilarating to me. I was warned by an English officer not to be
misled by so dazzling a spectacle, for that so much was due
to borrowed capital, and has ended and will end in countless
bankruptcies. But the real wealth of this country in cities,
steamboats, &c., must soon, like its population, equal that of
the land from which it draws so much borrowed money.
Colonel Coulson told me in the ' Acadia,' that he and other
officers could not help being annoyed in Canada, at witness-
ing the superior rate at which each new American city
founded on the shores of the great lakes, surpassed in a few
years in population, wealth, steamers, churches, and buildings, its British neighbour, enjoying all the same natural
advantages. I remember asking Mr. Mallet,[3] who rarely
fails to clear up my difficulties in political economy, to explain this, and he admitted that although the French Seig-
neuries, and other remains of feudal tenures, as well as a
colder climate, would account for some part of the contrast,
yet it would not do for large tracts in Upper Canada. Colonel
Coulson told me, that when he expressed wonder during the
rebellion, that land sold so high on the borders, he was

[3] John Lewis Mallet, Esq., son of the celebrated Mallet du Pan, for many
years Secretary of the Audit Board.

answered by a purchaser, that should Canada change hands, its selling price would immediately be doubled. I think I have found a complete explanation of this enigma.

The population in the United States is now increasing in the enormous ratio of 800,000 a year, of which about 100,000 only come from Europe. The great movement of colonisation westward is supplied mainly by New Englanders, all, whether *Locofocos* (Radicals) or Whigs (that is Conservatives), devotedly attached to their own institutions, and mostly anti-slavery politicians.

An estate in Ohio is worth double one of equally good land across the river in Kentucky, because the New Eng. landers will not colonise the slave state. So will they avoid British or foreign institutions in Canada, and people densely inferior land on their own side the border.

There is not so much drought here as farther east. Among the trees we are much struck with a fir tree called a hemlock, and with several kinds of walnuts, such as the butternut and hickory, and various kinds of oaks, and the sugar maple. I have seen one tree frog, and last night heard several catydids or grasshoppers, which croak as loud as frogs. There is a yellow flower—a *solidago*, which is as vulgar a weed as our Scotch weebow,[4] but to my eyes more elegant; a knotweed, which looks just like ours, is most abundant. Heaths are unknown, so is furze, and we miss our wee crimson-tipped daisy. The wild roses are past, but we have seen a few. We do not meet with many people of colour so far north, but without being an anti-abolitionist, I am sometimes half inclined to believe, that when the geological time arrives according to the system of progressive development for a being as much transcending the white man in intellect as the Caucasian race excels the chimpanzee, he will be puzzled in his work on natural history, when he comes to the order Primates, and has to decide whether the projecting os calcis and elongated ulna, and the more moderate share of intellectual endowment, must not force him to admit of two species, in spite of the innumerable crosses which we meet with here at every step.

Believe me ever your affectionate son-in-law,

CHARLES LYELL.

[4] Senecio Jacobæa.

To Leonard Horner, Esq.

Albany : September 21, 1841.

My dear Horner,—Mary has mentioned how much I was pleased with the Blossburg coal-field, which is the only formation except Silurian, Old Red, and Post Pliocene, that I have yet seen. Its analogy in fossils, white quartzose sandstones, bituminous shales, carbonate of iron, &c., with English coal-fields, is truly surprising. Dr. Saynisch[5] was an excellent guide. When Prince Maximilian of Neuwied was here, he travelled with him to the prairies, and the genuine descendants of John Bull were very intrusive in their desire to see a Prince, and used to march in without ceremony and ask the Doctor whether they could see the Prince. ' Yes, if you will pay twenty-five cents.' ' Nay, but I wish to see him.' Upon which the good-natured Max came out and said, ' Here I am,' and they went away disappointed to see a shortish man, in ordinary attire, instead of a magnificent hero in purple and gold. Dr. Saynisch practises gratuitously on the homœopathic system, and we saw him measure out some powders of infinitesimal quantities to a tall stout miner, adding injunctions about diet, in which I suspect lay the real force of the panacea. He likes the people, but told me some anecdotes in explanation of his not being in love with their institutions. He came with a strong predilection for the democratic party, but says they have cured him, not by homœopathic, but allopathic doses. He says it is the same with most German Liberals whom he knows. I have just been calling on the Governor of this State—his Excellency Mr. Seward—and expected to see a man of venerable age, but found him, as I do almost all men who are in active employment in this stirring country, a good deal younger than myself. He said he had already written to me, and he heard me very attentively on the subject of the Survey, and the best manner of getting out the final report with illustrations. General Dix, formerly Secretary of State, and originator of the Geological Survey, called on me, a gentlemanlike man of my own age, and talked on the same subject. I was glad to tell him how well I thought upon the whole

[5] President of the mine.

the plan had worked, especially in creating a set of practical
and scientific hammer-bearers. You see in many books on
the United States that the people are very good humoured
and full of jokes. The fact seems to be, that the great
throng of them are from fifteen to twenty years younger
than those you meet with playing the same active part in
Europe. Yesterday, as I stood in the bar-room of a wayside
tavern, on our return from Scholarie, the young proprietor
of a large pedlar's waggon marches up to the bar-keeper and
says, ' Can we have dinner ? ' ' Yes, for how many ? ' ' Eight
remarkably handsome young men.' ' Very well; in ten
minutes.' All day this sort of thing goes on as a matter of
course between perfect strangers.

I am very glad to hear that you have a geological paper
on trap dikes on the stocks; the width of some dikes in
Connecticut, and the manner in which they turned from
intrusive walls into intercalated and nearly horizontal sheets,
has interested me much, but the subdivisions of their Silurian
system have chiefly occupied me.

Philadelphia: September 27.—Here I am, working away in
quarries of greensand, and picking up belemnites and other
cretaceous fossils, with Conrad for my guide, whom I am
happy to find quite at my service, and the best-informed
paleontologist on this side the Atlantic.

You probably never heard of the Helderberg war, but as
it was the great talk while we were geologising among the
hills of that name in Albany County, New York, I must in-
form all you who live out of the world, that this is now the
third year's campaign. When old Van Renssalær died, the
last nobleman in the United States, his vast landed estate
was divided among three sons. A population of about
40,000 inhabited the share which fell to the eldest, who
endeavoured to get regular payment of his rent, besides
arrears, which his indulgent predecessor had allowed to
accumulate. Now the payment of rent being most unusual
in the United States, where every man farms his own land,
or what he calls his own, for it is often deeply mortgaged,
was voted quite an abuse of the old feudal times. They
thought they had paid long enough, and said with some
truth, that they had cleared a wilderness and created the

property themselves. So they would pay nothing. The laird appealed to the Sheriff, but they laughed at law processes, and at constables, till finally the Sheriff asked for troops from the Governor, who called on the Volunteers, who, as with us, proved, however constitutional a force, to be a most unfit one to interfere with temper and discipline. The farmers and peasantry turned out, erected barricades, and mounted some brass cannons, and have set the Government for three years at defiance, not one farthing of rent paid the while, and a party disguised as Indians maltreated a constable while we were there, and it was feared for a time that he was murdered. It is allowed that the landlord has mismanaged matters, and been far too uncompromising, but I must say, that after what the New Yorkers did in the Canada War, and since in Macleod's affair, I think there is a want either of energy in those who should enforce law, or of respect for it in the minds of the majority.[6]

With my love to all, believe me yours affectionately,

CHARLES LYELL.

To DR. MANTELL.

Boston : October 29, 1841.

My dear Mantell,—I was glad to hear of you from Dr. Silliman, who has probably told you of a visit which we paid to him and his very agreeable family at Newhaven. After staying two days with them we went by New York and the Hudson to Albany, where I began my explorings in the Silurian strata, and from thence I examined the valley of the Mohawk, in company with Mr. James Hall, who has been employed by the Government with four others, to survey the State of New York, which is about the size of our island. The Falls of Niagara were as beautiful as I expected, perhaps scarcely so grand, but in geological interest far beyond my most sanguine hopes. As I shall send a paper on the proofs of their recession to the Geological Society, I will not dwell on them now. After spending some time there, I examined seriatim all the Silurian groups in the Old Red and Coal on the borders of Pennsylvania. Returning to Albany

[6] See *Travels in North America*, by Charles Lyell, vol. i. p. 68.

I went south to Philadelphia, and spent four days in collecting in the different divisions of the Green sand in New Jersey, having Conrad as my guide. The analogy of the genera, and even of the species to the European chalk, is most striking. I went with Dr. Harlan [7] to see the great skeleton brought by a German, Koch, from the Missouri; a very large Mastodon which he calls the Missourium. He has turned the wonderfully huge tusks the wrong way— horizontally—has made the first pair of ribs into clavicles, and has intercalated several spurious dorsal and caudal vertebræ, and has placed the toe-bones wrong, to prove, what he really believes, that it was web-footed. 1 think he is a mixture of an enthusiast and an impostor, but more of the former, and amusingly ignorant. His mode of advertising is a thousand dollars reward for any one who will prove that the bones of his Missourium are made of wood. He is soon to take them to London, when you will have a great treat, and see a larger femur than that of the Iguanodon. Harlan is lost in admiration at the bones of this and other individuals, all belonging to the old Ohio Mastodon of Cuvier, from very young to very old individuals. He has also other fossils. Of my tour into the anthracite regions of Pennsylvania you will hear the results when the paper I sent Fitton on the Stigmaria clays is read. I like the people here very much, and have a most attentive class of about 2,000 both at my morning and evening lectures.

My lectures here will take me four weeks more, and my plan is then to run away from the winter so far south as to enable me to keep the field, examining especially the cretaceous and tertiary formations, and not to go northwards till the spring has fairly opened this fine country.

Believe me, my dear Mantell, ever most truly yours,

CHARLES LYELL.

[7] An eminent osteologist ; died in 1843.

CHAPTER XXIII.

JUNE 1842–DECEMBER 1843.

NIAGARA—NOVA SCOTIA—SUBTERRANEAN FOREST—UNFAIR STATEMENTS
ON THE UNITED STATES— RETURN TO ENGLAND — CONTEMPLATES
WRITING A JOURNAL OF HIS IMPRESSIONS OF AMERICA—MUCH GEOLO-
GICAL WORK IN PROSPECT—GIANT'S CAUSEWAY—LORD ROSSE'S TELE-
SCOPE—PRESCOTT'S ' MEXICO.'

CORRESPONDENCE.

To LEONARD HORNER, ESQ.

Lewiston, Canada : June 13, 1842.

My dear Horner,—We have come to this place from the
Great Falls to-day, having by the way enjoyed some excursions to the top and foot of the great precipice which
bounds the river, and I have found some additional evidence
of value to my mind, in favour of the recession of the Falls,
having traced the freshwater formation three miles and a half
down on the summit of the cliff, the old river bed.   I was
occupied a week in active exploration of the Niagara district,
and understood it much better.   The raised beaches of Lake
Erie are more difficult to explain than the Niagara ravine,
and I hope in descending the St. Lawrence to get some new
lights on these matters.   To-morrow we sail for Toronto.   I
shall now look back a little into my journal.   The contrast
of the disturbed and bent Coal, Devonian, and Silurian in the
Alleghanies, with the vast expanse of horizontal beds of the
same formations west of the immediate base of those mountains as far as we went in Ohio and borders of Kentucky,
is very interesting, and the magnitude of the coal-fields is
prodigious.   The new aspect worn by the same subdivisions
of the rocks below the coal in Ohio, as compared with what

I had previously seen of the same in New York, was very instructive. I continue to be struck with the general analogy of American and European geology, as regards the fossiliferous groups, which surprises me, it seems so much greater than between North and South Europe. I shall have a great stock of communications to make to the Geological Society on my return. We have been having remarkably cool weather at Niagara. There was even ice two nights during our stay there, but this was favourable to health and work. It has a strange effect when you have succeeded in obtaining some view of the Falls in which nothing appears but sky, wood, and water, and, when you are listening to the sound of the Falls, to be suddenly wakened out of your reverie by the loud whistle of a locomotive drawing a load of tourists and of merchants trafficking between the east and west, who discuss the Falls in three hours between two trains. Goat Island is the most perfect fairyland that I know. The views of the two Falls from it, and of the rapids which surround it, are delightful, and in the walks through its natural and aboriginal forest, you catch no view of the houses and mills which rise in the village on the United States side, bearing the ominous name of Manchester, and which may, I fear, ere ten years are past, extend some of its factories into the beautiful island. We purchased, on coming away, a daguerreotype of the Falls, and I think you will be surprised to see how well the sun has overcome the difficulty of the moving waters. What would I give for a daguerreotype of the scene as it was 4,000, and again 40,000 years ago! even four centuries would have been very important.

I am glad to have seen Ohio, as the finest example of rapid colonisation on record. The passage of manufactured goods and of the plough was at first from the Ohio northwards; but no sooner were steamers introduced into the lakes, than wealth and emigrants were landed on the northern shore from Lake Erie, and the central forests are now alone uncleared. It is supposed it might easily support a population equal to that of England and Scotland, but the movement westward prevents the filling up after the first occupation. The annual addition of people now made to the United States is just equal to that of the most flourishing State, Massa-

chusetts, exceeding 700,000, of which a small part, only a
seventh I am told, comes from Europe. It is a curious fact,
with which Mr. Ticknor surprised Lord Holland some years
ago, that the number of white persons now living in the United
States exceeds the total number of all who have been buried in
the United States from the landing of the Pilgrim Fathers,
and this may probably be true fifty years hence. I leave you
to work out the enigma. So long as the Pennsylvanian Germans and settlers coming from the Rhine predominated in
Ohio, no money could be obtained for public education, nor
for roads; but the moment the New Englanders had the
majority, a liberal tax for these purposes was assessed.
About a fifth of the whole population of Ohio is said to
speak German, but a much greater number are of German
origin. We found at Columbus, in the centre of the State
and the seat of Legislature, that there was a church with a
congregation of 300 to whom sermons are preached in the
Welsh language. In the Legislature of New Orleans, I
learn that speeches are made indifferently in French and
English. Is it not singular that the whole body of German
farmers, all proprietors of land, those anti-innovation men,
who would vote nothing for any improvements, are the most
ultra democrats, or rather they are blind tools in the hands
of demagogues? One principle avowed by their leaders is,
that schools and colleges naturally breed aristocrats, and
they therefore seize every opportunity of permanently voting
away those lands and funds which were originally set apart
by the central government for popular and university education. In various other States the same mischief has been
done, and a war waged against all corporation property and
State lands. In Georgia they have lately put the whole
State lands into a lottery, and every one drew for prizes,
and they could not give the usual grant to the Medical
College at Augusta. If some of the State debts are re-
pudiated, it will be the fruit of universal suffrage, for they
might all pay easily if they would tax themselves moderately.
Although I have been too busy with geology to have learnt
much in proportion to our tour respecting the politics and
institutions of the United States, I am already surprised at
the little that is known in England about matters here, where

there is so much worth imitating as well as avoiding. Part
of Ohio is called the Indian Reservation, and it is singular
that there is more wild game there than anywhere else, for
they will let no wild animals be disturbed except they want
them. You often see written up on the high road 'Mover's
House,' an empty wood building, in which emigrants may
pass the night. Another common inscription is 'Cash
Store,' which does not mean a bank, but that you cannot
buy in that shop for barter. The terms bakery and book-
bindery seem useful Teutonicisms. They have many mules
in Ohio, which I heard commended for their longevity, an
advantage I never thought of before. Ask Darwin if he is
aware of the hybrid having borrowed a portion of the ass's
length of days. What a perfectly intermediate creature !

Kingston, Upper Canada: *January* 19.—In looking over
part of this letter I see some inconsistency, and you might
ask how could the annulling of universal suffrage bring
matters right in Ohio, if nearly half the landed proprietors
are most ultra democrats. The fact is, that had the votes
of the low Irish and German Catholic emigrants who can
vote after five years, by a law which was an innovation on
the original charter of Washington and his colleagues ;
had these votes of the merest breakers of stone and hewers
of wood been cancelled at the last election, the returns
would have been entirely opposite, as these uneducated new-
comers, except a minority of Protestants, all go *en masse*
with the demagogues ; yet in spite of this and other evils,
I have not been in any country, where, if I was so unfortunate as to live out of my own, I could so well settle as in the
United States ; even in Ohio, as at Cincinnati. You would
find ample sympathy among the vast mass of more intelligent
Americans in your disapproval of all the more glaring faults
in their institutions ; nine-tenths of those with whom we
are thrown feel keenly the disgrace of any repudiation of
State debts, and had the New Englanders been left to
people the new States, without the influx of the dregs of
Europe, bringing with them violent anti-aristocratic and
Chartist feelings, great want of education, Irish sectarian
Catholicism, and other prejudices, their credit would now I
believe have stood as high as it does in the New England

States, where they are so thriving, in spite of universal suffrage. When once Americans see that you have a thorough respect for them and their country, they at once throw aside their optimism. For example, on my remarking to a merchant of Cincinnati that their Presidential elections gave rise to so much intrigue, bribery, and corruption, as to reconcile any one to monarchy where it was established, 'You are right, sir, and our Presidents have more power in the way of patronage and vetos than your kings ever had. The worst pages in English history are the wars of a disputed succession, but ours is one continued civil war of the same kind, and if the contending parties do not kill each other in the body, they do all they can with bribes and threats to do injury to their morals and principles.' You and I would hear more in good society here (in Canada) in one week, which we should consider narrow-minded and prejudiced and ungenerous to foreigners, in matters of politics, religion, and political economy, than we heard in nine months in the United States, for they have here all the Kleinstädterei of a colony and the enmity of the borderer, added to everything that you might disapprove of which they bring from home. They know very little of the United States, and do not wish to know more ; but of course there are many exceptions, and allowance must be made for the abominable interference of the uncontrolled American mobs in the late rebellion here, where the United States sympathisers brought cannon from the State arsenals, and the collision of the violent democratic party, of those born and bred in the new clearings, with the aristocratic feelings of capitalists and officials who come straight from Europe. But I must conclude, and with love to all,

Believe me ever most affectionately yours,

CHARLES LYELL.

To HIS SISTER.

Truro, Nova Scotia : July 30, 1842.

My dear Marianne,—We have just returned from an expedition of three days to the Strait which divides Nova Scotia from New Brunswick, whither I went to see a forest of fossil coal-trees—the most wonderful phenomenon perhaps that I have seen, so upright do the trees stand, or so perpendicular to the strata, in the ever-wasting cliffs, every year a new crop being brought into view, as the violent tides of the Bay of Fundy, and the intense frost of the winters here, combine to destroy, undermine, and sweep away the old one— trees twenty-five feet high, and some have been seen of forty feet, piercing the beds of sandstone and terminating downwards in the same beds, usually coal. This subterranean forest exceeds in extent and *quantity* of *timber* all that have been discovered in Europe put together. The new deposit of red mud of the numerous estuaries here affords me endless instruction. At this place, Truro, the tide is said to rise seventy-five feet. So we see the bottom of a deep saltwater sea, its rippled sands, shells, the holes of *Mya* and *Tellina* and their tracks, footmarks of birds and worms, the manner in which the clays crack and are marked by rain, and sometimes shells enclosed recently in solid nodules of claystone. I have also learnt more about the *geological* effects of drifting ice in the last ten days than in all the Canadian tour. The people here are quite loyal and monarchical, but their manners just as in the States. Mr. Lowell remarked to me one day, that nothing so much surprises the American readers of the ' Quarterly Review ' as to see so much which they know to be purely the result of circumstances, not political in the United States, attributed exclusively to the difference of their institutions from ours. There is so strong a conservative party among the rich and literary class in the United States, that the constant bitterness of the ' Quarterly Review,' ' Blackwood,' and other journals on the same side, and newspapers, against their country, which they resent from patriotic feelings, works precisely in the way which an ultra democrat, and still more a person opposed to an established church, would wish. When one reflects in how short a time, a time which even the present generation will live to see, the population of the United States will exceed ours, how short-sighted is all this! for the ideas of the United States must soon react much more powerfully on the English part of the Old World than they have done yet.

We have seen a great many woods, in which the low *Kalmia angustifolia* in flower purpled the ground almost as

much as the large heath in Hampshire, and reminded us of
the entire absence of heaths. The variety of wood here is very
great, but the trees not so large. The quantity of birch and
fir is the most novel feature. The quantity of rich marsh
land gained from the sea forms the chief wealth of the
region we have seen. You can no more let land here than
in the United States, for anyone may obtain acres of his
own nearly as cheap. This circumstance alone, besides the
absence of any hereditary aristocracy, and a well-endowed
established church, would assimilate this or any other genuine
English colony to one of the States of the Union.

Your affectionate brother,

CHARLES LYELL.

To GEORGE TICKNOR, ESQ.

Kinnordy: October 12, 1842.

My dear Mr. Ticknor,—When I wrote a few words to
you last week I was busy with the preparation of my paper
on the Niagara district. On reconsidering my journal, I find
a good many subjects of general interest to the naturalist as
well as to the geologist, which would not easily enter into any
of the twelve or more distinct subjects, such as Niagara,
fossil Mastodons, Alleghany Coal, Nova Scotia Coal, Martha's
Vineyard, &c., some of which have been, and others are to
be, read at our Geological Society, abstracts appearing in
their proceedings.

It struck me that in a journal of my tour a place could
be found for these matters, and a few of my impressions of
the country and people, omitting names of persons, except
my scientific fellow-labourers, and generally all things not
scientific. On making notes this last week, for this journal,
or 'personal narrative,' I soon found that there would be
danger of its growing to a book, which, as I gave all my
thoughts in the United States to my own science, would do
me no credit, though a bookseller would prefer it to the
scientific ballast which might be overcome by the specific
levity of the first part of the performance, and so float up
the vessel. Yet, I cannot bear not to take the opportunity,
not only of telling my scientific friends what route I took,

and how long I tarried, that they may test my opportunities of observation—but also of saying how I liked and what I thought of the people and country, as I ran through it on the railway, or the deck of the steamer. It may be said in a few words, thought I; so I made a few notes of my voyage and tour to Niagara, Blossburg, and Boston, with quotations from my journal, on the exhilarating effect of viewing the signs of such rapid progress in population, and the railways, churches, and school-houses among the stumps, &c. But I know by experience the remark which this would provoke here. 'It is very easy to go ahead with other people's money.' The answer is, that all the States of New England and New York, traversed in that first excursion, have either no debts, or have paid the interest of their debts, or have made railways with their own money, and their credit in trying times has stood high; yet their form of government is as popular as anywhere. But I must not shirk the 'Helderberg War,' but allude honestly to the only shady side of the picture forced on my sight in the first six weeks.

The New York people admitted it was a disgrace, just as we do when in Ireland the sheriff and soldiers fail year after year in distraining for rent, and ejectments are defied by the physical force of a mob, and are often permanently defeated. There are only two or three other difficult points which have yet struck me as requiring to be entered into, in order that I may not be addressed by the critics in the words with which Lord Lovelace met me soon after I came back: ' So, Lyell, I understand you have returned ipsis Americanis Americanior ! '

First, as to Repudiation: your letter on Sydney Smith only expressed what persons of opposite parties, and all whom we lived with, had said. As to gambling speculations, not guaranteed by States, they are like our bubbles of 1826, our late Australian bankruptcies, &c.; but the suspension of interest by Pennsylvania and other States, I attribute to universal suffrage and the votes of aliens. But I may add the similar extension of franchise would immediately cause the non-payment of the interest of our debt; therefore, the wonder is that with these institutions any States pay, and that the Central Government can still borrow. It seems to imply a higher standard of education or morality, or less

poverty. I would enter on this subject à *propos* to the great
distress I witnessed in Philadelphia, partly from the alarm
of the stoppage of the State dividends, partly the United
States bank. On this side of the water the sufferers think
that Brother Jonathan has come off a winner by their losses,
and they are surprised and softened when I tell them how
great has been the calamity on the other side of the Atlantic.
Perhaps I may say that, in proportion to the small capital
which the Americans have to invest permanently for incomes,
the ruin was as severe with them as with us. But then
comes the hardest task of all— the Southern States. No part
of my tour was more agreeable and instructive, geologically,
than that south of the Potomac, and it has much changed
my feelings towards the planters, however much I may
think of slavery as I did before. The domestic and farm
slaves whom we saw were a cheerful, often merry and light-
hearted set ; childlike, conceited, boastful, but not a suffering class, when compared with what may be witnessed in
Europe, and at home. They are uneducated, and not in the
way of being improved or raised in station. I had them
often with me for days, and neither saw nor heard of ill-
treatment. It is anything but an economical system, unless
where rice-grounds, sugar, or cotton, and crops which I did
not see, may make it desirable to have an animal that can
stand the climate. The evils to the whites are innumerable.
If poor, there is no place for them ; if rich, they have to
submit to the indolence and inefficiency of their slaves, to
doctor them when ill, support them when bedridden, guard
against their being excited by Abolitionists, &c. Their
children are corrupted by them, being made vain by flattery,
spoilt by power. If I give the favourable part of such a
picture, and enumerate the difficulty and danger of attempting any reform, because convinced that foreign and American
interference has been hitherto injudicious, I might throw my
mite into the wrong scale. When with the planters, seeing
their kindness to the slaves, and feeling that had I inherited
their estates, I should not well know what to do ; I could not
but feel that a London emancipation meeting, or a list of
advertisements by Dickens, raked out of newspapers from
all parts, would irritate and indispose me to exert myself

in forwarding the cause of emancipation. Sydney Smith said to me one day: 'But you should hold up the system to the reprobation of mankind.' I replied that it must be a work of time, sacrifices must be made, and the philanthropists ought to share them with the planters. Then comes the objection that an indemnity to purchase the liberty of all the slaves would exceed the means of the whole Union. This only proves that the Abolitionists ought to have confined their efforts within practicable limits. Might they not have begun attending to a few northern slave States, most ripe for a change to the better? As to our English meddling, it is like repeal meetings in America; and Lord Lansdowne said very well, when Lord Ashley's *exposé* of the collieries was made, they had better have been looking nearer home than speechifying about the miseries of the American blacks. A candid anti-slavery philanthropist from England in the midst of a black population, and asked by a planter what he would do if his British ancestors had bequeathed to him such an inheritance, would generally, I suspect, feel taken aback.

Perhaps by this time you are beginning to feel the same sort of surprise and alarm about my journal that I might feel were I to receive by next post a letter from you, announcing a project of a work on the literature, arts, and manners of Spain, Italy, &c., with your first impressions on the *geology* of those countries. But I beg you to consider that I am now setting before you all the most doubtful and difficult topics, which will only come in incidentally, if at all, in the journal devoted chiefly to remarks and speculations on natural history and science.

When they are placed in their niches in a large building, they will shrink, I hope, into their due proportions, or may be omitted many of them altogether. I should be glad to give vent to the sensation of freshness, cheerfulness, hope, and delight on first visiting America, and seeing such a glorious prospect of rapid progress in knowledge and civilisation, a feeling which I retained to the last; and my work will not be the worse received here, from my being able to declare how much I found of a kindred spirit in the state of society and institutions of our Queen's most loyal subjects in Nova Scotia. I could wish that what I draw up by-and-by,

when I have heard from you, should be shown to some candid Southern man. I do not think that these points of
difficulty would come once in twenty pages in that fourth or
fifth part of my volume which is not made up of purely
scientific memoirs. I shall have to talk of the sea-shells in
Boston harbour being in great part like European species, of
forest trees, humming birds, the absence of heaths, azaleas,
distinctness of plants and animals, tortoises, alligators, &c.
After all, I was so much in cities, with men who were not
mere naturalists, and have had since my return to talk right
or wrong on United States affairs, with those who know less of
them than me, that I should like to have my say. A manager of the Maryland Iron-works, when I stood up against the
tariff and for free trade, said: ' I grant all your political
economy, and you may be right, if to increase wealth and population was the only good; but I ask you, as a literary and
scientific man, whether you want to have all the United
States property represented by farmers? You have seen
them in New York, in Connecticut, in Pennsylvania. By a
protection to manufacturers you get a higher class into the
country. Iron rails will cost somewhat more, but do you wish
for that dull monotony of a nation of small landed proprietors? Here, in Maryland, if anyone does by chance accumulate a little capital, his luxury is litigation, and it soon evapo-
rates.' *À propos* to this I was told that if a Connecticut
farmer sold off, and invested all his money at six per cent.,
his average income would be no more than 80*l.* to 120*l.* a
year in English money. If so, they are below the average
farmers with us, who, though not proprietors, are worth
more. Mr. Jared Sparks told me, I think, that Philadelphia
would have been the capital of the United States, *vice* Washington, if the revolutionary soldiers in arrear of pay had not
been alarming to the Legislature.

Nothing astonishes me more in reading most books on the
United States, and hearing people talk, than to see how they
attribute to democracy and your institutions both the evil
and the good which existed under the monarchy. If in
deference to your judgment, or that of any other in whom I
place equal reliance, I were to say nothing of these topics, I
should at least be able in conversation to talk with more

effect on the United States affairs. So many people besides Sydney Smith have said that I ought to give them a book on what I saw and thought *de omnibus rebus*, that the presumption of saying a little in the course of the journal on moral and political matters seems less than it did. But, on the other hand, I feel that I ought to have seen schools, and attended legislative assemblies, and so forth, and that I might say, what is the truth, that I avoided giving any time to collect facts and observations on these heads, and therefore feel that I am not entitled to speak on them. Meanwhile, I shall try my hand at the ' Journal of a Naturalist and Geologist,' and see how it promises.

Believe me, ever most truly yours,
CHARLES LYELL.

To the REV. DR. FLEMING.

Kinnordy: October 17, 1842.

My dear Dr. Fleming,—We returned after a delightful tour in the United States in August last, in nine days and a half from Halifax, after a month's tour in Nova Scotia.

In all we were absent thirteen months, less than one of these being spent on the ocean, nearly ten in active geological field work, and a little more than two in cities, during which I gave by invitation some geological lectures to large and most patient audiences.

We leave this for London on the 25th instant, but shall be here till then, and, my father and mother desire me to say, shall all be most happy to see you and Mrs. Fleming at Kinnordy, if on such short notice you will do us the friendly act of accepting the invitation. I shall then be able to tell you *vivâ voce* what it would be in vain to attempt on a sheet of paper, of what I saw in the other world.

I shall also have some news to tell you of the Balruddery lobster, alias cherubim.[1] Suppose the ' blackberries ' which accompany it are the said lobster's eggs ?   But I must not indulge in these matters, but pray you to come, and with my wife's regards, believe me, ever most truly yours,

CHARLES LYELL.

[1] This lobster was the crustacean *Pterygotus*, and the ' blackberries ' called *Parka* are believed to be its eggs.

To PROFESSOR SEDGWICK.

Kinnordy, Kirriemuir: October 18, 1842.

My dear Sedgwick,—As we so rarely have the privilege of seeing you now in town, as of old, I shall not trust to meeting you in November, though I hope we may, but write to ask some news of your proceedings and your health. We had a very good muster of working geologists in May last, at the anniversary of the Geological Association at Boston, and they have many excellent men, as both the Brothers Rogers, Hitchcock, Vanuxem, Hall, Emmons, Conrad, and others. I have brought home three dozen boxes of specimens, the fruits of nearly ten months' field work, and I shall open my budget in November, with a paper on the lake ridges, as they are called, and the elevated beaches of the Canadian lakes, with new facts obtained during a second survey of the Niagara district, and bearing on the recession of the Falls, and on the drift of the valley of the St. Lawrence. I shall afterwards have to open ground on the tertiary, cretaceous, coal, and older rocks. I was very glad that you joined my friend John Carrick Moore in Galloway, and we heard sundry reports of certain fair ladies having rejoiced in the bad weather, which kept the philosopher indoors. So you see how these scientific news travel across the Atlantic. Your announcement of coal plants in the Rothliegendes at Whitehaven interested me much, and I should like to talk to you of the zoological fossils, if any, in the same beds. A lady sent us the important intelligence last week that Professor Sedgwick had found organic remains in granite, which her reverend informant had told her was very satisfactory, as overturning all received geological theories respecting the antiquity of the earth. This story has perhaps some foundation, and you have really hit upon some novelty in the old rocks this summer? We are here till the 24th instant, and hope on the 28th to be in Hart Street, at the old family mansion. Let us hear from you there or see you. My father and other members of the family here desire to be remembered to you, and with my wife's kind regards, believe me, my dear Sedgwick,

Ever most truly yours,
CHARLES LYELL.

P.S.—Remember us to the Master of Trinity College when you see him.

To SIR PHILIP EGERTON, BART.

December 21, 1842.

My dear Sir Philip,—We should have been very glad to pay you a visit, which we have really long looked forward to with pleasure, but must wait, and be a little stationary after such wanderings. To have examined the fish teeth with you would have been most improving, and just what I want, but I must really work in my cabinet to subdue my great mass of arrears. Lord Enniskillen will tell you that on Wednesday last I gave out nearly two hours of my American budget, and return to the charge next time, but I shall not open upon Big Bone Lick till after you are here. I send more than three hundred separate squaloids, but beg you will not bother yourself if you find the task tedious. I should be most glad if you would look at the Martha's Vineyard, marked ' M. V.,' because I am clear for this formation being post-eocene, whereas the geologists over the water have been disputing whether it was cretaceous or eocene. Most of the teeth are lettered.

Lonsdale writes in better spirits, and is collecting and observing the living polyparia.

Ever most truly yours,
CHARLES LYELL.

To HIS WIFE.

Bristol : August 13, 1843.

My dearest Mary,—My journey answered very well, as I had my back to the engine, and therefore escaped the air in my face, and the weather was delightful. I had some intelligent companions : one of them, a Bristol solicitor, told me that the twelve miles of rail from Bath to Bristol cost

80,000*l.* per mile, near a million sterling. It is a fine specimen of Brunel's stylish way of engineering—such viaducts
over the river and tunnels! We went fifty miles an hour for
several miles, and once for two miles more than fifty. Soon
after my arrival I was with Mr. Stutchbury, seeing coal
plants in the museum, then with him to Durdham Down,
where the ' Permian ' saurians [2] were found, &c., then home
by the Hotwells, Clifton. Next day with Stutchbury to
Bath on the rail, and then eight miles south to Radstock
coal-mines. Went down some hundred fathoms in an
iron bucket, and spent five and a half hours underground,
going miles in coal galleries; much delighted, as the mines
were dry and in a good state—upright trees, roofs full of
ferns, &c.

I hope to meet Ramsay at Bath, and may catch your
train at Bath, or more likely midway between Bath and
Bristol. I am grateful to think my eyes are keeping well in
spite of fossilising.

With love to all of you,

Your affectionate husband,

CHARLES LYELL.

To HIS SISTER.

Giant's Causeway : September 4, 1843.

My dear Marianne,—We have just returned from a walk
over the grand pavement, the effect of which was as picturesque as the evening sun and some white breakers
rolling and foaming over the black rocks could make it.
Much as I have been pleased with the sight, it strikes me
that there are parts of Staffa away from Fingal's cave, and
which travellers have seldom leisure to visit, which are even
finer in precisely the same style.

The geology of Antrim is very interesting, so many
formations, such as chalk, green sand, lias, new red, and the
coal being each represented by such distinctly characterised
and yet such thin sets of strata compared to the same groups
elsewhere, and then the grand trap or basaltic mass covering
and cutting through all. The people of this more northern

---

[2] These conglomerate beds and their remains are now considered at the
base of the new red sandstone.

region, like the heather-covered hills, are wonderfully more
like the Scotch than those of the southern counties—less handsome and to us less interesting. They are (I speak of the
humbler classes only) not so foreign, but like Scotch who
ought to be more thriving, who are content to be in rags
when you perceive that their circumstances ought to make
them above it. But we have been for the last two days very
much among a Catholic set, and whether these are more
degraded I know not. The begging continues in spite of
great poor-houses, which, by the way, are far better in an
architectural point of view than those in England. I think
the sordid penury of the Irish which Von Raumer described
must have been exaggerated, for he had only time to take a
superficial view, as we do. All agree, however, that the
habits of the people in regard to temperance have worked
wonders even since Raumer's time. I have much more
hopes of them after having seen the country. How fortunate
are we that we have not a black slave population to turn into
true and quiet subjects, instead of the quick, obliging, and
fine-looking natives of this green island, properly so called,
for the verdure is very remarkable ! I got a new light about
repeal from Lord Rosse. The farmers in Connaught had
become so annoyed at the exactions of the priests that a
strong feeling of resistance was gaining ground, which would
soon have spread, as the same spirit was manifested elsewhere, at Parsonstown for instance. O'Connell was able
much more easily to get the priests to join in the repeal
agitation, in order that by that greater excitement they
might be led to forget the other grievance. This has
answered for the time, but it shows that had Government
paid the Catholic priests, as they do the Irish Presbyterians,
the farmers would not have tolerated their Romanist church
dues, and the priests would have been forced to receive
support from the State, and would then probably have been
no longer rebellious. In short, until they are raised in the
scale there is no hope of their flocks improving, all which I
presume Sir R. Peel and the Duke have long known, but
they would never have carried any measures. One cannot
help fearing that the anti-English spirit has sunk deep into
the hearts of the millions here, for they read nothing but

O'Connell's newspapers, from which he artfully excludes, without appearing to them to do so, every other foreign or domestic topic of interest except repeal and Irish grievances, a great proportion of them now bygone. You will be glad to hear that even if the gigantic telescope of Lord Rosse should not succeed— of which, however, there seems no apprehension— the great one which we looked through has achieved grand results, not only reducing nebulæ into clusters of distinct stars, but showing that many of the regular geometric figures in which they presented themselves to Herschel, when viewed with a glass of less power, disappear and become very much like parts of the Milky Way. Lord Rosse showed me a model which he had made of one of the mountains in the moon, which I thought not so much like a volcano as one of the largest atolls, its sides being externally so steep and lofty and its crater sixty miles in diameter; but you must suppose the lagoon of enormous depth, and the ocean, of course, to be removed. During our call on the Archbishop of Dublin he said, among other things, that he thought that brutes used language for communicating ideas as we do, but never, like us, as an instrument of thought. We were much diverted with a story told us of one of his eccentric ways of taking exercise. He used to go to the seashore with Dr. Dickinson, the late Bishop of Meath, and they would go on for three hours together throwing up pebbles into the air, each trying to hit the other's pebble. I have no doubt that Whately was sometimes speculating on the doctrine of chances, and calculating how many misses went to one hit; but his chaplains were in several cases obliged to interfere, and succeeded in representing to him that some of his gymnastics were not in Dublin thought quite dignified enough in an archbishop. Our scientific meeting at Cork went off pleasantly and very well as far as concerns the muster of scientific men, but the money received was not near half the usual sum, owing entirely to the gentry of the neighbourhood and county holding off, except the Lord Lieutenant and a few others who had originally joined the town in inviting us. The reason of this was that the townspeople, comprising many rich merchants and most of the tradesmen, were repealers, and the agitation having occurred since we were invited, the opposite parties could never in Ireland act or pull together. In addition to this, however, the gentry in Ireland, at least in the south, seem very much behind those in England in interest for any scientific matters. We are struck with the similarity of the common flowers and plants here to those of England, Scotland, and France. Certainly the naturalist has a great additional pleasure in touring through Canada or the United States in the surprising novelty of all the wild plants and animals. We had an amusing scene in starting from Limerick, the coach being already full, and a petitioner addressing the guard very eagerly with, ' Will I come up ? that's the proposition.' To which the reply was, ' Where can you sit ? that's the query.' The other saying he would go on the baggage at the top, he was told to lie flat, or his head would be taken off going under the arch, and by way of comfort he was asked, when in a great fright, if his life was insured. ' No.' ' So much the worse for your wife ! ' All this passed as quick as thought. We have seen many remarkable round towers, as at Kildare, like the Brechin tower but taller ; also many ruins of monasteries. The school-houses, often very neat, pretty buildings, put us in mind of the United States, and we are told that the Catholics and Protestants get on well together in these schools, and are each allowed their own version of the Scriptures.

Since I began this we have seen more of the Causeway in delightful weather, but I shall leave Mary to tell you of this. With love to all at Kinnordy,

Believe me, your affectionate brother,

CHARLES LYELL.

To JOHN CARRICK MOORE, ESQ., CORSWALL, STRANRAER.

Kinnordy : September 17, 1843.

My dear Moore,—We made out our tour very successfully, seeing the Ballantrae section, and going the old road by Alloway, so as to see the witches' bridge and the ruined church where they danced, and Burns' monument, and the cottage where he was born, and still just reached the five o'clock Glasgow railway train, so that we got a good long night's rest before starting the next morning by six o'clock. We found the two parcels had come safe by the ' Maid of Galloway,' and were already in the hotel. Had we not been free of these incumbrances, and had not the cars ordered by you been all ready, we could not have accomplished all this by five o'clock at Ayr. The geology at Ballantrae [3] was quite what you had given me to expect—a small dike of serpentinous trap about one foot broad, and running magnetic N. and S., being the only fact which you seem to have missed. I certainly think the red sandstone is newer, as your white and variously coloured sandstones of Sloughnagarry are older, than the conglomerate with fragments of graywacke. As to the said conglomerate and the red sandstone, we can say nothing more of its age. I saw ripple marks and casts of cracks, and perhaps fucoids (?) as you did. As to crustacea I could not detect any good markings to confirm that idea. But I saw distinct rain drops. Study the recent ones on the shores of Lough Ryan in all states, after a short shower and a hard one, and after hail, and when half effaced by tide, &c. In short, that Lough is a grand magazine of geological analogies—tidal, littoral, conchological, sedimentary, &c.—which I envy you having at your door. My father saw here a splendid meteor, and other neighbours did at the same moment, as large as a man's head, more splendid than the moon, on Wednesday the 6th, at a quarter past eleven o'clock P.M. One person thought he saw it divide. The Milmans arrived here a few hours after us yesterday, all well, and in fine weather, their three sons with them.

My wife joins in kind remembrances to your whole party, and we are glad to have finished our tour with the recollection of so agreeable a visit.

Pray tell Mrs. Moore they have had the good taste to put the sculptured Tam o'Shanter and Sutor Johnnie in an apartment separate from the monument to Burns.

Yours very sincerely,

CHARLES LYELL.

---

[3] On the coast of Ayrshire.

To SIR PHILIP EGERTON, BART.

Kinnordy, Kirriemuir, N.B.: September 22, 1843.

My dear Sir Philip,—I was in hopes of learning from the Milmans, who have just left us after a visit of five days, in what part of the world you and Lady Egerton might be at present, whether to the north or south of us, but as I could get no intelligence I shall send this to Cheshire to be forwarded if you are in the Highlands. We are only a week returned from Ireland, having gone from Killarney to the Giant's Causeway after the Cork meeting, visited Lord Rosse on the way, and seen his telescope, besides seeing other Irish friends. My Auvergne tour was also very successful, and I shall be glad to talk with you on the wonderful collection of mammalia, of one hundred and sixty five extinct species of all periods from the eocene to the pliocene, which the Abbé Croizet and Bravard have disinterred from the old fresh-water marks and the volcanic alluviums, &c., of all ages. But I have only time in this letter to beg you to let me know whether you shall be returning from the Highlands to the south before the first week of November, or going northwards from England, as in that case I should like to send you a special invitation from my father to stay here some days on your way. The only visit we have to pay is to Aberdeen, and I might regulate the time of that by your movements, if you would write immediately. We are enjoying delightful summer weather here, after having had already two summers, one in France and one in Ireland.

Ever most truly yours,

CHARLES LYELL.

To HIS SISTER.

December 7, 1843.

My dear Marianne,—We went yesterday to Mr. Curtis, who named for me my N. American insects. The fire-fly is a *Lampyris*, very like the Italian one which I saw in 1818 in the rice-fields near Padua. I am going to-day to visit poor Mantell, who is very ill. I am getting on very steadily with

my geological travels in North America, and am particularly
working at the fossils, which can only be done by getting
the assistance of many others in different branches. Edward
Forbes has helped me for my chalk fossils of New Jersey, and
I have packed up some twenty packages of American corals
from different rocks, and sent them off by Lonsdale's invitation to Falmouth. Having seen in the United States that
there was no one who understood the corals, and only two men
in France, I willingly let these specimens run the risk of
another 600 miles of travelling, though I wish my naturalist
was nearer. To class and pack them was no small work for me
and my new aid, young Sowerby ; I miss Hall[4] in a hundred
little arrangements. Three Institutions at Manchester
clubbed together, to invite me to give twelve lectures at each,
repeating the same. Mr. James Heywood was deputed to
negotiate with me. I told him that, had my occupations permitted it, the average sum per lecture was about one-fifth of
the fee for which alone I could think of lecturing anywhere,
or had done so, in the last three years. The fact is that
there, as everywhere in England, magnificent subscriptions
which might have endowed lectures equal to the Lowell
Institute, have been all spent in building, and it is the same
whether 50*l.* or 150,000*l.* are subscribed in this country ;
and as in the case of the London University and King's
College here, the buildings, which cripple them with debt,
remain unfinished. If anyone should have the wisdom
of Mr. Lowell, and forbid a farthing to be spent on building,
an institution with half the funds of the K. C. or L. U.
would have the pick of all the first teachers, and leave the
rest behind.

　　　　　　Believe me, your affectionate brother,

　　　　　　　　　　　　　　CHARLES LYELL.

To GEORGE TICKNOR, ESQ.

　　　　　　　　　　　　London : December 27, 1843.

My dear Mr. Ticknor,—We are reading Mr. Prescott's
' Mexico ' with great delight. I have not seen Milman's

---

[4] His clerk, who had been eighteen years with him, and who died of con-
sumption the previous month.

review in the 'Quarterly,' but he has been so much overworked lately with parochial and other professional duties,
by which his health has suffered, that I shall be relieved
to find the review worthy of Milman and the subject;
at least the reviewer was much pleased with his task. The
parts on which Mr. Hallam, when he talked of it to me the
other day, dwelt with high commendations, were the essays
in the beginning, and the appendix on the Aztecs, the abori-
gines, &c. Everybody seems to be reading it. I was much
pleased with the comparison between the human sacrifices of
the Spanish Inquisition and the Mexican priests, and the
justice done to Montezuma for not becoming an apostate.
I am very much obliged to you for your information, and that
sent by Mr. Sparks, and had I not been very busy in preparing for the Geological Society papers on the plumbago of
Worcester and the green sand of New Jersey, I should have
digested the matter you sent me, and read up to it by aid of
your references. It strikes me that it may do good to
remind our ultras here (who could forgive the Americans
more easily for dispensing with a King than for doing without a church establishment) that in the monarchical times
of the States all sects were, as now, on an equality.

If I can state this broadly it would astonish many. It
was new to me, and people are amazed when I tell them that
now, in Nova Scotia, all sects are equal by the charter.
What struck me was the advantage in a country open to
colonisation of finding no other sect dominant, and one's
own politically and socially inferior.

It is not *ungenteel* to be a Baptist or Unitarian. One may
know that the orthodox of other creeds believe that one has
no hope of salvation in the next world, but one is not irritated
by the thought of being despised and thought to hold a low
position in this. But was there not once in some part of
New England a general provision, afterwards abandoned, for
all religious sects, a tax levied equally from all, and paid, as in
France, to different religious communities, according to their
numbers? Was there nothing of this kind anywhere in the
United States? It is my *beau idéal* of a *mezzo termine* between
the Establishment and the voluntary system, and I believe
we have lately begun this in Australia, but I must inquire.

More than half a million have lately seceded in Scotland from
the Establishment, and as it is very possible that the leaders
of the Catholics and Protestant dissenters may in some
future election bury their religious animosities, and make
common cause, the time may be nearer than some think when
we shall have all sects endowed, which I trust will happen,
instead of none being so. But, at all events, I abhor the
political disaffection created in Ireland, Scotland, and
England by the exclusive privileges of Church of England
ascendency.

It is really the power which is oppressive here, and not
the monarchy, nor the aristocracy. Perhaps I feel it too
sensitively as a scientific man, since our Puseyites have excluded physical science from Oxford. They are wise in their
generation. The abject deference to authority advocated
conscientiously by them can never survive a sound philosophical education.

Ever most faithfully yours,
CHARLES LYELL.

CHAPTER XXIV.

JANUARY, 1844—DECEMBER, 1846.

LETTER TO MR. TICKNOR ON AMERICA — FOSSIL BOTANY—GOES WITH
FARADAY ON A COMMISSION OF INQUIRY ON ACCIDENTS IN THE COL-
LIERIES—BLANCO WHITE'S BIOGRAPHY—NEWFOUNDLAND—ICEBERG—
POLITICS IN MAINE — VEGETATION IN GEORGIA—SLAVERY — CHEIRO-
THERIUM—ICEBERGS—LETTER TO EDWARD FORBES, AND REPLY—DINNER
AT MILMAN'S—VISIT TO BOWOOD.

[In the autumn of 1844, a disastrous explosion took place at
Haswell Colliery, Durham ; on which the Government instituted an
inquiry, and at the request of Sir Robert Peel, Mr. Lyell and Mr.
Faraday went as Commissioners, and prepared a report on the causes
of the accident.

He attended the British Association at York this year, and in
1845 he published his 'Travels in North America, with Geological
Observations.' In September he returned to the United States, and
had a still more extensive tour in the countries mentioned afterwards
in his ' Second Visit,' some of the geological fruits of which have also
appeared in separate scientific memoirs. He was nine months absent from England, returning in June, 1846. He attended the British
Association at Southampton, and in September moved to 11 Harley
Street, after spending fourteen happy years in Hart Street, Bloomsbury, but his collections demanded more space.

His travels in 1847 were confined to England and Scotland, and
he attended the British Association at Oxford.

In 1848 he visited Southampton and other places in England, and
went to Scotland. From Kinnordy he rode over the hills by Clova
and Loch-na-gar to Balmoral, when he had the honour of being
knighted by the Queen, in September.]

CORRESPONDENCE.

To George Ticknor, Esq.

16 Hart Street, Bloomsbury, London : January 7, 1844.

My dear Mr. Ticknor,—I found your letter on our return from Cornwall. Your information respecting the slaves, and your speculations, were most interesting to me. I had arrived at the notion, from reading Gurney's book on the West Indies, that in the most Southern states, and wherever rice and sugar are cultivated, the emancipated negro might stand the competition of the white labourer.

But in all the higher parts, and many of the low grounds of Virginia, Maryland, the Carolinas, and Georgia which we saw, I should expect the free black to give place, like the Indian, before the white immigrant.

You mention Massachusetts having refused magistrates and use of prisons to those who followed runaway slaves. I should have thought it good policy to give some indemnification in those cases, not the whole value perhaps, and to aim always at sharing the sacrifice required by humanity with the planters, and ultimately forcing the Central Government to incur a debt to set free one state and then another. Mr. Everett told me that black men did vote for the governor of Massachusetts. De Tocqueville says that the free blacks do not dare to vote. He also says that they die out much faster than slaves. If so, must it not be for want of an adequate poor-law system ?

I thank you much for the admirable pamphlet 'The State Debts,' which it is painful to read, and reminds me of the state of indignation which I was in at Naples, when associating with a set of proscribed literary and scientific men, who seemed to me very moderate in their political opinions, but who were so far in advance of their own countrymen and government as to be martyrs to their patriotism.

There was an absolute power over which they had no control, and which had no sympathies in common with them.

I should do injustice no doubt to the legislature at Harrisburg were I to compare them to the corrupt government of Naples ; still I could not help feeling that the men of finer intellects in Philadelphia were under the control of men of coarser clay, delegates of the Pennsylvanian rural population. On them I cannot help fearing that the manly and spirited and touching appeal of your friend will be thrown away, yet I hope most sincerely that I am mistaken. I had an argument the other day with one of Lord Ashburton's suite (in 1842) about the cause of repudiation, I attributing it to the ignorance of the lowest and most numerous class of voters, especially those newly arrived, or who do not advance with the rest, because of a different language, &c., he maintaining that it was simply the sharing of the shame among so many which alone prevents communities from dishonesty, and which no ministry or moderate-sized constituency could endure or brave. Surely it is not a question whether popular governments can be virtuous and honourable, but whether the democratic principle has not been carried out so far as to subject the educated to the uneducated, those who have independence of fortune and leisure to those who have neither. Is it Utopian to try to raise the franchise and exclude those not born in the United States? When were settlers of five years' standing first admitted to vote? When that passed, they should all have been received into colleges and educated for the first five years, and not allowed to clear the forest, &c., till afterwards.

As an admirer and well-wisher of the United States I look upon this particular breach of faith, with all the loss and misery occasioned by it on both sides of the water, as a trifling matter compared to the symptoms of some inherent disorder in the general constitution of the body politic. If the opinion of the more highly-educated, wealthier, and middle classes can have so little control, or act so tardily on the vast majority below, an impolitic or absurd war, or some measure as serious as the disunion of the States, or any. other mischief, might as readily occur, in direct opposition to the almost unanimous sense of those who, for the good of the whole, ought to govern.

I am getting as prosy as if I were upon our Anti-Corn-

law League, or the Irish state trials. But I hope you will indulge me with a like 'expectoration,' as a German scholar
said to our great botanist Robert Brown, meaning that he
should unbosom himself.

Ever most sincerely yours,
CHARLES LYELL.

To GEORGE TICKNOR, ESQ.

London : April 2, 1844.

My dear Mr. Ticknor,—Have you seen Godley's book?
He agrees so entirely with the opinions we formed on almost
all points when his *Catholic* Church notions do not interfere,
that I am much pleased with the work. It is a great contrast in its tone of feeling to the ordinary English portraits
of America and the Americans, and even his Puseyism will
be a bait to precisely that class of readers who from their
anti-democratic, high church principles, are the most prejudiced against the United States.

I have heard some of this school say of late that there
is one and one only redeeming point about the United States
—the recent progress of the Episcopal Church. How far is
it true that the German professorial system is almost universal instead of our tutorial system, which, for my part, I
think a very bad one in the universities, as it does not inspire
the pupils nor the teacher with a love of what they learn or
teach? I hope it is true that the physical sciences have as
large a share of attention in the United States colleges and
schools as this thorough Oxonian complains of their enjoying. He says that the Americans will admit any sect to
be Christian and Protestant which receives the Christian
scriptures, and any doctrinal scheme which it thinks may be
deduced therefrom, so long as it allows all others to be
equally right who do the same; also, that the American
mind has a natural repugnance to anything which affects an
exclusive or dogmatic character. This is his heaviest charge
against you. I wish I felt as sure as he does of its truth.
One great evil which I complain of in our system of education, both in schools and colleges, is the monopoly of all professorial and tutorial places, masterships and usherships in

schools, &c., by the clergy.   Their *Catholic* propensities have
led them to seize upon professorships of astronomy (Armagh),
geology (Cambridge and Oxford), botany (Cambridge), mineralogy (Oxford and Cambridge), natural philosophy (King's
College, London), engineering (ditto), political economy
(ditto, Prof. Jones), and I could give you a long list of others.
In divinity, church history, Hebrew, and many others we would
not grudge them exclusive possession, but the evil now is,
that they not only regard the sciences they teach as subordinate to professional duties, but are liable, if eminent in
science, to be rewarded by church preferment, and immediately stopped in one career of usefulness.   I was told of a
college in Albany, I think, newly established on Oxford principles, in which all the teachers seem to be Episcopalian
divines, and no physical science or natural history to be
taught.   We are struck with the fact that amongst the
many *families* we visited at Boston, New York, Philadelphia,
and other places in south and west, altogether more than a
hundred, not one was residing in a boarding-house, and yet
this is represented as the American mode of living.

Ever most truly yours,

CHARLES LYELL.

To GEORGE TICKNOR, ESQ.

16 Hart Street, Bloomsbury, London : June 12, 1844.

My dear Mr. Ticknor,—I have two of your letters before
me unanswered, and we are just stepping into a steamer,
which in forty hours is to carry us nearly five hundred miles
northward, so I must thank you first for what you told me
of your universities and toleration, on the excess of which
Mr. Godley had been so eloquent.   I was very glad to see by
your letter to your French friend, which I read and forwarded immediately, that you had such hopes of the country,
about which I am sufficiently sanguine myself, if time be
allowed.   For what is fifty years in the history of a nation, or
any great experiment in politics, and the art of governing
men under entirely new circumstances?   If the institutions
are in arrear of the point of civilisation, morals, and knowledge to which the people have advanced, I doubt not that

they will rally and put things right; but as it is a costly machinery which requires such a stir as you are now making, in order to remedy abuses, I hope some improvements will be made to cheapen and simplify it, all which will come, and I hope in our time. Your public men ought to be more upright and independent than ours, because, owing to the wider distribution of the means of living, and of education and other causes, people whom they govern are as a whole superior. If the constituencies or voters as a whole are not better than ours, this must be and will be in time improved, first by excluding foreigners, and then by a property test. I cannot imagine that universal suffrage can ever be good for any community, and where it is so easy to acquire some property, and implies such a want of industry to be penniless, it must be still less unjust than here to exclude all who are not tolerably well off. As the suffrage has undergone so many modifications, surely the idea of restricting it within less wide bounds is not Utopian.

Ever most truly yours,
CHARLES LYELL.

To CHARLES BUNBURY, ESQ.

1844.

My dear Bunbury,—I was very glad to hear from you, though I have not time to say half the things which your letter has set me thinking upon. When my friend Babbage heard you were upon fossil plants, he said it was a good thing, as in such cases ' two and two made more than four,' and I hope together we shall strike out some improved view of the carboniferous era so far as relates to its chief feature, the flora of its strata. The identity of the *Pteris aquilina* with some impressions of coal plants so far as leaves are concerned was new to me, though to a superficial eye the resemblance is striking. May we be led by this to doubt whether the European and American coal-ferns, which seem equally identical, may really have been distinct in species? I think not. Some of the genera of shells and corals of the carboniferous period are the same as living ones, but not the species in any instance. The Virginian morasses allow, under a hot summer sun, great accumulations of black vege-

table matter, nearly like peat, and which might make coal.
The shade of *Cupressus distichi, Thuya,* and water-oaks, &c.,
shut out the sun, and ferns and mosses grew in the damp air
beneath, while the heat causes evaporation, and evaporation
cold.   One swamp is forty miles long by twenty, which I
saw.   Thousands of prostrate trees in the peat.   I hold Ad.
Brongniart's atmosphere of carbonic acid in the coal period to
be apocryphal, and if you can relieve us from intense heat,
so much the better.   The corals and shells of the then existing seas in northern latitudes show a warm state of things
far from the line.

Believe me, ever affectionately yours,

CHARLES LYELL.

To CHARLES BABBAGE, ESQ.

October 7, 1844.

My dear Babbage,—Mr. Phillipps called here to-day, and
told me he had felt at liberty to lay your letter before Sir
Robert Peel, Sir James Graham not having come up; and
that Sir Robert had asked him, Mr. P., to go and ask
me if I would go down with some chemist, Faraday, if possible.   I told him that I thought I should not carry down
the sort of knowledge most appropriate; that a lawyer, or one
like you, of general scientific knowledge, and special in
mechanical affairs, and to whom he, P., had first applied,
would be better.   He replied that Sir R., having seen me
mentioned in your letter as being in town, and thinking that
a geologist who had had to do with coal-mines would, combined with a chemist, do best for them, had asked him to
invite me.   I said I was very busy with my American book.
He, P., said 'it would be fully as inconvenient or worse for
Mr. Babbage just now, as he assured me.'   I said, 'A coal-
viewer and Faraday would do better.'   He replied, 'No; the
Government will empower you in your credentials to employ
any coal surveyor you like, but they want scientific men
known in your and Mr. Faraday's line to be there.'   I accordingly agreed, and Faraday has done so too, as I hear from
Phillipps since.   I observe that in all the Parliamentary
reports of 'children in collieries,' 'accidents in collieries,' &c.,

1835 and 1841, the geology of the district and position of the coal seams is laid down with minuteness, maps and sections. I therefore perceive that there is a geological side to the commission, and if I get knowledge or credit I have to thank your letter for it, and only regret we do not go together. I hope still to be back so as to see you, if disengaged, with Sir David Brewster on Sunday evening next, when, at all events, my wife will be *at home*. I hope Phillipps will write to you, as he said he should, about our long conference, and acknowledgments for your prompt aid to the Government.

Sincerely yours,
CHARLES LYELL.

To SIR PHILIP EGERTON.

16 Hart Street, Bloomsbury : October 22, 1844.

My dear Sir Philip,—I should have answered your kind letter sooner if I had not been very unexpectedly, and just when I thought I was settled in for the winter, and at my American book, induced by a direct invitation from Sir R. Peel to go with Faraday and attend the Haswell inquest, and see into the cause of the accident and suggest a remedy in future. We have only just drawn up our report to Sir J. Graham, in which the geology is a small part, the chemistry a large, and if I mistake not, a most valuable one, practically speaking.

I am very much obliged to you for offering to compare the fish, and will send them soon, and wish my engagements would have allowed of our accompanying them, but I regret to say this is not the case.

I shall write again soon, when I send the fish, and reply to you a little more fully.

Ever most truly yours,
CHARLES LYELL.

To GEORGE TICKNOR, ESQ.

London : March 1, 1845.

My dear Mr. Ticknor,—I am trying to negotiate with Mr. Lowell for a course of lectures from the celebrated Agassiz, the ichthyologist and Swiss naturalist and writer on glaciers, for 1845–46, but, perhaps, all are filled up. Ch. Buonaparte, Prince of Canino, has offered to take him to the United States, as he visits it with his son this year. I am sure Mr. Lowell will do it if he can, as I have answered for his English being passable. You will be much pleased with Agassiz, and his visit will be most useful, as it always is to us when he comes here. The British Association has thrice voted him sums of money to describe our fossils.

The marriage of Ward, the Oxford Puseyite martyr, is delightful, as illustrating their Romanist zeal for the celibacy of the clergy. Milman wonders if he will sign the marriage articles, as well as the thirty-nine, in a ' non-natural' sense, and has always maintained that the ladies and the rich endowments of the Anglican Church will keep the majority from Rome. He was joking in this way with an intimate Puseyite friend, a very thin man, Mr. Manning, the other day, and said how fortunate it was there were so many sleek incumbents of livings, and looking at Manning, he said :

> Yon Cassius has a lean and hungry look,
> He thinks too much—such men are dangerous.

Believe me, ever most truly yours,
CHARLES LYELL.

To LEONARD HORNER, ESQ.

Kinnordy : July 27, 1845.

My dear Horner,—I was vexed to hear you had so much work thrown on you at the Geological Society. I regret it the more as I am much struck with the value of the first and second numbers of the Journal, and wish it was generally known how much we owe them to you. But next year the machine will work better. I forgot to thank you for having

sent the ' Spectator,' the first critique I saw.[1] I mean to
send them to Mr. Ticknor, or take them to him, because
they all show how willingly our press welcomes anything
favourable to the United States, though I suppose some of
the bitter anti-American magazines will act otherwise.

Blanco White's book keeps up its interest to me, and
certainly it should teach every scientific man to modify his
opinions, and never to contend for doctrines, because he may
have once favoured them, as soon as new discoveries, facts,
and reasonings, require their modification or abandonment.
The little suffering or annoyance that he, or his self-love,
have to endure, is so insignificant in amount in comparison
to the penalty which the theological professor must pay for
relinquishing a little of any one of the numerous dogmas or
forms of interpreting scripture, which he has pledged himself
to adopt for life, that he must feel ashamed if he hesitates
for a moment to recant, after reading the confessions of
Saint Blanco the Martyr. For the sake of this moral, which
I hope many a philosopher will draw from it, I am glad that
all his sufferings are given in full length.

Dr. Falconer showed me the Dinotherium tooth from
Perrin Island, just like one I saw from the marine *faluns* of
Touraine, and confirming me in the view I had previously
expressed and printed in the Asiatic Society's proceedings,
that the formation was Miocene.

Yours affectionately,

CHARLES LYELL.

To CHARLES BUNBURY, ESQ.

Kinnordy : July 28, 1845.

My dear Bunbury,—I have been re-reading your letter to
me on fossil botany with no small interest, as well as your
excellent abstracts of Ad. Brongniart on *Sigillaria, Stigmaria,*
&c. Now, pray give me the benefit of your last ideas on the
climate of the carboniferous period, as deduced from the
plants, and refer to the last edition of my ' Principles ' about
the minimum of light, and criticise this. The near resem-

---

[1] On his travels in North America.

blance of the pteroid ferns is, as you say, a warning not to assume extraordinary heat, and you ask if in old formations we find some genera of animals very like living ones, mixed with very anomalous and unknown forms. Certainly we do. In regard to shells we have carboniferous *Orthocerata*, with nucula and many other shells very like living ones, and I might accumulate cases in other classes, so here the flora and fauna yield analogous results. In one letter you allude to the idea of a continuation of Lindley and Hutton's ' Fossil Flora,' but probably you will think better of the plan, and either give us first a carboniferous fossil flora, to be followed, as circumstances may lead you, by other floras, or some other independent work. The fact is, that the science will have got so far beyond the point where Lindley left off, that even in treating to an English reader of the very species and genera which Lindley spoke of, you would be able to write articles sufficiently original, and if you give your time gratis, and do not make money, or very little by such a work, you would do wrong, both for your own sake and that of science, not to have a work quite distinct from any other, and without any associate, for there is great disadvantage in joint authorships, partly because no one knows who is answerable. To make your knowledge useful your name must acquire authority, so that independently of laudable ambition, you should take care to get known by the title of the book. Probably a ' British Carboniferous Flora,' like Owen's ' Mammalia,' would sell best, and whatever foreign knowledge you obtained of American, or other carboniferous plants, would still tell upon it. I would not have you delay more than a year from this time in organising some plan. I persuaded Owen to change from monthly to every two months, which he rejoices in, but the frequent coming out of new parts keeps the public interest alive, and leave of absence for a foreign tour can be obtained by getting a few parts ahead.

With love to all, ever affectionately yours,

CHARLES LYELL.

To GEORGE TICKNOR, ESQ.

Kinnordy, Kirriemuir, N.B. : August, 1845.

My dear Mr. Ticknor,—I have been finishing the auto-biography of St. Blanco the Martyr. I wish he had said more about Ferdinand and Isabella. You remember my argument with you, that our friend (Prescott) would have done well if he had pointed the moral differently, or more strongly, to show how the result of the energy and talents of his heroine, considered as a monarch, was to place the force of an established monarchy and a great standing army at the command of that new church inquisition, which was to crush the free spirit both in political and ecclesiastical matters of Aragon. In short, her superstition, combined with her virtues, was laying the foundation of the future degradation of her country. The historian might have liked the woman more, for trusting with so much faith to her father confessor, and yet dwelt with pain on the degree in which it unfitted her most fatally to be the founder of the political constitution of a great country.

I told you what I had heard of certain letters in Blanco White's biography. I admired those of Channing very much ; and that concluding passage, vol. iii., p. 312, where he compares the undertone of truth, in spite of the occasional false-ness of particular views, to that general solemn roar of the sea which is quite distinct from the dashing of the separate waves, is as just and well applied as it is poetic. Mr. Norton's letters also are excellent, and the whole work is painfully interesting.

Remember me most kindly to Mrs. Ticknor and your daughters.

Believe me, ever most truly yours,

CHARLES LYELL.

To HIS SISTER.

Off Trepassey Bay, Newfoundland : September 15, 1845.

My dear Marianne,—Here we are within three and a half miles of the coast of Newfoundland, which, although it lies in a direct line to Halifax from Cape Clear, we did not approach in the 'Acadia' in 1841. The first three days after taking leave of Mr. Horner at Liverpool, from which we sailed on September 4, we had delightful weather and fair wind, making, though heavy with our full freight of coal, 220 to 240 miles a day. Then came our adverse winds and two regular equinoctial gales, which I am glad now they are over to have seen, but they made us both ill. Yet I have had much pleasant and instructive conversation with Mr. Everett, also with Mr. Ward, the American agent of the Barings' house in Boston, well acquainted with all United States affairs. The first equinoctial lasted twelve hours, the second on the 14th, when we were nearing the Great Bank, lasted twenty-three hours, and part of the time a perfect hurricane. But the engineer never missed a stroke, and they have a very full and well-disciplined crew.

When we were near the Great Bank, and the day before our grand gale, we saw an iceberg 200 feet high eight miles south of us. It was too far off to be a distinct object to my eye, though white and visible with a glass. They are very rare this month. This was the day before our heavy squall, and when the wind struck us in the night we thought we were running against an iceberg.

Your affectionate brother,

CHARLES LYELL.

To HIS SISTER.

Portland, Maine, U.S.A. : September 27, 1845.

My dear Marianne,—We have just returned from a visit to two of the most eminent medical men of this town, the largest in Maine, 15,000 inhabitants. Dr. Mighels showed us a fine collection of recent American shells, and some fossils, from a formation which interests me, as being a continuation of that of which I have spoken so much in my travels in Canada, which appears here under some new aspects. It is the old Uddevalla affair, with recent Arctic shells ; but I was pleased at recognising among the novelties the tusk of a walrus, which my discovery of one in Martha's Vineyard, figured in my book, enabled me to make out. We little dreamt when we left Boston that we should go full

200 miles in a north-easterly direction to Augusta, in Maine, to which I was taken by Squire Allen, as our driver termed the lawyer who was our host. I went to the State house, a handsome building of granite, and saw their collection, made during the geological survey. I was introduced to the Governor of the State, and when you think of 'those wild people of Maine,' of whom Lord Palmerston spoke in those terms during the border feud (an expressionthey will never forget), and learn that this chief magistrate, a Unitarian, has been re-elected several successive years by the democratic party, you might imagine that no great satisfaction was to be derived from the interview; but I found him a quiet, sensible, well-mannered man, who told me he was very desirous to resume the geological survey, which the legislature had suspended from economical grounds, and he drew from me all the utilitarian arguments he could which my travels in the States could furnish, in enabling him to talk over the party who might be averse to the outlay. He was once a Senator at Washington. One of their late governors was a Roman Catholic,—a proof that these New Englanders do not mix up religion with their political divisions.

We have now been in all the six New England states. The constitution of Maine is singularly democratic, members of both houses being elected by universal suffrage, and re-elected annually, and precisely the same qualifications in both for the electors and the elected. The judges, however, are appointed by the governor and council; the extreme in some states is where they are elective. I asked Mr. Allen whether Mr. Gardiner (a rich relation, and one of the most cultivated men in the country, who lives in a handsome house in a park in the English style) had ever been in the legislature or political life. He answered, 'Oh, no; his landed property is much too large : besides that, he derives much wealth from other sources.' I remarked that in Massachusetts I had understood that the reason the electors objected to rich men was that they were inaccessible to them, too much above them, could not be troubled to listen to all their wants. He answered that this motive really operated strongly, for in regard to envy and jealousy, they often felt that as keenly or more so, when they made one of

their equals their representative. Lately, in choosing
railway directors at Portland, they objected to the first
list because they were too rich and aristocratic, and the
numerous petty shareholders wanted to confer with them
on equal and familiar terms. Mr. Allen admits that the extent to which the democratic spirit throughout the United
States succeeds in making wealth a disqualification for
political influence is annoying to many of the richer
citizens, but he says that their rights and property are perfectly safe, and all the interests in the country were never so
flourishing as at this moment. After all, I suspect that
property, especially land, governs here as elsewhere, but
ninety-nine hundredths of the land is divided into a countless
number of small fractions, and the large proprietors, who
own the remaining hundredth, may very naturally go for nothing in the balance. That the acquisition of large estates
and fortunes has had the same charm as with us, appears
from the eagerness with which they toil to obtain them. A
grand railway is planned from this town (Portland) to Montreal, and they hope to get the English steamers to land
here, where there is a very fine harbour, and will shorten
the road to the British provinces very much. It would be
natural, too, that the English mail and passengers should
land here, and go 110 miles by railroad to Boston by a rail
which carried us at the rate of thirty miles an hour, instead
of having so much more sea. All the money, more than
two millions sterling, is subscribed, and the road surveyed
and the Acts passed both in Canada and Maine. The
Kennebec is navigable for some forty miles from its mouth
to Augusta, to which the tide goes up. At Bath we saw
numerous ships building, destined to carry cotton from the
Southern States to England. Streets of new houses rising
there, as at Gardiner, Hallowell, and Augusta. New large
cotton factories with steam-mills in progress at all these
places, new saw-mills for the timber which is floated down
the river in rafts, three or four huge steamboats carrying
300 to 400 passengers each going up and down, brigs and
schooners laden with hay for Alabama and Louisiana, that
the horses in New Orleans and Mobile may enjoy the sweet
herbage of Maine, so much better for them than the rank

grass of the rich soil of the western prairies. They are to
return with cotton in exchange for the hay. What an
advantage to have all this free trade between such distant
regions in the same country ! On our way here we passed
the Wenham Pond, or rather lake, for it is nine miles
round, with its pure sandy bottom and fed by clear springs
which when frozen over in winter yields the ice sent to
London. And now, dearest Marny, I must finish, and before
starting for the White Mountains, which they say are already
tipped with snow, I will put this into the post at once,
though my tale is half told, especially our Portsmouth visit
to Mr. Hayes, who drove me in a gig through some green
lanes in a delightful forest where the *Coleas* were swarming.
Under the trees a kind of gale smelling like our bog myrtle
(*Myrica cerifera*) was abundant, and a beautiful *Gerrardia*
quite new to me. We collected fossils in this wood. But I
must end, with our love to all, and hopes to receive good
news at Albany.

Ever your affectionate brother,

CHARLES LYELL.

To LEONARD HORNER, ESQ.

Hopetown, near Darien, Georgia: January 9, 1846.

My dear Horner,—I have been visiting the site of three
Megatheriums near Savannah, and the places where four or
five others, some entire skeletons, were dug up here. I can
prove them by their position relatively to beds of marine
shells, to be a shade more modern than the elephant and
rhinoceros of the valley of the Thames. They co-existed
here with our *Elephas primigenius*, the American Mastodon,
&c. The modern or Post Pliocene series of changes before
and after the Megatherium in this low country are very
interesting. Tell Darwin I have quite a counterpart to his
Patagonian steps, or successive cliffs cut out of the Tertiary.
The botany here, in spite of the season, which is now free
at last from frost, and sunshiny, is very striking. The
three palmettos—*Chamærops* one of them—a tree actually
forty feet high at Savannah, and having all the aspect of a
palm tree—are very abundant. The tall cabbage palm in

the seaboard islands, requiring the salt sea-air, the saw palmetto abounding under the long-leaved pines, and blue palmetto in the clay swamps. There is a common bay that grows very high, *Laurus Caroliniensis*, with large leaves, having the odour of our bog myrtle. Magnolias in abundance, and three kinds of holly with red berries. Nothing has pleased me more than to witness the improvement that is going on in the condition and education of the slaves, both in the rice plantations and in the small farms, Sunday Schools being general, oral instruction, though reading feared, lest they get hold of exciting pamphlets. Many, however, do learn to read, and receive presents of Bibles, &c., from their masters. I have seen a large steam engine which has been entirely trusted for fifteen years to a thorough black; all the carpenters here, blacksmiths, &c., are Africans or mulattos. Their masters talk to them as much as we should to Irish labourers, and they of the new generation of blacks calling themselves country born, feel and talk of their superiority to the preceding generation their parents, whom they style *Africanians*, as the young Irish educated in the schools at Boston feel towards their unlettered emigrant sires. Their labour is calculated for eight hours in all the rice, cotton, and sugar plantations about here. They often finish it in seven or less, and it is frequently all over by two o'clock, or twelve or even eleven in the day, after which they usually take some hours' sleep, because it is their pleasure to sit up half the night, gossiping, singing, or listening to some favourite black Baptist preacher. I am convinced that people are too apt to forget the very low platform of civilisation from which the African starts. One-fourth of the five hundred negroes now collected round this house were born in Africa, and to bring them up, or elevate them to the grade of the lowest Irish, is a step far beyond turning the said Irish into the average American labourer's standard. When I see how much has been done in so short a time I begin to be more hopeful than in my last tour, for unless the fanatical party of the North force on a collision, the next generation will be just as much beyond the blacks we see as they are above the Africans, and the treatment of them is necessarily regulated by the position and intellectual and moral condition to which they have attained. It is too large a subject to enter now upon, but I am glad to find in the South, slave-holders who will speak out upon all the worst evils of the system and who are convinced that it must wear out, though it would have done so much faster if the course of improvement had not been arrested by interference and insults which have hurt the temper of many Southerners. Were it not for this unfortunate question, the free-trade feelings of these Southern States would be so strong towards England, that a war would be impossible. They have no love for the Northern ultra democracy, to which they ally themselves merely for protection against an apprehended interference for some scheme of speedy emancipation for which the blacks are not prepared; because the race cannot yet be reckoned upon for habits of continuous labour, which every savage, and his immediate descendants, has so great a repugnance to. The Northerners would not believe how attached many families here are to their negroes, and how it prevents them selling estates, when to do so would be both for their interests and tastes. I have looked into the statistics of crimes and punishments on this estate for thirty years, and wish that an average English parish could return as favourable results. Besides being clothed and very well fed, they are by nature most peaceful, so that they hardly ever fight, and the contrast of some Irish labourers who came here to dig a canal (to which, by the way, we owe the discovery of the Megatherium), would really be laughable if it were not such a serious evil. Mr. Hamilton Couper has written for the 'Geological Journal' from the beginning, and will be a subscriber for life. Others here will gradually come in, but it must be the work of time.

Believe me, my dear Horner, with love to all, ever your affectionate son-in-law,

CHARLES LYELL.

To His Father.

New Orleans : February 25, 1846.

My dear Father,—I have just been sending off a long letter, or paper, on the geology of Alabama, to the President, to read to the Geological Society.

I had to rough it a good deal in Alabama, but was well rewarded, both by the Coal-region and the Tertiary. I am now entering upon a new field, the great valley. I find everywhere much encouragement in the way of fellow-labourers, most of whom have either been educated in the more northern universities, or have migrated from the New England States on account of weak lungs. The unsettled migratory disposition of the inhabitants of a new country is quite curious. You meet with men, who with their wives have lived twenty years on one farm—improved it, grown rich on it, created it out of the wilderness, men without children, who are going to Texas. They have never gone first to see the promised land. If you enquire where shall you settle in that vast country, they reply, I shall go with my wife and negroes to Houston or Nachitochus, and then *look out.* ' Who never is, but always to be blest,' should be the motto of such landowners. I have seen nothing to alter my views of the condition of the slaves. If emancipated, they will suffer very much more than they will gain. They have separate houses, give parties, at which turkeys and all sorts of cakes are served up. They marry far more than our servants—eat pork—the women exempted from work a full month after childbirth, corporal punishment excessively rare ; they do so much less bodily work than the whites in the North, that the Southern planters will not believe in the stories of the former. The other day an Alabama brick-layer returned from a New England apprenticeship and reported at Tuscaloosa that he had earned two and a half dollars a day by laying 3,000 bricks daily. As the strongest negroes are only required to lay 1,000, it appeared to the planter incredible.

Ever, in haste, your affectionate son,

CHARLES LYELL.

Philadelphia : April 27, 1846.

My dear Horner,—The news which we received yesterday
have given us so much grief on dear Harry's [2] account, and
I feel so miserably our distance from him (which would be
as great, however, if we were in Europe instead of here), that
I have had great difficulty to-day to fix my thoughts on anything. I had some thoughts before my arrival here, to draw
up a paper on what I learnt and observed of the important
geological question respecting the proofs alleged to have
been found of the existence of mammalia, birds, and reptiles in the Pennsylvanian Coal Field. But I cannot do this
now, and until I have in town the slabs of sandstone which
are on their way, and have had more time to make sections
and maps, I cannot do the subject justice. Meanwhile, I
shall not be sorry to give the Geological Society the news in
any way you please. In a few words, I have satisfied myself
that Dr. King is right in believing that he has discovered in
the middle of the coal formation, the foot tracks of a large
reptilian quadruped or animal allied to the Cheirotherium.
These occur in one locality, and no others have yet been
found in the same place, nor under similar circumstances
elsewhere. The locality, which I visited, of the supposed
bird tracks is simply a ledge of white coal-grit or sandstone,
sculptured by the Indians, who have, I doubt not, intended
to represent dogs (or wolves), birds, and other animals. Dr.
King was a beginner in geology when he first found, and to
his credit appreciated duly, the importance of the Cheirotherium tracks. He is a man of thirty years of age, and in an
extensive medical practice, who has suffered some persecution, professionally and socially, for believing the world to be
more than 6,000 years old, and avowing this at a Lyceum.
He has been held up as an infidel by the President of a
Catholic College, by some German Calvinists, &c. I have
met with other proofs of similar illiberality from persons of

---

[2] His brother, Captain Henry Lyell, ' wounded severely ' at the battle of
Sobraon.

all sects, lay and clerical, in the United States, where the subject is much in the same state as in Europe.

The Cheirotherium tracks occur in pairs, each pair consisting of a hind and fore foot. There are two rows of these which are parallel, or have been formed the one by the right fore and hind feet, the other by the left, the toes turning one set to the right and the other to the left, and the distance between the successive footsteps being about the same throughout.

Few geologists will now be prepared to believe that this single species or genus of reptile, or that one class only of vertebrated animal, had possession of the islands and continents on which so widely extended and magnificent a vegetation flourished.

Tell Dr. Falconer to look out for some splendid teeth of Texas elephants which are going from New Orleans to London for sale.

Believe me, ever your affectionate son-in-law,

CHARLES LYELL.

To HIS SISTER.

[3] Steamship ' Britannia ': June 11, 1846.

Dearest Carry,—The Captain declares we were at least three ships' length off the great iceberg which appeared suddenly out of the fog in the night of Friday, June 8. As we were going at the rate of ten miles an hour, the moon obscured by clouds and a thick fog, you may believe that Captain Elliot as a naval man, and better able than the rest of us to estimate the danger, was not a little excited at what he considers the rashness and inexcusable folly of not slacking the ship's speed. All these mails run on in this way, and will do it till the Admiralty interferes. Going three or four knots they could always turn in time for their broadside to hit the ice, and only receive a moderate bumping. As we passed fifty bergs or more in daylight, the Captain (one of the most cautious of the whole set) does not doubt that we were within pistol-shot of many which were not seen in the fog. They keep a bright look out, officers and

[3] Return voyage from America.

men, and are aware of the risk. But what folly, when they
would only lose six hours of the night at most! One iceberg,
almost the only one before dark which came close to us when
I was below, had a large rock twelve feet square on the top,
and much gravel and dark sand on its side. They were from
50 to 400 feet in height; pyramidal, pinnacled, dome-
shaped, single-peaked, double-peaked, flat-topped, and of
every form and most picturesque, and only a quarter of a mile
off us, and numbers more distant.

Ever affectionately yours,
CHARLES LYELL.

To SIR PHILIP EGERTON.

11 Harley Street, London : September 26, 1846.

My dear Sir Philip,—I am very glad to know that Agassiz
has really sailed at last, and when I knew of the ' Great
Britain' going ashore, I could not help rejoicing that I refused him even the three or four days' grace he prayed so
hard for in Southampton. The work he did when with you
was really gigantic. All the while he was looking over the
Connecticut fish you had named, he exclaimed how rejoiced
he was to see you had named and separated them as he
would have done, and that 'you will have in England while
I am away, one to whom you may safely refer.' Of the
homocercal (lias?) fish, I have duplicates, of which you
may depend on having the first sight and pickings. C.
Bunbury is drawing up an account of the accompanying
fossil plants (oolitic forms) like those of the Whitby (oolite)
coal in Yorkshire. I hope to get a description of the single
species of fish from you, the only perfect one, and anything
that you may have to say of his associate, the Tetragonolepis,
of which unfortunately there are only scales in patches.

As to the Durham (Connecticut) fish which you examined,
they have no plants associated, but I am very much interested to hear from Agassiz that the fish resemble most
nearly those of *Autun,* a place I visited in 1843, and got
some plants (psarrolites and ferns) and some few fish, rather
fine ones though few, which I will show you. Now I begin
most strongly to suspect that the fish called Paleonisci, but

which Agassiz agrees with you in thinking not of that genus, belong to a formation almost as old as the coal, if not upper coal measures, whereas the Richmond coal plants and fish are triassic. This will be the more interesting as you remember the bird's footsteps of Hitchcock belong to the same period as the so-called Paleonisci and Catopteri. By the way, I have a duplicate copy of Hitchcock's 'Geology of Massachusetts,' 2 vols., 4to., with plates of the bird's footsteps, which I could give you if you have room for the same in your library.

Believe me ever very truly yours,

CHARLES LYELL.

To LEONARD HORNER, ESQ.

London : October 9, 1846.

My dear Horner,—I mean to refer to your last anniversary address in my sixth chapter of 'Principles,' for the best *résumé* of the present state of the facts and theory of the coal. I shall send you copies of letters which I am writing to E. Forbes and Ramsay on their excellent papers. I think E. Forbes's reply to you about not citing my climatal theory sufficient. But as I claim to have first in my 'Principles' laid down the foundations of explaining the difficulties experienced in reconciling the doctrine of *specific centres* with the apparent anomalous position of many living plants and animals, viz. by showing that *since* the existing species came into being the land and sea had so changed, chains of islands disappeared, shallow seas turned into land, or into deep seas, this might naturally or ought to have been adverted to. Still more, my proofs referred to by Owen, in his Introduction, of the isthmus which in the Miocene period joined Dover and Calais. But do not suppose I am going to enact the character of the 'injured man.' I am in great good humour with E. Forbes's beautiful paper, and even his numerous citations of me, though he has missed my two most important claims, which I shall be glad of the President of the Geological Society setting right, and so will Forbes himself. I am very desirous of reading the Richmond papers when you are in the chair. What I learnt of the fish from Agassiz will add to its importance very much—it is oolitic coal—but I have also learnt and shall treat of it in the same paper, that my fine collection of so-called New Red or Connecticut fish, are not oolitic, but much older, and they throw new light on the age of the bird's footsteps, if not of the true carboniferous reptilian. But I write enigmas, as I cannot enter into details.

Ever your affectionate

CHARLES LYELL.

To EDWARD FORBES, ESQ.

11 Harley Street: October 14, 1846.

My dear Forbes,—I have been reading your beautiful essay on the glacial epoch, and the origin of the existing fauna and flora of Great Britain, with the highest interest and pleasure, and think it one of the most original treatises, and so far as I have yet studied it (for I mean to read it again), one of the soundest as well as boldest that I have ever had the pleasure of perusing. Among many views which I never ventured to promulgate, from want of sufficient knowledge of the Mediterranean and British mollusca now living, but what I had nevertheless more than suspected from what I knew, was the probable contemporaneousness of the newest Sicilian Pliocene, and the drift; as I thought some of the northern species which were in the Sicilian bed indicated a chill in the sea there, though no erratics had reached so far. I am very glad to see you come out with that, and many other novelties which I am sure are new. Horner, I find, had remarked to you, before I had read a word of your paper, that he should have liked you to have made more references to my 'Principles of Geology.' I am not aware that they are mentioned, though you have cited me for many other papers most fully. I am perfectly aware that there is no one who is in the habit of doing more justice to any claim I may have in our science than yourself, and therefore I attribute to mere accident any omission you may have made in citing me, especially as I have no doubt you think the publicity of a work which has gone through six editions is enough to satisfy any man, and supersede

tedious historical details. I may, however, mention on this
head, that on the Continent I gain no priority for any original
views or facts which have only appeared in my ' Principles '
and ' Elements.' When the Geological Society of France
voted a sum of money to Archiac, to draw up a report on
the progress of geology for ten years (1835 to 1845 I believe),
he wrote to me to say that all treatises on geology were
left out of such reports, as they were presumed to be com-
pilations, authors taking care to take date for their discoveries in scientific journals, but as my book was an
exception to such rules, he wished me to send him an exact
list of all my original theories and facts, and their dates,
which owing to their numerous editions no one could make
out, and which he must neglect without such aid. I have
not yet complied with this requisition, nor do I think I shall,
but it makes it the more agreeable to me, when any able
writer like Owen ('British Mammals ') cites me, and gives
the date of the earliest editions when my own views first
appeared. I will now allude to some parts of my ' Principles,' and some of my separate memoirs, to which I think
it would have been natural for you to refer, that you may
take an opportunity on some future occasion, if you think
fit, to acknowledge my labours.

   To say nothing of my theory of the cause of fluctuation
in climate, which although no eminent men have controverted it, none except Darwin have avowed their belief in,
I will come at once to a grand truth of which your paper is
a splendid illustration, viz. that ' zoologists and botanists
were unable to refer the distribution of species to any deter-
minate principles because they have usually speculated on the
phenomena on the assumption that the physical geography
of the globe has undergone no alteration since the introduction of species now living,' &c. (' Principles,' 1st Edition, vol.
ii. Ch. II.) I explained De Candolle's difficulty of mountain
chains sometimes being barriers to the migration of plants
and sometimes not. I insisted on the fact that in Sicily the
species of plants, shells, &c., the whole flora and fauna, were
older than their *stations*, that they existed before the seas,
hills, rivers, lakes, mountains, &c., were formed. I need
not remind you of the entire chapters devoted to this kind

of argument, which so far as Darwin's and my reading goes, was new. I certainly did not borrow it, and no one had dealt with it systematically. As I assumed 'specific centres,' as you do, and endeavoured to remove many anomalies which beset the subject and explained how species may have migrated 'where islands since destroyed and submerged existed,' &c., and as I had so recently opened up this new line of research connecting geology and natural history, I think some brief notice, in a prefatory sentence, of my treatise would have formed a natural beginning to your grand working out of the problem. But I will now come to some of my papers which enter more specially into your own arena. I will first take one of my papers on Touraine, out of the order of date I believe, for I am writing this, as you will perceive, off-hand, in which I have endeavoured to prove, after carefully studying on the spot the fossil shells of the Norfolk and Suffolk Crag, and those of the Faluns, and compared them with each other, that a tract of land must have intervened in the Miocene period, between the regions of the Loire, &c., on the south, and the Orwell, Yare, &c., on the north. Owen has referred to this paper in his Introduction, when speculating on the isthmus between Dover and Calais. Murchison in an anniversary address alluded with wonder to my identifying the groups of fossil shells in Norfolk and Touraine as contemporaneous, when nearly every species was distinct. I had accounted for this difference in the species, not only by a geographical barrier like the isthmus of Suez, but also by the meeting of two different marine faunas in the same sea, one from higher and the other from lower latitudes, as Darwin had shown in South America. (See my paper, 'Magazine of Natural History,' July 1839.) In that paper the comparative northern aspect of the Crag shells is reasoned upon. But this subject is taken up more fully in my comparison of the Faluns and Crag, published in 1841 ('Geological Proceedings,' vol. iii. p. 437), in which memoir there are many passages which I think called for some acknowledgment, and among others, my proof from the terrestrial mammalia of adjacent land in the neighbourhood of the Faluns. The geologist might wish to know whether you doubted my Miocene east and

west barrier, or if not, whether plants would not have
migrated by it into the south of England at an earlier period
than that of your West of Ireland flora, No. 1. You may
reply that such land may have had no connection with
Spain, &c. I had made, as you know, the finest collection
of Touraine shells, and had been aided by Searles Wood, as
well as yourself and George Sowerby, in regard to my
Miocene shells, besides getting the corals and fish, teeth, &c.,
determined. I traced the real Suffolk Crag to its southern
limit in Normandy ('Geological Proceedings,' vol. iii. p. 438).

In my paper, read 1840 ('Geological Proceedings,' vol. iii.
p. 171), on the Norfolk drift, I refer the disturbed stratification of the drift to the agency of ice, and I identify in age
certain freshwater beds with part of the drift period, giving
insects and fish and plants of that age. The fish are afterwards described, and I state that we must look northward
for analogous types. Some allusion might have been made to
these attempts to work out some of the dates of the modern
fauna contemporary with *recent* shells.

In my 'Elements of Geology' I entered copiously on the
phenomena of the glacial period, and insisted on the arctic
character of the fauna, in which I may have been anticipated
by Smith of Jordan Hill, though I had drawn the inference
from Canada and Sweden before he published. In the list
you give of the shells of the glacial epoch, you have forgotten
my list of Canada shells (Quebec) given in my American
Travels; thus, for example, *Mya arenaria, Saxicava rugosa,
Cardium Groenlandicum, Scalaria Groenlandica, Littorina
palliata*, are shells common at Quebec, and given in my list,
to which you have not appended Canada as a good drift
locality. I well know, however, how difficult it is to recollect all the scattered notices of any author, and quite impossible to find them in time for reference, even if you do
remember them, and I imagine that none of us ever publish
without giving some one the power of pointing out such
omissions.

You speak in your paper of the mingling of deep and
shallow water species of shells, a fact to which Dr. Fleming
had often called my attention, as the effect of what he calls a
ground swell, throwing up both kinds of testacea, littoral and

pelagic, on one shore. You hint at icebergs and northern waves. The former has no doubt had its influence, and when icebergs turn over, or fall to pieces, huge waves are caused not merely *from* the north. But it has always seemed to me that much more influence ought to be attributed to simple denudation where beds of loose sand, gravel, or mud were upheaved, and sometimes alternately depressed and upraised in an open sea. The exposure of such destructible materials must have led to the confusion you allude to, but much less so where the beds were protected in fiords, &c. The broken fossils found in these strata would agree with my denudation hypothesis, which is I think strengthened by the frequent regular re-stratification of the beds containing the deep and shallow water species.

I think your paper and Owen's Introduction have come out at a peculiarly opportune period, when Agassiz, Alcide D'Orbigny, and their followers, are trying to make out sudden revolutions in organic life in support of equally hypothetical catastrophes in the physical geography of the globe. Do not imagine that I shall in my new edition of the ' Principles,' or elsewhere in print, ever allude to the slightest regret on my part about any of the omissions of reference in your book which I have here enumerated. On the contrary, I shall take every opportunity of expressing the real admiration I feel for your paper, and of citing it in my first, second, and third books.

Believe me ever most truly yours,

CHARLES LYELL.

From EDWARD FORBES, ESQ.—*Reply to the foregoing Letter.*

Bala, North Wales : October 18, 1846.

My dear Sir,—Many thanks for your letter of the 14th, with your remarks on my essay in the Survey Reports. The kind spirit in which you have written assures me that you cannot consider I have sent forth that essay with any intention of being unjust, or of overlooking your great claims in any part of it.

My paper was written under very disadvantageous circumstances. It had to be prepared for the press by a certain

time, and no sooner had I commenced the preparation of it
than the severe illness which laid me up last winter came on,
and two months, January and February, were thus entirely
lost. During those months I was confined to my bed.
March only remained for the composition, and during that
month, whilst very weak and by no means fit for work of
any kind, I had not only to write the paper as printed, but
also to prepare my course of lectures on Distribution for the
London Institution, and to get my botanical lectures in
order, for their commencement in April. The superintendence of Spratt's and my Travels in Lycia, now at length
ready to come out, was also on my hands. Under such circumstances many references I should have liked to quote
were lost sight of and omitted. I could not go about to
libraries, and my own books are mostly still lying packed up
in various places, as it is useless taking them out until I find
myself in some more settled position. My only course under
such circumstances was to write out my argument in my
own way, and to back it by such references as were at hand,
or as I had already taken notes of, when backing seemed to
me absolutely necessary. The plan of my essay was purely
inductive, and consequently did not admit of more citation
than was absolutely necessary for my purpose. I had proposed to myself a *special* problem (the origin of the *British*
fauna and flora), and in the working of it out had neither
time nor space to draw up an *exhaustive* essay. I had to
compress my argument into as small a space as possible. It
was my wish to have appended notes and extracts in an
appendix, and to have given a bibliography of the scientific
literature bearing upon the subjects treated of. In this way
I could have done the fullest justice to everyone who had
touched upon them; but illness and the printers prevented
the fulfilment of the intention. As you were away when I
was at work, I have thought it necessary to offer these preliminary statements before replying to the special points in
your letter. I shall now take them seriatim.

1st. You remark on my not alluding to the ' Principles.'
There cannot be a warmer admirer or more grateful pupil of
your ' Principles ' than myself. I have read every edition,
and most often over; and much that I have done has grown

up from the seed sown by you. My not quoting them
oftener, however, is not an accident, but because I did not
consider allusions to them necessary. I wrote for naturalists
and geologists—who are, or ought to be, as familiar with
them as myself. In all such essays as mine, it is impossible
to refer except in very special cases or when the book or
paper is supposed not to be familiar to the readers. What
you say respecting the French does not affect this. It is
part and parcel of the conceit of that nation to assume they
have a right to put admitted truths, or sources of truth,
aside. The idea of leaving out references to treatises on
geology in a report on its progress is preposterous. In
geology, as in the other natural history sciences, the last
general treatise of acknowledged authority must always be
the starting point in special essays, and its contents taken
for granted, or else citations would be endless. At the same
time the references you quote from the ' Principles,' especially
that respecting the age of the plants and animals of Sicily,
might have been most advantageously (to me) noticed—and
would have been, had they not escaped my memory. I have
not the book by me here, but when I get back will refer to
them most carefully. I still look forward to (and am preparing) an Essay on the Origin of the Flora of the Levant, in
which the history of the whole Mediterranean flora will have
to be commented on. This will give me an opportunity of
showing the value of your views on these points. On this
subject I should like to have much conversation.

Now as to the more special papers. Your references to
the Touraine and Crag papers are of the greatest consequence, and the quoting of them would have been strong
supports to my argument. It would have made all those
parts of my essay relating to the Crags much clearer and
more satisfactory. But that part of the subject was kept as
brief as possible, subservient to the main questions about the
drift, &c. I had neither time nor sufficient *practical* knowledge of the Crags to warrant my adventuring farther than I
have done, but look forward to future work in that direction.
The same may be said with respect to your Norfolk drift
papers. I could not have entered critically on them or the
Touraine papers without doing so in the case of many other

valuable essays, by various authors, British and foreign—too
many to cite consistent with my plan. This was also the
case with respect to questions about action of ice, &c. As
to the mingling of deep and shallow water species of shells,
Dr. Fleming's explanation does not appear to me to apply in
the cases referred to. When we meet, I will explain why.

The references to the Canada list of Pleistocene fossils
were omitted in the table at page 380 by an oversight, but
you will find them all inserted in the full list forming the
appendix, and at the end of the appendix your name given
as my authority.

One important point I have omitted to allude to, viz.
your climatal views. I·look on my paper in a great measure
as a contribution towards this confirmation. But to have
expressed a positive opinion on my part would have been of
too little weight in the discussion.

You are pleased to compliment my paper on its *originality*.
Any praise from you must ever be among the greatest grati-
fications to me, and to any honest labourer in the great field
of nature. But I had rather hear the views I have set forward be proved *not original* than the contrary. It seems to
me that the surest proof of the *truth* of such conclusions as
I have summed up at the end of my essay is the fact of their
not being *original* so far as *one* person is concerned, and of
their having become manifest to *more than one* mind, either
about the same time, or successively, without communication.
I believe laws discover themselves to individuals, and not
that individuals discover laws. If a law have truth in it,
many will see it about the same time. Hence to me an-
nouncements of anticipations are confirmations of the truth
of one's opinions, and welcome accordingly. In this spirit
I have gladly perused your letter, and in the same spirit
I feel sure you will regard whatever I have written or may
write, worthy to interest you.

Ever, dear Mr. Lyell, most sincerely yours,

EDWARD FORBES.

To HIS FATHER.

11 Harley Street : November 26, 1846.

My dear Father,—I have got through half, and by far the most difficult half, of my new edition, and it gives me an interest in reading up my arrears of new geological papers which reconciles me to the deferring of my American second Travels. We went this morning to buy a dozen of knives for our dinner party on Monday. The old cutler in Oxford Street said, ' No. 11 Harley Street ; I remember well the house, eleven doors on the right hand from Cavendish Square, for I went there with knives to the Duke of Wellington, then Sir Arthur Wellesley (thirty eight years ago), for he lived there.' Old Mr. Rogers, who with the Milmans dine with us on Monday, to meet Charles Bunbury and Frances, will, I daresay, be able to tell us of Sir Arthur in this quarter. Macaulay was most entertaining at Milman's last dinner, giving and taking, and not overpowering. He is hard at work with his ' History of England.' I asked him if he had read 'Constantinople,' in the last 'Quarterly Review.' He said, ' No, but all about St. Chrysostom is got out of the edition of his works, which I read at Calcutta, and ended by liking the old saint, which is more than one can say of most of the old Fathers.' Milman remarked, that at Oxford such high prices are no longer obtained for editions of the Fathers or Puseyite mediæval books, but they are selling at Cambridge. A few days before, Herman Merivale told me he had heard the same, and that there was an extraordinary spread of scepticism and rationalism at Oxford. In large parties, men holding forth that as a high admiration of the beauty of form was the characteristic of the Greeks as a nation, so the Jews had the religious instinct very largely developed, and hence they developed Judaism, Christianity, &c. To get back to Dean's Yard, Milman was talking of the fortune he could have made if he had had the gift of prophecy for five years, as, when he came to Westminster, whole streets of houses were offered him for a fifth of what they let for, when railway companies were bidding for offices near the Houses of Parliament, &c. On which Macaulay, recurring to the former talk about Chrysostom said, ' But think if one could have bought up the Fathers at their value in 1800 (when they were fairly appreciated), and sold them at the Oxford price of 1840!' Some one at the other end of the table, where there was a dish of larks, was talking of the destruction of life, such small birds, when Macaulay said, ' On that principle you ought to feed on blubber.' Would not old Dr. Johnson have just said that, if Boz had been sentimental? Lord Lansdowne was in good spirits, but the great affair of the Education Bill being put on him, evidently makes him feel as the old man. Kay Shuttleworth, who, dined with us, is his right-hand man in that, and to him we shall be, I hope, indebted for some good plan of national education. George Lewis, the much-abused poor-law commissioner, was very agreeable. Ever since his article on serfdom in the ' Edinburgh,' last spring, I have thought more highly of his talents, which we always knew to be great. Lewis was talking of the ancient Britons not trying to convert the Anglo-Saxons because they despised them so much, and thought they might as well go to the devil their own way. Macaulay said, ' Would the Saxons have gained much, being such respectable Pagans, and the Welsh such superstitious Christians ? ' Henry Drummond (the saint), when Cunningham (of Harrow) tried to convert Byron, said, ' It was better to be an infidel such as Byron, than such a churchman as Cunningham.' Milman talked of Talfourd's dismay when his ' Ion ' was published entire in a Calcutta newspaper, and called ' John.' Alison's ' History ' was spoken of, and his blunder of calling the French ' impôt des timbres,' the timber duties. I wish the accounts from Kinnordy could make me feel more cheerful about my mother.

With our joint love, believe me, my dear father, your affectionate son,

CHARLES LYELL.

To His Father.

11 Harley Street: December 16, 1846.

My dear Father,—Eleanor's letter to Mary, and yours to me, arrived at very unusual hours of the day, and I suppose the

railway punctuality has been disturbed by the snow. I
assure you I expected no answers to my letters, and if they
ever give any amusement to the party at Kinnordy, it is a
pleasure to us, and a very poor contribution towards our
share of cheering the winter there,[4] where so many are
proving themselves such excellent nurses. I can hardly
remember enough of Mary's letter to be sure I may not make
some repetitions of our three very agreeable days at Bowood,
but I will add a few of my remembrances. Nothing made
the whole visit more agreeable than the perfectly equal part
that most of the guests played in conversation, which was
lively and well worth listening to. It is too often the case
in such society, that some one gets an undue monopoly, not
so much from a desire to shine, as from exuberance of ideas
and spirits, as was remarkably the case with Sydney Smith,
and even Rogers in his better days when I first knew him,
and is now with Macaulay and Lord Jeffrey, and occasionally
even with Hallam, but never with Milman, though his talk
is of the very best. One day Lord Lansdowne asked Lord
Clarendon what the Pope was originally. He answered, 'He
wanted to go into the imperial guard and be a soldier, but
they would not have him because he had been epileptic.'
'Then,' said Milman, 'he may make, perhaps, the fourth
great epileptic hero;' Julius Cæsar, Mahomet, and Napoleon
being the three others. They were talking of Sydney Smith,
and how he never talked of his mother, whereas his brother,
Bobus Smith, was so fond of dilating on the excellences and
extraordinary beauty of their mother. On this Lord Lansdowne said one day, he (Bobus) bored Talleyrand so much
with that theme in London, that Talleyrand said, 'C'étoit
donc monsieur votre père qui étoit si laid.'

Mr. Greville had learnt from Harness about my reptilian
footsteps from the Pennsylvanian coal, and as his brother-in-
law, Lord Ellesmere (late Lord F. Gower), had been recently
taking up slabs of Cheirotherium footsteps in the New Red
on his property, he (Mr. Greville) was curious about them,
and wanted to know the antiquity I ascribed to them as
compared to the Megatherium. After the ladies had retired
they fell to in earnest on the whole subject, and after I had

[4] His mother and sister being laid up with illness.

given my views, Milman explained how writers of research and reputed orthodoxy in the English church, had adopted, some the Hebrew, others the Septuagint, others the Samaritan chronology, and how that of St. Paul differed from that of Moses, and on the whole, how they varied 2,000 years from each other. After his very clear exposition of the case, Lord Clarendon, who had never, I think, read or thought much on the point before, observed, ' It seems then, that the geologists are blamed for not making their notions of the world's age agree exactly with a chronology (supposed to be Scriptural), which we really know nothing about.'

One day at dinner I was telling Milman that Pius IX. had not only encouraged the scientific professors of the Roman States to attend this year's congress of ' savans ' in Italy, but another (the next I believe) was to be held at Rome. ' Yes,' he said, ' and the British Association is to meet at Oxford with Sir R. Inglis as president. Inglis and the Pope ; no wonder people are fearing that Oxford and Rome are approximating too nearly. And both of them, after opposing railways, are adopting them.' Lord Clarendon told a good story of the last papal election, but I cannot remember the names. Cardinal A., who was ambitious of mounting the papal chair, asked Cardinal B. who he thought would succeed. B. replied, ' If, as we are given to understand, the Holy Ghost directs our decisions, I think it will be either Cardinal Chigi or Ferretti (the present Pope), but if, as some think, the Devil meddles with the election, I shall not be surprised if the choice should fall to your Eminence or me.' Within five or six miles from Chippenham, through Calne or Bowood Park, to the summit of the chalk downs, which are visible from Bowood, there is a beautiful geological series of formations exposed, ranging from the Bath Oolite to the chalk. By going one way and returning another, I was able in a few hours to give a good field lesson to a party of three ; they were apt scholars, and I was very glad to explore the country myself.

Ever your affectionate son,
Charles Lyell.

To George Ticknor, Esq.

11 Harley Street, London : December 26, 1846.

My dear Ticknor,—I shall begin this letter some days
before the next mail, that I may thank you for your last,
which interested us much, especially your account of the
inhaling the ether, of which many successful experiments
have already been made in the hospitals here, and about
which everyone was anxious to know, and knew not how
much to believe. Sir J. Clark and Dr. Boott learnt from
your statement, which Mary gave them when she met them
one day, the first precise account of the longest time (seventeen minutes) that the effect had been prolonged without
injury. We have been reading Dana's 'Two Years before
the Mast,' a delightful book; it is so rare to get a clear insight from an educated mind of what life really is in a
humbler grade among the working classes. Is he getting
on as a lawyer, and has his health never suffered from his
voyage? When you introduced him to me he seemed quite
well. I believe I mentioned to you in my last letter that we
were going to pay a visit at Bowood, at Lord Lansdowne's, in
the country (Wiltshire) of a few days. It turned out a very
agreeable holiday. The party were Mr. and Mrs. Milman,
and Mr. Harness, Lord and Lady Clarendon (he is now
President of the Board of Trade), the Chancellor of the
Exchequer, Mr. Wood and Lady Mary Wood (she is a daughter
of Lord Grey, and a very pleasant person), Mr. Charles
Greville, author of an excellent work on ' Ireland,' Lord
Devon. I had much talk with the President of the Board of
Trade on American affairs, on which he is well informed,
and has liberal and statesmanlike notions, without prejudice.
Indeed the freedom from that upon this subject is, I think,
singularly in proportion to the more elevated station in
society which men occupy here, which is more than some
would have expected between a democracy and aristocracy.

Mary showed the daguerreotype of Mr. Prescott's bust to
Lord Lansdowne, and they think that none have been as well
executed here, and Lord Lansdowne wished Mr. Prescott
would make haste and come to England, in which the others
joined. The Milmans were in good health and spirits. Tom

Moore was asked, but did not come. One day Lady Lansdowne gave me an Irish car to myself, and I gave Milman, Lord Clarendon, and the Chancellor of Exchequer a regular field lecture, taking them from ' the oolite to the chalk downs,' into which they entered with great spirit, especially the two ministers, who wanted to forget the state of Ireland, growing as it does every day worse and worse. I send you a letter I was induced to send to the ' Times ' [5] (the first time I was ever guilty of such familiar intercourse with the public press) in answer to innumerable inquiries on the subject, some of them from your side of the water. Dr. Gould will, I think, be glad to see it when you have read it. I am glad to hear that Agassiz saw so much of Gould, and appreciated, as he could not fail to do, his profound knowledge of zoology.

Believe me, dear Ticknor, ever most truly yours,

CHARLES LYELL.

[5] On the alleged co-existence of man and the Megatherium.

CHAPTER XXV.

FEBRUARY 1847—SEPTEMBER 1848.

ANNIVERSARY OF GEOLOGICAL SOCIETY — EDITORSHIP OF 'EDINBURGH REVIEW'—CALL ON ROGERS—HIS REMARKS ON HISTORIANS—BREAKFAST WITH HIM—CHAPTER ON UNIVERSITIES IN SECOND TRAVELS—RARE BOOKS—BRITISH ASSOCIATION, OXFORD—LINNÆA BOREALIS—DECAY OF THE DRAMA—HUGH MILLER'S 'FIRST IMPRESSIONS OF ENGLAND' —VISIT TO BOWOOD—PARTY AT SIR ROBERT PEEL'S—MEETING AT ROYAL SOCIETY—KNIGHTHOOD—DINNER WITH THE ARCHBISHOP OF DUBLIN.

To His Father.

11 Harley Street: February 21, 1847.

My dear Father,—I have been so much occupied since I got my book off my hands with arrears which had accumulated, that I have not given you any account of our scientific anniversary of the Geological Society when the first, and only copy yet out of *the book*, was laid on the table as a present to the Geological Society, and really looks better than you might expect a condensation of three into one to do, the print being respectably large. Horner, to whom as President of the Geological Society I have dedicated my seventh edition, got great credit, as he deserved, for his two years' services to the Geological Society. Sir H. De la Bêche, the new President, was supported by Lord Morpeth, who spoke with his usual eloquence, and after him Bancroft made a very good speech, ending with saying he had obtained leave to give my health. So I was called up very early, and told them my last news from De Verneuil and Agassiz, both foreign members of the Geological Society, and who having both visited the United States since I returned from America last spring, had given us as flattering an account of their reception, and the number of congenial enthusiastic minds they had been in contact with, as I could have done. I amused them by narrating what De Verneuil had said to me, when I asked him if he was not struck with the reigning idea which cheered one in travelling in that country: progress, progress. 'Yes,' he said, 'your information is quite antiquated; it is six months old. You have only seen the railroad penetrating westward 300 miles, through forests and morasses, but I have seen the electric telegraph extending beyond the termination of the railway; the wires attached to the trunks of trees from which the boughs had just been lopped off, and the news carried with the speed of lightning to villages and towns which have never yet seen a locomotive engine.' The Bishop of Norwich spoke well, though some might have been less pleased than we geologists to hear him wish that 'they who differ but in a little in theological matters, would only live in as great harmony with each other as he saw scientific men do, who differed so widely in their theoretical opinions on many points, which they could discuss, as we did, without quarrelling, and cherishing, as became Christians, the unity of spirit and the bond of peace.' Poor man! he has enough of it in his diocese, although he has not, I believe, been of late personally embroiled in the ecclesiastical disputes that divide many of his clergy. We are glad to have the Bunburys with us. 'The fresh woods and pastures new,' which you congratulated me upon now enjoying, consist of my joint paper with Charles Bunbury, on the more modern or Oolitic Coal and Coal Plants of Chesterfield County, near Richmond, Virginia.

Ever your affectionate son,

CHARLES LYELL.

To GEORGE TICKNOR, ESQ.

11 Harley Street: February 23, 1847.

My dear Ticknor,—We were glad to receive your kind letters from Park Street, with one, long looked for, from the Hamilton Coupers from Georgia, and others from Mac-Ilvaine, Miss Wadsworth, and other friends, which took us back very pleasantly to many of the points where we spent our time most agreeably a year ago. At a dinner at Hallam's the great subject was the vacant editorship of the

'Edinburgh Review.' Macvey Napier having died, Hallam and Milman had for years thought they had better bring the 'Edinburgh' up to London, as most of the contributors in these days live south of the Tweed. Longman the bookseller had been closeted with Macaulay for hours, consulting what could be done, the pecuniary and political stake being considerable. They had at first thought seriously of a Mr. Rogers (no relation of the poet, and I believe a clergyman), a contributor to the 'Review' of many able articles of late, on 'Puseyism,' 'Leibnitz,' and in the last, 'Pascal.' Lord Jeffrey was much opposed to the editorship leaving 'Auld Reekie;' but yet they talk now of his son-in-law Empson being editor, who has a professorship of law in the East India College, near London. He wrote the long article on 'David Hume' in the last. The article on 'Ragged Schools' in the 'Quarterly Review,' of which you inquire, is by Lord Ashley, who has gone practically and personally into the investigation. Milman wrote the other, on 'Popular Education,' which I am glad you approve of as much as we do. The article on 'Centralisation' in the 'Edinburgh' is by John Austin: they tell me it is 'good, but dry.' I remember when Lord Melbourne was considering the best way of dispersing a mob which they were anticipating, Sydney Smith recommended him to get John Austin to go and read them a chapter out of his 'Jurisprudence,' then just published.

We were discussing the other day the effect which it would have here in modifying religious intolerance, if the social and educational equality of the different sects were greater, which made me wish to know from you, as a matter of fact, whether any literary man or woman in Boston, whether of the Episcopal, Presbyterian (Calvinist), Romanist (if any there), Methodist, &c. &c., lived through the days of Channing's greatest celebrity as a preacher and writer, without, once in a way, and many of them oftener, going to hear him preach. I apprehend that the effect of such occasional interchange of churches, even where rare, is not only to make part of a congregation see that there is some good in the leading men of other sects (at all events that they have not horns and cloven feet), but that the controversialists who teach a flock, who have access to their opponents' arguments orally delivered, are more cautious and exact in what they say themselves. Bancroft made a good speech at our geological anniversary.

The education scheme here is crippled by the bottomless pit of Ireland, but is ripening and expanding the teaching of *schoolmasters*, more and more determined on as indispensable. But they are deplorably behind Boston.

Believe me ever most truly yours,

CHARLES LYELL.

To HIS FATHER.

11 Harley Street : March 5, 1847.

My dear Father,—I hope you will get this letter on your birthday, of which Mary and I wish you many happy returns; and beg you to accept from us a print of Paris, which recalled to our memory in a bird's-eye view all the leading features, and streets and gardens and boulevards, of that beautiful city most vividly.

I am busy writing a paper on the more modern Coal Field near Richmond in Virginia, and spent a good day yesterday in my museums here with Sir Philip Egerton, who has named my fossil fish for me, on which subject he is quite an authority, founding new genera and species which are generally adopted in Europe. We had a pleasant call at Mr. Rogers', whose sister is recovering from her fall. We found on the old man's table a speech of Charles II. to his Parliament, printed in 1661, in bad English, which he observed could never have been shown to Clarendon. Alluding to Macaulay, he said ' he had found him once writing a review with five folios open, each on separate chairs, but unfortunately, though conscious that the article would be known to be his, he was writing with that confidence and rapidity which if he had had to sign his name at length to the pages, he would not have presumed to do. Such was the unfortunate tendency of anonymous historical literature.' He then repeated, what I had often heard him declare, that Hallam wrote history as a judge and Macaulay as an advocate, and he blamed the latter for giving a set-down to Charles Fox's 'Life of James II.,' for which Samuel Rogers stood up manfully, taking the book down from his shelf, and, without spectacles, pointed to three or four of his favourite passages.

Mary sends her love and best wishes; and with love to all the family, believe me, my dear father,

Your affectionate son,

CHARLES LYELL.

To GEORGE TICKNOR, ESQ.

London: April 2, 1847.

My dear Ticknor,—We had a very agreeable breakfast the other morning with Mr. Rogers, in his eighty-fifth year, and though he is now often tired at dinner after the society of the morning, I never knew his mind more vigorous than it was. The party, besides ourselves, were the Bishop of Norwich and his wife Mrs. Stanley, both original and pleasant people, and the Bancrofts. The old poet took pains, as the Bancrofts were there for the first time, to do the honours of the house and pictures well, and often as I have heard him, it was most of it new to me. He has a very spirited reduced picture by Haydon, of Napoleon on the rock overhanging the ocean at St. Helena, the back turned towards you. I like it better than the one which I have seen at Peel's, at Tamworth (Drayton), which is as large as life; too large. This gives us all the sentiment, which was much heightened to us by Rogers, who you know is an excellent reciter, repeating some lines of a poem by Lyte, a Devonshire clergyman, written when the French first asked leave to remove the ashes of the Emperor from St. Helena.

> Disturb him not, he slumbers well
> On his rock 'mid the western deep!
> Where the broad blue waters round him swell,
> And the tempests o'er him sweep.
> Disturb him not, though bleak and bare,
> That spot is all his own,
> And greater homage was paid him there
> Than on his hard-won throne.
> Earth's mightiest monarchs there at bay
> The caged lion kept,
> For they knew with dread that his iron tread
> Waked earthquakes where he stept.

On my asking if he could show me the whole poem of Lyte's, he took an album from a drawer, where it was pasted in,

cut out of a newspaper, and as he gave it me, he said something in his caustic way which I cannot report correctly.
' When men have hit on a dozen good lines, they must
always dilute them with a hundred weak ones.'   In fact I
found a long set of verses, from which, not all from one place,
he had skilfully culled the best.   After he had told the story
of Chantrey having come as a workman at five shillings a
day to receive orders for the ornamented woodwork of the
table in the middle of the drawing-room, Mary happened to
say she had never seen Addison's table, on which he wrote
the ' Spectator.'   S. R. went for it, for it is too plain and
simple to accord with the rest of the furniture.   Opening a
drawer, he took out Addison's works, and read his dedication to Craggs, where he speaks of his desire that the
memory of his writings should not outlive that of their
friendship. 'A compliment,' said Rogers, ' which is the more
pleasing when we recollect, which few do, that three days
after he penned the dedication Addison died.'   S. R. was
describing some picture to Mr. Bancroft and Mary, while
Bancroft was still at Addison's table with the book.   I asked
him, as he had been so often at the Abbey lately, and in
Poets' Corner, if he could help me to the last two verses of
Pope's Epitaph on Craggs.   It is nearly thus, said I :

> He missed no public, served no private end,
> He gained no title, and he lost no friend.

But ' missed ' is certainly not the word. Afterwards, when at
home, I had turned up Pope, and read 'He broke no promise,'
&c.   There was some talk of duelling in the United States,
and the usual explanation that the North is not the South,
and that John Bull must try and learn to draw a line between
the New Englander and the Southern chivalry.   Upon which
Rogers told a story which was the more laughed at because
the ladies hardly knew whether they ought to laugh, and
which I think the old gentleman told expressly to Mrs. Bancroft to try whether she had any of the reputed prudery of
the Northern lady.   He remembered an acquaintance going
out to fight one Dr. Humphrey Howard, and seeing that his
antagonist had no garments on, sent his second for an explanation.   H. H. replied, that being a surgeon he well

knew that a bit of cloth carried into the flesh by a bullet
would make it fester. The other declined shooting at a naked
man, and so it ended. 'And if ever,' said S. R., 'I am called
out, I am determined to adopt the same costume.' There
was some talk of prose translations of poetry; Rogers in
favour of them, wishing the 'Iliad' had been well done into
prose, 'though I provoked Milman by saying so, and he declared he had made a vow never again to read any prose
version of any poem.' 'You make no exception?' said I. 'None
whatever.' 'Then we shall never hear you again read the
Book of Job, or the Psalms of David, in our ordinary version,'
said I. Empson has undertaken the 'Edinburgh Review'
after all. They talked of Rogers. 'Is he not a dissenting
minister at Birmingham?' 'Yes, and evidently a man of
first-rate talent; but how do you know that they did not
mean me, when they talked of Rogers as editor?' 'You
have written reviews?' 'A few. One on Carey's "Dante,"
and Moxon the publisher said I caused the sale of 2,000
copies immediately. Another on "Joanna Baillie," when I
called her several times "he" and "him" in the course of
the article.'

I am much obliged to you for what you told me in your
letter about Channing's works, and their sale, and Dewey's
'History,' and we read in consequence with still greater interest Channing's powerful discourse on the occasion of Jared
Sparkes' ordination. There is nothing which people here
so overrate as the relative proportion of the Unitarians as
compared to the population of all other sects in Boston, and
still more in New England, while they equally underrate, I
suspect, the influence which the writings of the American
Unitarians have exerted and do exert both in the United
States and in this country.

As to Mexico, one cannot but feel what Madame de Staël
said about *nations*, 'that nations always deserve their fate,
whatever that fate may be,' although it is no apology for
those who think it is 'our destiny' to be conquerors and the
instruments of fate. Few here have any idea how much
colonisation will be promoted in California, and what great
results must come of this movement.

My University chapter in my 'Travels' has been much

read here the last few weeks, and cited as authority in pamphlets, there being a stir for reform in higher as well as lower departments. Whewell has prepared a scheme for Cambridge reform which I have not seen yet. The other day Baron Stockmar, who is so much about the Prince (now Chancellor of Cambridge by a small majority on the poll!!), told me at a party 'that Bunsen had set him to read my *temperate* and *manly* article on the "Universities," by which he and Bunsen had learnt with surprise that abuses which they had attributed to Romanist times dated chiefly from Protestant ages, and were therefore less excusable.' Nothing, I find, surprises our University men themselves so much as to learn, *inter alia*, that whereas under Roman Catholic rule the principal teachers of academical youth married and settled at Oxford and Cambridge, all those who now really and efficiently engross the educational function are enjoined to celibacy.

I am making it a matter of conscience to get out what Lockhart calls my 'Greywacke,' and which he and the public generally skip, before I indulge in other matters; but if you can tell me anything about the pay and station of your schoolmasters of schools for the many, by which I can compare them to ours, I shall be glad. I know how many complaints there are in New England of teachers being underpaid, yet I fancy they are in affluence when contrasted with ours. I should also be glad to know about the foundation and endowment of the new chairs lately founded, or according to Gould about to be created, for science in Harvard University. Whewell has replied to my chapter on our University system, and he has also written me a private letter of remonstrance the other day. I am going to take a day or two to write a reply in full, of which I will send you a copy, which I shall be glad that Everett should read, as some of the statements by Whewell had made some impression on him; but he will see that I was right. Indeed public opinion is rapidly strengthening. There is a move now in the right direction; but the clerical influence arrayed against all progressive sciences, whether physical or literary, is too powerful to be easily overcome. My University chapter has been praised of late in quarters where three years ago it would have acted as an exclusion from society or good fellowship.

The moment I mentioned to Hallam, some weeks or a month ago, that England was the only country that had not conferred some public honour on Prescott, he said they usually waited till some friend took it up. Shortly after he told me, that he and Lord Mahon had immediately proposed him at the Antiquaries, the oldest body, after the Royal, which is for ' natural knowledge ' exclusively. In due time he will be elected, and will have the privilege, as explained in one of the old farces, of appending A.S.S. to his name as being of the Antiquarian Society a Socius.

Dr. Gould's letter of enthusiasm about Agassiz is very refreshing, and shows that Channing need not have been so severe on the atmosphere of Boston, as repressing all such feelings, or the expression of them.

As to Agassiz saying the negro's brain is like a child's fourteen or sixteen years old, if I am not greatly in error, Owen says the same of the adult male stolid and uneducated agricultural labourer. Tell Agassiz this, and see if it is new to him. On our recent fast day, the charity children, the youngest down to four years old, had to go to church, in many parishes three times; service two hours, and an hour and a half. Is there any persecution of the young perpetrated in New England on Sundays or other days of this kind?

But I must conclude. Believe me, dear Ticknor,

Ever most truly yours,
CHARLES LYELL.

To GEORGE TICKNOR, ESQ.

London: May 3, 1847.

My dear Ticknor,—Nothing which you have told me gives me greater hopes of the future than your account of Mr. Hillard's lectures, *and his class.* Pray tell me all such news. Mr. Hallam, when we breakfasted with him, was giving Mr. Winthrop an account of the sums paid for rare books by the Americans, one of whom bought at a sale the Mazarine Bible for 500*l.* (sterling) the other day. Sir Thomas Phillips told this to Hallam, who remarked that according to the old proverb, there must have been two fools at that auction. ' Yes,' said Sir Thomas, ' I was the other ; I bid against him.' Milman said that Sir Thomas could not have paid, so perhaps he was not such a fool after all. You see that Sir Walter Scott is dead. So ends that title, which one would have liked to have seen hereditary, if any. Lockhart looks ill.

Read Hugh Miller's ' First Impressions of England,' &c. I hope to send you a copy of my letter to Whewell by this post. Beg Mr. Everett to read it, and tell him I have an answer from Whewell which is quite a knock-under as to every disputed fact, and extremely pacific.

With my love to Mrs. Ticknor and her daughters,

Believe me most truly yours,

Charles Lyell.

To His Father.

11 Harley Street, London : June 25, 1847.

My dear Father,—I went down on Wednesday morning by the 10.0 express train to Oxford,[1] most of the time at the rate of a mile a minute, and our party, consisting of Milman, Bancroft, Grove, and myself, conversed easily, and it was certainly a magical transfer from place to place. It seemed strange to find Richards rector. I first encountered my old Swedish friend, of Lund in Scania, Professor Nilsson, who told me ' he was very unhappy, having lost his coffer ' (alias portmanteau), which with his broken English he could not recover. So I made a stir, and found it at an inn. He then showed me what he considers conclusive evidence of man having co-existed with *Ursus spelæus* and other extinct mammalia in Scania. Though not quite satisfactory to me, it is the best proof I have seen; he had his fossil teeth, bones, arrow-heads, &c., with him. I then met in the Theatre Whewell, Peacock, Conybeare, Buckland, Adams, Faraday, Owen, C. Darwin, Bishop of Norwich, Chevalier Bunsen, Sir R. Inglis, Murchison, De la Bêche, Sedgwick, Sir J. Richardson, Wheatstone, and many others with whom I spoke. I called on Mr. Cary, but found him overwhelmed

[1] Where the British Association was held.

with law business in the Vice-Chancellor's Court. He spoke of your present of your 'Dante,' &c. As Darwin was lodging with the Jacobsons, I called with him, knowing Mr. Jacobson, who is Vice-Principal, I think, of Magdalen Hall. After I was gone, Darwin, seeing my 'Travels' on Jacobson's shelf, asked him how my University chapter had been taken at Oxford. He said, 'Extremely well; much read, and considered very temperate, and full of proofs of my having thought deeply on their system, and that he had heard many who differed from it *in toto* praise it.' I found on my table many cards of Exeter men, now become fellows, &c. The first day I dined with Baden Powell, who has married my friend Captain W. H. Smyth's daughter, to meet Sir John Herschel and Sir James and Lady Ross. I learnt much of the Antarctic expedition; some very wonderful new facts about the temperature of the very deep parts of the sea. In the evening we had an immense party at the Botanic Garden, at Dr. Daubeny's, where I was introduced by Sir Thomas Acland to his son, now Professor of Comparative Anatomy, and who takes all my views of the reform required in their system. Trevelyan, who was there with his wife, tells me he is one of the cleverest of the scientific Oxonians, and will be very distinguished. Young Buckland had a young bear, dressed up as a student of Christ Church, with cap and gown, whom he formally introduced to me, and successively to the Prince of Canino (Charles Buonaparte), Milne-Edwards, member of the French Institute, and Sir T. Acland. The bear sucked all our hands, and was very caressing. Amid our shouts of laughter in the garden by moonlight, it was diverting to see two or three of the dons, who were very shy, not knowing how far their dignity was compromised. Next morning I breakfasted at New College with Mr. Philip Duncan, a large party in Hall, most of whom I knew. The Dean of Winchester and some I had seen at Southampton in September were there. I then went and spoke at some length at the Geological Section, on what I had seen of 'raised beaches' in North America, on the Canadian Lakes, and in the valley of the Seine, subjects on which Robert Chambers (author of the 'Vestiges') read a paper. Sedgwick spoke well, and Professor James Forbes of Edinburgh, and Darwin. Dalby was with us in the omnibus for a stage, and he and the sub-rector and my friends Darwin and Major Clarke had had a luncheon dinner together in the Common Room at Exeter before starting.

I left at six o'clock. With me in the coach was Dr. Twiss, who had been active on Cardwell's committee, and whose indignation at the conduct of even the more moderate party at Oxford was at the boiling-point. How extremes meet! I really might have thought that some one of my New England Whig friends was complaining of the ultra democrats reducing them to mere delegates, and destroying all free agency by requiring so many pledges as to future votes. It was not enough to go into all pending questions, but they put several hypothetical queries as to possible measures of future Government; as to educating or endowing Roman Catholics; and declared the answers not specific enough, and so drove him to retire; and now Round is said to have the best chance, and Maynooth and the Corn Bill, which the Cardwell party had hoped were dead and buried, and in the tomb of the Capulets, are to make the University, as Milman says, 'totus teres atque *rotundus.*' Inglis (who is quite a star in the House compared to Round) being so sleek, fat, and good-humoured, and on such good terms personally with all, even those who regard him as the parliamentary representative of the prejudices of the age, is truly the *totus teres.* Out of twenty-four heads of houses, only four at Oxford to receive the Association! But it will go off the better by the absence of the lukewarm or the hostile. I spoke to Hallam, Ehrenberg, and a great many others. Mary will tell you, I hope, of the Lansdowne House party; my eyes are so much better now that I enjoyed it much. I was glad at Oxford to see more of Ruskin, who was secretary of our Geological Section. I like him very much. If you have not seen his book on ' Painting,' you will be able to do so at Kinnordy. I have borrowed it of Susan.

Believe me, my dear father, your affectionate son,

CHARLES LYELL.

To CHARLES BUNBURY, ESQ.

Kinnordy: July 27, 1847.

My dear Bunbury,—The prospect of the Cape Journal coming to light is to me a great source of pleasure. I found your *Linnæa borealis* with more than *a hundred flowers* in full blow in the first week of this month. Some shoots, eight inches further from the centre of the patch than last year, and I think I have reckoned backwards that thirty years ago it might have been sown by a bird, but is probably older. Under a hot sun, the almondy scent of the flowers is very strong and delicious; after rain there is no odour at all. The day I last wrote to you I was going up the Clare Hill. I saw there, and took, the northern butterfly *Polyommatus Artaxerxes*, which I had never seen (Lepidopterist as I am) on the wing before. Our fairy ring is of larger diameter annually.

In haste, and with love to Frances, believe me, my dear Bunbury, ever affectionately yours,

CHARLES LYELL.

To GEORGE TICKNOR, ESQ.

Kinnordy: August 25, 1847.

My dear Ticknor,—I am now fairly through my principal geological papers, one of which, of some length, on the Coal Field near Richmond, is just printed, with illustrations of the fossils I procured there. In looking over my notes nothing surprises me so much as the activity and progress, and the intellectual excitement and promise for the future, which I saw in so many parts of the Union, and of which there is here such an inadequate appreciation, and a kind of dogged incredulity which nothing but an array of statistics will disturb, coupled with the evidence of an eye-witness. In spite of the multiplication of steamboats and newspapers, and complimentary speeches to Winthrop and Bancroft at public and private dinners, John Bull is in general in a state of wilful blindness as to what is doing on the west side of

the Atlantic, and he will wake up some fine morning, and
hardly believe his eyes when he beholds the *white* population equalling his own, and the number of those who can
and do read and write twice as great.   I remember when
first the French Chambers began to discuss the first principles of a representative government, the English were
astonished to see how many things even the highest Tories
had been accustomed to take for granted, and how much
they had inherited as a prejudice which they of the old
régime in France deemed revolutionary.   Pray let me have
letters from you on the evidence of the advance of the
educational and literary and scientific cause when the spirit
moves you.   When I was first at Boston all theatres were
shut up; now you have several.   Are any steps taken to
render them less objectionable in the eyes of those people
who in England regard them as corrupting and loose in their
morality, from the bad selection of pieces acted and the
character of the performances?   I fear that circumstances
do not favour in New England what one would wish to see—
the converting the theatre into a place of refined intellectual
culture, holding the same rank in that line which a philosophical lecture-room does in science, or a church in matters
of religion.   Is not this decay of the drama a strange anomaly?   If the mere music and show, and costly scenery of
our modern melodrama, to which crowds go, had been mediæval, and the acting of 'Macbeth,' and the writing of it had
belonged to this age, would not this have seemed to you
more in the natural order of things?   Song charms the sense,
&c.   De Verneuil told me, and I repeated it at a public
dinner in London, much to the edification of my scientific
friends, that when they had extended the electric telegraph
to the extent of the railroad, they carried it on from one tree
to another in the aboriginal forest on trees from which the
boughs had been lopped off.   It gives a lively image of going
ahead, which no Frenchman could have invented, least of all
our quiet friend de Verneuil.   Could you learn from one of
your Boston law friends, whether in old colonial times the
business of barrister and attorney was separated in New
England, as it has always been in this country, or whether
the union of the two grades was made to suit a more demo-

cratic constitution. We suffer, in my opinion, from this marked line of demarcation.

Ever, my dear Ticknor, yours most truly,

CHARLES LYELL.

To GEORGE TICKNOR, ESQ.

11 Harley Street, London : September 26, 1847.

My dear Ticknor,—Perhaps you have seen some notices in the ' Athenæum ' of a book by Hugh Miller called ' First Impressions of England and its People.' Do read it, as we have been doing. He is the editor of the ' Witness,' Free Kirk newspaper, has risen from being a common quarryman to be one of our best writers. If you are now and then startled at the inconsistency of his highly philosophical views on the connection of science and religion being held in the same mind with ultra-covenantic views on other points, you must remember that he is a sincere enthusiast in the Free Kirk line. I have been re-reading with care your letters on educational matters. It is clear that you have reached what I can never live to see accomplished in England—a successful raising up of a body of secular or lay teachers as schoolmasters and mistresses, having on an average the same scale of pay, and a similar position in society to the clergy. The average pay of the latter in New England it would, I suppose, be more difficult to ascertain, as there are so many sects ; and as they get presents, and have fees for marriages, burials, and christenings apart from their fixed salaries. If the most eminent lay teacher in England could hope to equal in salary the Archbishop of Canterbury, or if the latter could be levelled down to the standard, if something could be done to place secular and clerical teachers on the same social footing, we might get on and be emancipated from our trammels, and education proceed and move on with the spirit of the age, which it now lags behind. If it were equally gentlemanlike, men would take less and be lay teachers, so many of the best have scruples in undertaking the clerical office, such as it is with us. Making the Bible a school-book here, and setting poor children to read *Deuteronomy*, is a proof that our Church teaching is not meant to open their minds. People will have education, so they manage thus by sham instruction to evade what they dread, *i.e.* the making them capable of thinking and reasoning. I believe the Bible is not a school-book with you. The success of the 'ragged schools' in Edinburgh has been lamentably checked by the Presbyterians and their clergy insisting on the Bible, and this making the Catholic priests object to the Romanist children attending, to whom it would have been of peculiar use. I am afraid that our clergy think they should lose influence if lay teachers were raised in schools. I imagine, on the contrary, that all the higher minds at least would gain. Here it is notorious that if you have three sons, and determine that they shall belong to the legal, medical, and clerical profes-sions, you would select the least talented as the safest for the *Church.* I hope the progress of popular education may one day change this.

Ever most truly yours,
CHARLES LYELL.

To HIS FATHER.

Bowood : January 13, 1848.

My dear Father,—I presented your 'Dante's Lyrics,'[2] as you suggested, to Lord Lansdowne as from yourself this morning, and gave him also a nicely bound copy with the Italian on one side which is always much liked.

Lord L. remembered your essay, not only because he reads and remembers all that comes out, but because Mr. Davenport used to talk for ever to him of Rossetti's views, and Lord L. had thought that Rossetti had gone too far. Lord L. also asked me what you thought of Ugo Foscolo's performance, which he had read and commented on. He skimmed through your last edition, and said the analysis prefixed to the Sonnets was most useful and necessary. He thought there was so much in what is now going on in Italy like the old contests of Guelphs and Ghibellines, that it was calculated to keep alive the interest of Dante's writ-ings. He begged me particularly to send his thanks for the books.

---

[2] *The Canzoniere of Dante Alighieri, and the Vita Nuova and Convito,* translated by Charles Lyell, Senior, of Kinnordy, N.B.

We found, what is always most pleasant, a small party, Lord Auckland and Macaulay having left to attend a Privy or Cabinet Council, the number at dinner being ten—our host and hostess, the Milmans, Mrs. Norton and her boy, a gentlemanlike young Etonian, Mr. Charles Austin, famed for having made more money in four years at the law than anyone ever did before, some say 40,000*l.* a year, by railways, and Mr. Charles Buller, M.P., who has now undertaken the poor law, and to be chief of the department. The conversation was very agreeable and without effort; Milman in good spirits though not strong, Buller and Austin both very original and literary. Fielding's novels were discussed *con amore,* Dickens and Thackeray rather run down too much, till young Norton said he was ' boiling;' Chaucer and Dryden; and then Mrs. Norton, who is really almost as handsome as ever, very ladylike and clever, had an argument with Milman, who maintained that translations of poems should be in verse and not ' stilted prose ' as he called all prose translations.

Our party at Mr. Rogers' on Monday was brilliant, and no one engrossed too much. Mr. Empson, now editor of the ' Edinburgh Review,' and Mrs. Empson (Miss Jeffrey), Hallam, Babbage, Eastlake, and Mr. Luttrell; the latter, though oldish now, came in now and then with his witty sayings. Lord Campbell's ' Chancellors,' in which a letter of Lady Philip Francis, acknowledging her husband to be Junius, is given, brought up that old controversy, and Rogers confessed the truth of the tale, that when he was set on at Holland House to ask Sir Philip Francis if he might put a question to him, Sir Philip replied ' At your peril ! ' in so forbidding a tone, that Rogers retreated to the rest, and said ' If he is Junius, it is Junius *Brutus.*' On some one calling in question the great superiority of Junius, Rogers cited in support of it an able passage on the difference between injuries and insults; but Hallam said, ' After all, there is nothing in Junius so powerful as the comment of Dr. Johnson on it, when he said " that some people mistake the venom of the shaft for the vigour of the bow." ' When Luttrell complained of cold, Hallam said, ' Don't let Rogers hear you, for his maxim is that no man can be cold except he be a fool or a beggar.'

The party are all going for a walk, so I must join them. Perhaps Mary will write to-morrow.

Believe me, my dear father, your affectionate son,

CHARLES LYELL.

To CHARLES BUNBURY, ESQ. (*who was travelling in Italy*).

Bowood : January 14, 1848.

My dear Bunbury,—Your letter was most agreeable to me,[3] especially as knowing something of that region of calcareous rocks in the south, and remembering the general aspect of the vegetation without having made out why it looked as it does, which your description of the part played by the different species of trees, shrubs, and plants, many of which I well remember, so well explains. I believe most of the limestone which we used to consider as part of the cretaceous series, with its nummulites, is now voted by many to be intermediate between Cretaceous and Eocene, but I should like to know more about that. I am very sorry that Susan's drawing-book was lost, as I am sure it contained many valuable memoranda.[4] I am much pleased with her picture of Mary, and think the last touches improved it much. Eastlake says it is very well painted, much beyond any former work of hers, and as to likeness he added, 'Unhappy as I have often thought the lot of the portrait painter, yet I do not believe that their plight is quite so miserable as that people should not be satisfied with such an excellent resemblance to the original as this.' My father has read your 'Botany of Southern France' with much delight. The chief geological news is the arrival of eight hundred specimens of bones of fossil birds, in the most perfect state of preservation, from clefts and caves of New Zealand, buried in loose volcanic sand. They belong to ten or more species of Moa, or Dinornis, one ten to twelve feet high. The skulls and eggs are found. The largest would have made a

[3] Treating of the botany of the South of France.

[4] His sister-in-law, Miss Horner, who had been in Rome with Mr. and Mrs. Bunbury, and who suffered shipwreck in the 'Ariel,' off the Vado rocks, near Leghorn, and lost her drawings among other things.

more gigantic footprint than the biggest of Hitchcock's Connecticut marks. Eastlake, whom I sat next to at Rogers' dinner, was, I find, much taken with Owen's article on ' Paleontology' (on Broderip) in the last ' Quarterly Review.' Sir John Herschel's, on ' Humboldt's Cosmos,' in the last ' Edinburgh,' is also much thought of.

Ever affectionately yours,

CHARLES LYELL.

To HIS SISTER.

11 Harley Street : February 7, 1848.

My dearest Eleanor,—When we read your letter this morning, and saw how much you and the rest, including my dear mother, had been thinking of *our lecture*,[5] in spite of so many of you being invalids, we were quite sorry that we did not some of us send a note on Saturday to say that it came off much to my satisfaction. The consciousness that no one else either in Europe or the United States could from actual observation have given the same account of the nature of the evidence, and what was genuine and what spurious, respecting the proofs of the first quadruped or air-breathing reptile ever found in such ancient rocks as the coal-strata, gave me confidence and spirit, as I knew it would be of interest to all the geologists present. It was, as Dr. Fitton remarked to Mary before I began, a glorious audience, and the full benches made me perform with much more ease to myself. Before lecture, Faraday pleased me by some hints, and showing me his private memoranda for lectures, and telling me what I suspect he hardly ever did to anyone, part of his mechanism for timing the different parts. It is wonderful how much I had come by practice to the same results. He has given fifty or a hundred lectures to my one, and is the chief master of the art in England, to my mind. The number present, as noted down at the entrance, was 398, but it may have been more, as Mr. Barlow does not quite trust the new porter.

With love to all, ever your affectionate brother,

CHARLES LYELL.

[5] Lecture at the Royal Institution, on a Reptile in the Coal-formation of the Alleghanies.

To His Sister.

Sunday, February 27, 1848.

Dearest Marianne,—I must give you a few words on my party yesterday, as, if I lose any time, there is so much going on in these eventful times that I shall forget what passed. When I got to Sir R. Peel's, Bunsen was the only one arrived, so after making my bow to Lady Peel and her daughter, and shaking hands with my host, I heard the Prussian Ambassador give us a full account of a member of his embassy who had left Paris on the Thursday of the Revolution passing over barricades and having some hairbreadth escapes. Bunsen told us that there are some 30,000 communists in Paris who are for property in common and no marriage, and who are much to be feared by those who have aught to lose. He (Bunsen) blamed Louis Philippe for not having seen some months ago that a moderate extension of the suffrage could alone defer some catastrophe. The party which next came in were Faraday, Sir Stratford Canning, Buckland, Lord St. Germans, Owen, De la Bêche, Sir Benjamin Brodie, Mr. Hamilton, Hallam, Sir James Graham, Dr. Mantell; making with me and a young son of Sir Robert's seventeen; the eighteenth or missing one was Lord Aberdeen, whose excuse came in after I got there. I sat between Buckland and Faraday, F. being next Peel, on whose other side was Hamilton, and next him Owen, and we had much pleasant talk, both *tête-à-tête* and we six all together. After the ladies retired, I expected little more than a continuance of conversation with next neighbours; but Sir Robert remaining in the same place, asked Bunsen to tell them his latest Paris news, which he did from near the top of the table, all questioning him. Then Sir Robert gave an account of what the French Ambassador had told him just two hours before, and I do not think I ever before remember fifteen persons exchanging so many ideas, chiefly because most of them were in different ways public men, and not afraid to hear the sound of their own voices, and partly from so many being well acquainted with nearly all the others. Sir S. Canning was asked what he thought would be the effect on

Switzerland, from which he had just returned. Hallam talked a good deal with Bunsen, Graham, Peel, and Canning on the merits and demerits of Lamartine's new book on the French Revolutionary period, the effect of which some of them regard as very mischievous, especially now. If Louis Philippe, even at the last hour, had on giving up Guizot chosen Thiers or some other eminent reformer, Barrot for example (neither of whom are republicans), instead of a conservative like Molé, his throne might perhaps have been saved, and if not to him, at least to his dynasty. I was struck with the little feeling the whole party had for him, though they were of course hoping he would escape. Hallam, who was intimate with Guizot when he was in London, and since at Paris, told me as he took me home in his carriage, how anxious he was about him and his safety. He said he could not defend his late career, but he liked him much.

Sir Robert told us very clearly, at Brodie's request, two stories of his having respited two different criminals, condemned by the Judges on circumstantial evidence, both of whom, when Secretary for the Home Department, he proved to be guiltless.

Believe me your affectionate brother,

CHARLES LYELL.

To HIS SISTER.

March 16, 1848.

At the Romillys' party yesterday, we had a specimen of the monetary panic of Paris transferred to London; not a few in the room, especially those connected with Geneva, having considerable investments in the French funds, canal shares, railroads, and having their political misgivings greatly heightened by their sensitiveness about the loss of their incomes. Mr. Strutt, M.P., Mr. Wickham, of the Stamp Office, and their wives, Mr. Marcet, Dr. Roget, Mr. and Mrs. Kay Shuttleworth, Mr. Parker, of the Treasury. Shuttleworth told us that Mr. Jones Loyd, the great banker, had just informed him that his foreign broker had failed to negotiate a good bill on Vienna, and told him that at that moment there were only two important cities in all Europe, Hamburgh and Amsterdam, on which he was able to negotiate bills. So general a commercial disturbance was never known. Lord Normanby had just sent for 500 sovereigns to get on with, until specie was forthcoming in Paris. The general opinion seemed to be, that long before the 900 meet, or even before the elections begin, there will be a collision of parties. The day I dined with Sir R. Peel, when the Republic was just proclaimed, Sir Robert said there will be a financial crisis. But many here think that might have been avoided but for the measures of the Provisional Government, which frightened capitalists by increasing the burdens by 25,000 additional troops; increased rations for sailors in the navy; a promise to find work and wages for any one of the 35,000,000 that might want it; one pound a day for every one of the 900 until they had made a constitution; paying one per cent. more for all money in the savings bank; and no retrenchment except of the civil employés. Against which we have to set the giving up the tax on the newspaper stamps, and a promise to give up the *octroi* of Paris. Mr. Empson, now editor of the 'Edinburgh Review,' remarked to me that either the whole world, writers on political economy, statesmen and all, have been in their dotage up to this hour, and Louis Blanc has made the grandest discovery of the age, or they are the most arrant fools for promising to find work and wages, to give labourers the profit of capital. There are not a few who hope that after some extraordinary changes the country will right itself, for they have a plentiful harvest in prospect and vast resources, but it is an awful experiment and risk.

Ever affectionately yours,
CHARLES LYELL.

To HIS SISTER.

London: April 2, 1848.

We went yesterday to see the studio of Lough the sculptor, who is a man, I think, of considerable genius, and has some spacious rooms filled with models, many of them colossal, such as 'Milton's Satan,' and some which he made for the House of Lords, but did not succeed in his competi-

tion. I think his model for the Nelson column, which was
nearly chosen, for he missed it only by two votes, would have
been very superior to that which gained the prize. A large
statue of Nelson, without the cocked hat which now disfigures
it, not placed out of sight in the clouds, with four statues of
sailors at the four corners of the pedestal, and on the sides of
this alto-relievos of sea-fights, &c. Among the company
viewing the sculpture was Madame Bunsen, who told me
that her husband had just received the official news of the
fight in the streets of Berlin, and that instead of the newspaper story of 2,000 soldiers killed, there were only 140, and
only four officers instead of forty, and they presume of the
people in the same proportion. She flatters herself that the
Government is now firmly established in Prussia. We
had a pleasant party at Sir Edward Ryan's;[6] he in good
spirits, and a young friend of his, about thirty, Mr. Gibbs,
a lawyer, fellow of Trinity College, Cambridge, well informed and lively; Dr. Fitton in great force. I got him to
refresh my memory with his story of a Dublin professor
who said to his class, 'Gentlemen, the Hon. Mr. Boyle
was a great man; he was the father of chemistry, and
uncle to the Earl of Cork;' from which, says Fitton, his
pupils worked out the conclusion that chemistry and the
Earl of Cork were first cousins. At the Zoological Gardens
to-day, we saw Broadhead, the Secretary of Legation,
who told me there is no doubt that the Duke and Duchess
of Montpensier went off in a huff. They called suddenly
and unexpectedly on the Queen, not *incog.* but in their
public capacity, when she was in a delicate state of health,
and not knowing exactly what to do, and fearing she
might commit herself, she very naturally sent off to Lord
Palmerston to come directly, but before he got there, they
had waited and waited till they got out of humour, and
set off in about thirty hours afterwards, on their way to
Rotterdam, whence they are said to have taken ship to St.

[6] Late Chief Justice in Calcutta, from whence he returned in 1843. He was
made a member of the Privy Council and Civil Service Commissioner, and for
his sound judgment and sense, his advice was sought by statesmen and others,
and his genial nature was a delight to all who had the privilege of knowing
him.

Sebastian, and if Spain be in the state reported, they will soon wish themselves back in London. The other day I had a short talk with the Bishop of Oxford, who said that Guizot had the merit not only of writing history, but of making it. The first day that news came of the French having proclaimed their Republic, and among other things the Provisional Government having ordered troops to the frontiers, I met the Bishop of Llandaff at the Athenæum, and he asked me what 1 thought of the affair. I said it reminded me of the old cycle of political events, which the ancients were so fond of believing, upon which he repeated the lines : ' erunt etiam altera bella; Atque iterum ad Trojam magnus mittetur Achilles ; ' and he added, ' I wonder who will be their great Achilles this time.' I told him probably some ' petit caporal ' of whom none of us have yet heard. The Swiss consul, John Prevost, remarked to me to-day, that one result is clear; most of the French capitalists and bankers are ruined.

On Sir Edward Ryan's table were several handsome pieces of plate, after classical models, worked by native Hindoos, and with inscriptions intimating that they were given from gratitude to him (Sir Edward) for his exertions to schools and colleges for the education of the natives. Sir Edward told us when we were talking of modern orators, that he heard Lord Lansdowne speak for an hour on the sending an ambassador to Rome, and although it read well in the papers, it was far better in the delivery, as he was very animated and his action and manner so good. A petition is getting up for a royal commission to go to Oxford and Cambridge, and the signatures of fellows of colleges, and even of the clergy, are already most numerous. I have obtained a good many, and a pamphlet is just come out, quite in the spirit of my chapter on the ' Universities,' by the Rev. Mr. Jowett and Mr. Arthur Stanley, son of the Bishop ot Norwich, which is excellent. A considerable reform at Cambridge is much agitated, but a little pressure from without is desirable, to aid the reformers within. A strong wish to revive the utility of Professor Smyth's Historical Professorship at Cambridge, is showing itself in the University.

Ever affectionately,
CHARLES LYELL.

To HIS SISTER.

April, 1848.

When I got to the council-room,[7] I found Sir Robert Inglis and his party mustering pretty strong. After De la Bêche had spoken in favour of scientific Presidents and a short term of office (two years), I seconded his motion, and as an argument against going on with aristocratic Presidents, as we have been doing so long, I pointed out that of forty-eight members of the Upper House who now sign F.R.S. to their names, and who may be taken as that part of our higher aristocracy who care most for science, not one ever wrote a single paper in the 'Philosophical Transactions' except Brougham, thirty-three years before he was made a peer. I then mentioned four or five exceptions, of men who like Lord Northampton, then in the chair, had a real feeling for science. I said I admired our peers as a body for their talent, but that made their neglect of science the more marked. It was owing to our University system, which also governed the great schools; even Cambridge, while she put a powerful instrument into the hands of a student, gave him no taste for its application, and, like Oxford, shunned the progressive sciences. I have no wish to disparage our aristocracy as a political body, nor wish them to have less social rank, but why do homage to a body that care for and know so little about science? Inglis and Hopkins spoke at full length on the other side, and it ended in our agreeing not to divide. We gave up insisting on two years' limitation, while they agreed not to insist on permanency; that science should have its turn at least, which as we have a majority they could not help; and finally that Colonel Sabine and Mr. Lyell be appointed a deputation to go down to Sir John Herschel's, in Kent, to try and persuade him to be President. Whether we succeed or not, it is a point gained that it should be offered to science, and if he declines, we will try some other. Faraday's health makes him out of the question. A great many of us are for Robert Brown.

CHARLES LYELL.

[7] Of the Royal Society.

To CHARLES BUNBURY, ESQ.

May, 1848.

My dear Bunbury,—If I have not written to you for some time, it is because the affairs of the Royal Society, still in a state of transition, have occupied my leisure moments. Trying in vain to get Herschel and Faraday to be Presidents, then to stir up Robert Brown's courage to let us assert a principle in nominating him, then ending with getting Lord Rosse to say he would stand, then choosing the fifteen out of all the candidates for the year, then a new secretary (Grove, I hope), and various reforms against a set of obstructives, compared to whom Metternich was, I presume, a progressive animal. When Sismonda, who has made a beautiful map of Sardinia, came here, I immediately asked him if he would lay before you all the evidence from fossil plants of the supposed occurrence of carboniferous species in the lias. He told me he wished much to do so, that it was high time that Adolphe Brongniart's first dicta thereupon should be reconsidered, and that he had the only complete collection, and all of them should be submitted to you. He thinks the problem, as I do, the most puzzling and curious of all the theoretical points yet undecided in Geology, and worth a journey to Italy by any competent botanist like you, whose papers he knows, as he takes in our Journal. He is not quite satisfied that Brongniart has gone critically into the affair, though I suspect you found that he had, yet Sismonda has got together great materials since that.

You will be glad to hear that Owen, after seeing Goldfuss's published account, entirely confirms all that Falconer told us of the true reptilian character of those fossils in the coal of Saarbrück, for the age of which V. Decken vouches.

So now, besides my Pennsylvanian footsteps we have about three species recognised by their bones, skull vertebræ and all, of an order which according to the theory of progressive development came in only with the Permian. But they will merely shift their starting point, and never take warning. Owen told me yesterday, that Bunyeep now turns out to be a

foal and not a calf, and I believe I shall have to make out
the sea-serpent to be a shark.

Ever your affectionate brother-in-law,

CHARLES LYELL.

To CHARLES BUNBURY, ESQ.

Kinnordy: August 2, 1848.

My dear Bunbury,—We have been following you with
much interest in your journey through Tuscany and the land
of musquitoes and robbers, and I was very glad to receive
here your letter to me, in which, among other matters, you
mention an entomological confirmation of a story I have often
told, till the wonder it excited began to make me doubt it myself. I took a large dragon-fly (if I am not mistaken, *Œshna
juncœa*), and having carefully eviscerated it, put a straw, *secun-
dum artem*, inside the abdomen to keep it distended. Before I
impaled it, the creature, which had been stunned by a blow
(in the thorax, I presume), recovered, and flew out of the
window. I ran out and watched it to the roof of the house,
when the wind taking it, it sailed off in grand style, the
straw seeming just to balance the lost weight of viscera.
But I admire your cicada much more for doing without a
make-weight. You inquire about Joseph Hooker. The last
piece of news is enough to make your mouth water. He
had found *three* new species of Magnolia in the Himalaya
mountains in one day. His father has printed a letter in
the 'Botanical Magazine,' with amusing gossip about the
journey through Egypt with Lord Dalhousie, Aden, and
Madras, where they were received (the Governor-General and
suite) in state, and a few passing notes on the plants and
fossil forest in the desert.

I have walked with Mary to the site (still the only one)
of the *Linnœa borealis*, which now measures 13 feet 3 inches
in a north and south direction, and 19 feet east and west.
By-and-by I shall compare it with its size in former years,
and we will see if we can calculate back to the day when
some grouse or emberiza, driven down by cold from the
Grampians, and burrowing for blackberries in the snow,
rubbed off one of the minute seeds from their feathers in the

spot where you were destined to see it so soon after you had dismounted from your steed in 1844. It was in full flower in the end of June, when Katharine [8] paid it a visit, and picked some of it. At Sir Philip Egerton's we saw a large sheet of water entirely covered with the leaves and yellow flowers of *Menyanthes nymphœfolia.* I am glad you saw Targioni, and hope you will have found Sismonda at home. I am glad you saw the Brazilian ferns of Raddi, as I have no doubt you gained new ideas. My father received your letter from Florence, which we all read with much pleasure. The foraminifera of the Sienna marls are innumerable as I am well aware, and I have seen recent marl from the Mediterranean off the delta of the Nile, equally full of recent species.

August 3.—I have just been reading an extract in the last 'Edinburgh Review' (art. on 'Sharpe's Egypt') from Miss Martineau's 'Egypt,' certainly one of the most eloquent in the English language. If Harriet should be excommunicated for certain doctrines boldly put forth in this the most able work she has produced, it is certainly better worth being a martyr for than her mesmerism. Joseph Hooker was shown in Malta some of the huge Miocene fossil shark's teeth as having been those of St. Paul. If the Apostles had a succession of teeth through life, like the Squalidæ, without a proportional quantity of toothache, from which I have been suffering for the last week, I envy them their gifts. I shall be glad if you return by the Simplon, as, having seen it twice, I should like to see it often again, whereas I am quite satisfied with the Mont Cenis, although the Italian ascent of that pass is very interesting. As we shall be talking of new 'Quarterlies' and 'Edinburghs' when you return, I may as well mention that Milman wrote the article on 'Pope Ganganelli' in the last 'Quarterly Review,' with which my father is much delighted, and Guizot that on the 'State of Religion in France,' translated by Mrs. Austin, with which I have been much interested. Carlyle is said to have written the entertaining article on 'Goldsmith' in the 'Edinburgh,' and I know that Francis Newman wrote that on 'Tests,' and Peacock that on 'Herschel.'

[8] His sister-in-law, Mrs. H. Lyell.

With my love to Frances, believe me, my dear Bunbury, ever your affectionate brother-in-law,

CHARLES LYELL.

To LEONARD HORNER, ESQ.

Kinnordy : September 11, 1848.

My dear Horner,—As there is nothing in this morning's (Friday's) Gazette, there may be in that of Tuesday. The delay has I presume been caused by Lord Lansdowne's having left town for Bowood. I send you an extract from Lord Lansdowne's letter, that you may understand how the matter stands : ' I have thought it on consideration so fit in the distinguished situation you occupy, and with your scientific reputation, that you should receive the distinction of knight-hood, that I took it upon myself to mention the subject to the Queen, and I have her Majesty's authority to state that she will most willingly confer it upon you, and she under-stands that it is without any solicitation on your part. Had you been near London, I should have proposed to you to attend the Council on Monday for that purpose, but it can be gazetted without your attendance.'

Ever yours affectionately,
CHARLES LYELL.

To GIDEON MANTELL, ESQ.

Kinnordy, Kirriemuir, N.B. : September 24, 1848.

My dear Mantell,—Although at the southern extremity of the island, you are the first person, excepting a member of my own family, who has addressed a letter to me under my new denomination, and I thank you sincerely for your congratulations. The manner in which the honour was con-ferred, both on the part of the Ministers and the Queen, has been such as every scientific friend would approve of, and I had a most agreeable geological exploring on the banks of the Dee, into which Prince Albert entered with much spirit. I am sure your lectures in the Isle of Wight will have sown some good seed, for the Milmans, to whom I lent your book, have been profiting by it in their examinations of Purbeck, and speak of it most approvingly. I am glad to hear of the

new discoveries, and that the two papers for the Royal
Society are getting ready. By staying in town, I and a few
others flatter ourselves we have put at the end of the session
the printing and referring of papers on a better footing for
the future, though we have still much to do. The Royal
Society council work is no sinecure to those who have a reforming spirit. I have hitherto missed no meeting since I
was elected.

Believe me, with my wife's kind remembrances, ever, my
dear Mantell, yours most truly,

CHARLES LYELL.

To HIS SISTER.

London, 1848.

My dear Marianne,—Mary has given you the journal of
our late proceedings, but has not had room for any particulars, and I have not time, as I should like to have, were
it only to refresh my memory, to tell you of some of the
people we have met and the things they said.

Few dinners have pleased me more than that at Dr.
Whately's. Fortunately two M.P.'s, I know not who they
were, did not come, but sent their excuses just before we sat
down, it being the second night of the debate on Church
Rates and of the division. This reduced the party to six, a
number who could all talk together, and could not talk *party*
politics, an excellent representation of France and of
America, and the Archbishop, a strange compound of an
Oxford Churchman grafted on Ireland, and full of information about all that is going on there, which he views with
interest more as a political economist than in any other light,
as far as I could judge.

Mr. Dewer is a lawyer of considerable practice in New
York, now deputed to arrange some banking concerns between the United States and England. He wrote the first
good digest of American law. His good temper, I would
almost say magnanimity, in talking to De Beaumont on the
tender question of slavery in America, considering what a
satire 'Marie' contains, struck me much. We talked
of Miss Martineau's book, parts of which must be very

clever. The Archbishop said, that if women ever became
invested with political rights here, it might be well to have
two Houses, and let the women *speak* in one and the men
vote in the other, for since the Irish members have got in, he
saw no other way of economising time.   Dr. Whately is a
great philologist.   When on such subjects he said, 'De
Beaumont, you have no word for " home." '   De Beaumont
said, ' No, perhaps because we have less of the *thing* than
you have.   We have said of late, " mon chez moi," " son chez
soi," but that is very clumsy; and then you have another
word, "job," which we cannot translate; it is a sublime
word that—God knows we have the *thing*.'   The Archbishop
was philosophising on the cause of their not having the word
'job,' and said that their representative form of government was so new, and in a pure monarchy there were fewer
true jobs.   De Beaumont said, ' Certainly there are no jobs
under an absolute despotism ; because it is all *one great job,*
and there is no room for small ones.'   De Beaumont gave us
an amusing account of his wonder at the aristocratic manners
of society in England as contrasted with France, and its
effects on social intercourse, in which he saw clearly enough
what was absurd and ridiculous.   Dr. Whately explained to
him why Englishmen, being in fact socially disposed, were
obliged to be stiff to strangers, because all were trying to
hang upon and be pulled up by the skirts of those above
them.   When De Beaumont asked how many grades there
were in society, the Archbishop said, ' I cannot say how low
it goes, but the other day some chimney-sweepers presented
a petition to the Lord Mayor against others who had intruded themselves into their privilege of dancing, &c., on
May day, and in this petition they said " that certain dustmen and other *low fellows, pretending to be chimney-sweepers,*"
&c., so the degrees of rank probably descend even below the
dustmen.'   A very interesting discussion arose about the
effects of the French, American, and English laws of the
division of property as contrasted with each other.   De Beaumont defended the system of division so far as agriculture
was concerned, which Dr. Whately was most afraid of.   I
then asked De Beaumont whether he did not feel already that
the small fortunes in France enabled a wealthy Government

to bribe the majority, not only of the electors, but even of those elected.　He said, 'Yes; but then we must greatly increase the number of electors, and we may do that safely when millions become landed proprietors, no matter how small their estates.'　I objected that still the Government could offer irresistible bribes to the majority of the deputies, which was impossible when there were so many large fortunes as in England.　When I had thus driven him home, he began a sort of confession, having previously seemed determined to be an optimist on French affairs.　He said, 'I grant you that it is equality rather than liberty that we are tending to; the wide distribution of property is raising the *morale* of the great mass of our population above yours, and it is producing a centralisation force in the Government; but this must end, I fear, in the central power being so great there is an end to that liberty which I believe nothing but an aristocracy can preserve.'　The Archbishop brought out a pamphlet to prove that in one district near Paris the average property of eleven thousand landed proprietors was the quarter of an English acre each, and he began imagining, when the division had gone farther, a question of law arising as to whether a huntsman had committed a trespass by clearing his neighbour's estate at one leap.　Dewer denied that the equal division in America was producing small properties, which he supposed must arise from the rapidity with which large fortunes were daily made, and the new lands obtained in the new settlements.

At our party at Mr. Whitmore's yesterday, the Archbishop told us one of the late Lord Norbury's puns.　The two curates of St. George's Chapel, Dublin, were then, and are now, Mr. Short and Mr. Bridge—and very long-winded preachers.　Lord N., on being asked how he liked them, said 'he could wish that Bridge was shortened, and Short abridged.'

With love to all, your affectionate brother,

CHARLES LYELL.

CHAPTER XXVI.

JANUARY 1849–JULY 1851.

DR. HOOKER IN INDIA—MACAULAY'S 'HISTORY OF ENGLAND'—TOUR ON DEE-SIDE --VISIT TO SIR JAMES CLARK—BALMORAL—TICKNOR'S 'HISTORY OF SPANISH LITERATURE'—VISIT TO GERMANY—BOTANY OF THE HARTZ—GEOLOGY OF SAXONY—LETTERS OF IGNATIUS—PREJUDICE AGAINST SCIENCE—TRACTARIANISM—LECTURE AT IPSWICH—EXHIBITION IN HYDE PARK—RAFFAEL'S CARTOONS.

[He was re-elected President of the Geological Society in February 1849, and during his term of office communicated two important papers, ' On Craters of Denudation, with Observations on the Structure and Growth of Volcanic Cones,' and 'On Fossil Rain-Marks of the recent Triassic and Carboniferous Periods.'   He published his ' Second Visit to the United States of North America ' this year, and he presided over the Geological Section at the British Association at Birmingham.

In 1850 he became one of the Royal Commissioners for the first Great Exhibition, which was opened in Hyde Park on May 1, 1851. Sir Charles Lyell gave several lectures at the Royal Institution and elsewhere during these years.]

CORRESPONDENCE.

To His Father.

11 Harley Street : January 11, 1849.

My dear Father, —Mary has, I believe, told you that I was clever enough to scald my foot, and was getting well, but walked so many miles yesterday that I am now a prisoner for a day.   I put a finishing stroke by walking to St. James' Street to attend a meeting of the Graphic Society, to which a committee of artists had invited me.   There I found Ruskin, Eastlake, Boxall, and other artists and amateurs, Hallam, Babbage, Sir R. Inglis, Bishop of Oxford, and saw some very beautiful drawings by Turner, and by one

Knowles, of the Parthenon, and heard good talk about Phidias, and how he would be astonished to see the black stains caused by smoking chimneys of cottages, and red stains of iron on the columns, instead of his azure and gold; and how, if Lord Elgin had not bribed the Pasha, the French Consul would, and had both left the sculpture, time would have destroyed more than half ere this. I am getting on steadily with my work, and we have finished vol. i. of Macaulay, though frequently referring back to it, to read again favourite passages. Longman is stereotyping it, encouraged by the rapid sale. There is a higher moral tone in it, and less party feeling, than in any of his reviews, and it is the result of an amount of reading and of remembering what he has read, which is wonderful to think of.

Sir William Hooker has been corresponding with Charles Bunbury, and tells him that Joseph Hooker was going under escort of a guard from the Rajah of Nepaul to ascend the highest mountain, more than 28,000 feet high, and that he tells him that since he left the plains of India, no less than *four* of his (J. H.'s) intimate friends have died, and among them Mr. Williams the geologist. This intelligence comes home to me, for when I drew up, after several communications with Lord Auckland, my address to the East India Board on sending out a geologist to India, I pointed out that every one of five geologists whom they had sent out had been cut off in the prime of life, for want of aid in assistants, elephants, steamers, &c., which could alone enable them safely and effectively to perform their mission; and I protested, with De la Bêche, against the best of his practical men (the said Williams) being sent out on a forlorn hope. He has done his business, poor fellow, well, put them in the way of working rich mines of coal, and is now left like his predecessors to die in a ditch. Heaven grant that poor Joseph Hooker may be spared, but I dread Assam and Borneo, and would rather have the work on the Antarctic regions and still unpublished notes than all the magnolias in the Himalaya. His last paper on fossil botany is much thought of. With love to all at Kinnordy,

Believe me, my dear father, your ever affectionate son,

CHARLES LYELL.

To GEORGE TICKNOR, ESQ.

February 7, 1849.

My dear Ticknor,—Agassiz' lectures on 'Embryology' interested us much, and the sale of them astonishes our naturalists. We have read Macaulay with much delight. It is a great proof of his genius that he makes one feel as if present at the scenes he describes, and as if one had lived in those times, almost as much as in reading Pepys' 'Diary.' There is a want of generosity I think in regard to Penn and some others, not that one wishes to have their delinquencies concealed, but the characters of such men, and of Marlborough, ought to be allowed to unfold themselves, and they ought not to be drawn before the events are related for which they are so unsparingly condemned. I had an argument with Macaulay about the atrocious crimes which he says God commanded the Israelites to perpetrate. All he could say was that he was on 'the orthodox' side, and that according to my view 'the God of the Old Testament differed from the God of the New,' and that if I merely thought that the Israelites *believed* they were acting under God's special command when they hewed Agag in pieces, and that they were mistaken, I should be obliged to renounce their history and the prophets.

That he himself believed it he neither said nor the contrary, but certainly he need not have written that passage about the Puritans in such strong terms. As Francis Newman, the brother of the great Tractarian who has gone over to Rome, takes the lead here of our rationalists, so the brother of the Puseyite saint, Mr. Froude of Oriel, who died and left some curious memoirs, has come out with a book called the 'Nemesis of Faith,' not very short of Theodore Parker's opinions, and written not without ability. As he is a lay fellow of Exeter College, Oxford, it remains to be seen whether they will or can expel him.

I have been very busy with my inauguration dinner as President of the Geological Society, and succeeded in getting the Archbishop of Canterbury (Dr. Sumner, author of 'Records of Creation,' a geologico-theological work), Sir

Robert Peel, Van de Weyer, and a great many M.P.'s and notabilities to come, so that the speaking is allowed to be the most brilliant we ever had at any anniversary. Sedgwick spoke very eloquently, and Peel; and the Archbishop made a straightforward and manly speech. I have been returning the Archbishop's call to-day, and was much struck with the magnificent appearance which the new houses of the Legislature, or the ' Palace of Westminster,' make from the Lambeth bank of the river. When will you come and see the new creation, in the old florid gothic style, grouping beautifully with the towers of Westminster Abbey? Ever most truly yours,

CHARLES LYELL.

To LEONARD HORNER, ESQ.

Kinnordy : July 29, 1849.

My dear Horner,—I return you Empson's letter with many thanks. His commendations and Lord Jeffrey's are a reward for taking pains with the book, and if, like an *achromatic* telescope, it has presented objects as they are, without imparting to them any colouring of its own, I may feel gratified by that compliment. I fear, however, that the strong anti-slavery movement in New England, from a natural desire at this moment to stop the spread and accession of new Slave territory, is causing them so to misrepresent the real state of the South, and the negroes, that they will write against that part of my work. I have just finished reading this morning your first anniversary address, which is excellent. Your criticism of Agassiz where he disputes all identification of recent and fossil species is well put. You ought to have seen him at Southampton on our dredging expedition. E. Forbes showed him some of the varieties of common shells, which occur also fossil in the red crag, and which Agassiz had declared were not varieties, but extinct species. When he was convinced that the same varieties flourished now, he said ' if Lyell would give up half his percentage doctrines, he would give up half his objections,' or something to that effect.

Believe me ever affectionately yours,

CHARLES LYELL.

To LEONARD HORNER, ESQ.

Birkhall: September 3, 1849.

My dear Horner,—I have been spending my time in a very enjoyable and by no means uninstructive way here, and have felt all my visiting at Balmoral, in the whole course of which there has not been a single drawback, not only pleasant, but I hope useful, as helping forward that taste for natural history and science which is intended to form a part of the education of the young people, and expected to prove a resource to them.   What Van de Weyer said of the steady development of Prince Albert's mind, in a great variety of directions, I had been able more to appreciate.   His German reading on serious subjects makes him an improving companion to one who is not versed in what is going on in that world, and I had much good talk with him alone, on a variety of grave subjects, as well as on the different *insects* which belong to Switzerland, the Isle of Wight, and Scotland respectively.   That he knew so much about these was quite a new light to me.   I had two long walks with Mr. Birch and his pupil.   Lord John Russell made himself very agreeable to me, and I like the rest of the household.   There reigns a strong wish to live here in privacy, which is most wholesome to the mind, after such a life of representation. The Prince has just finished Macaulay, and his discussion of it with me was one of many fruitful subjects for comparing notes and opinions.

With love to all at Rivermede, ever affectionately yours,

CHARLES LYELL.

To HIS SISTER-IN-LAW, MISS HORNER.

Kinnordy: September 5, 1849.

My dearest Susan,—I told your father that I would send you a little gossip of our late tour on Deeside, during which I had some good geological work on some remarkable hills composed entirely of serpentine in Glen Muick, and some botanising, with much useful and healthy exercise, in addition to agreeable companionship with the Clarks, *père*, *mère*, and *fils*, and parts of three days which I spent at Balmoral.

The day I went to dine there, Saturday last, I had first a long walk—Sir James Clark and I—with Mr. Birch [9] and his pupil, a pleasing, lively boy, whose animated description of the conjuror, or Wizard of the North, whom they had seen a few days before, was very amusing. 'He (the wizard) had cut to pieces mamma's pocket-handkerchief, then darned it and ironed it, so that it was as entire as ever; he had fired a pistol, and caused five or six watches to go through Gibb's (one of their footmen) head, and all were tied to a chair on Gibb's other side,' and so forth; 'but papa (Prince Albert) knows how all these things are done, and had the watches really gone through Gibb's head he could hardly have looked so well, though he was confounded.' Sometimes I walked alone with the child, who asked me the names of plants, and to let him see spiders, &c., through my magnifying glass, sometimes with the tutor, whom I continue to like more as I become better acquainted. After our ramble of two hours and a half through some wild scenery, I was sent for to join another party, when I found the Queen, Prince, and Lord John, by a deep pool on the river Dee, fishing for trout and salmon.

On the way home we visited the quarry with the trap dike cutting through limestone, and producing garnets, which I found out last year, and I talked a little about it. After the Queen had entered the Castle, the Prince kept me so long, and we kept one another so late, talking on all kinds of subjects, that a messenger came from her Majesty saying it was only a quarter of an hour to dinner-time.

After the ladies had gone to the drawing-room, we had much lively talk, which the Prince promoted greatly, telling some amusing stories himself, and encouraging others by laughing at theirs. Lord John Russell, and Lord Portman and myself, had nearly the whole talk to ourselves, but anyone who chose could cut in. Sir J. Clark was not well, and could not dine with us. I played whist in the evening with Colonel Gordon, Mr. Anson, and Mr. Farquharson of Invercauld.

Next day I went to church, and we heard a good practical sermon on good works by Mr. Anderson, who has none of that shyness attributed to him by the 'Times' correspond-

---

[9] Tutor to the Prince of Wales, who was then not eight years old.

ence. The prayer for the parish, magistracy, Queen and royal family, judges, ministers of religion, parliament, and whole nation, was just such as you would have liked, and in excellent taste, with nothing which a republican jealous of equality could, I think, have objected to, and which I believe our sovereign and her husband would thoroughly appreciate the simplicity of. They shoved the box on the end of a long pole to Queen and Prince and maids of honour, as to all the rest of the congregation, and each dropped in their piece of coin. After church I had much conversation alone with Prince Albert, whose mind is in full activity on a variety of grave subjects while he is invigorating his body with field sports. Charles Bunbury has just summoned me to take a geological walk with him. Caroline has just come in, in great glee, to announce the discovery, for the first time, of *Orthotrichum Lyellii*[1] in fruit, which Charles Bunbury admits is a grand find.

Believe me, with love to all,

Ever your affectionate brother,

CHARLES LYELL.

To CHARLES BUNBURY, ESQ.

11 Harley Street, London : January 17, 1850.

My dear Bunbury,—I have been so busy with my anniversary speech that I have had no time to thank you for your letter. There was a good discussion, and Darwin adopts my views as to Mauritius, St. Jago, and so-called elevation craters, which he has examined, and was puzzled with.

I am much pleased to hear that you have paid my ' Second Visit' a compliment which I fear I rarely pay any modern book, that of re-reading it. I believe if you visited the United States, and stayed long enough, above all, if you re-visited it after a lapse of five years, you would feel exhilarated, as I did, at the rapid rate of general progress in wealth, order, intelligence, education, and rational (or at least more rational) ideas of government in so brief a period. I am told by my correspondents that I should now find things greatly improved

---

[1] A moss named after Mr. Lyell, senior.

since 1846, even in literature. For example, my friend Ticknor has come out with three large thick octavo volumes on the ' History of the Literature of Spain,' the work of his life, of which Hallam, Milman, Ford, and Lord Mahon speak in very high terms, and which I have read scraps of with much pleasure. Murray says that such a book could scarcely here do more than pay its expenses, 750 copies being as much as he dares reprint. Now three different booksellers offered Ticknor for the use, for one year, of his stereotype plates, 5,000 dollars, and Harper, of New York, at last gave 5,250, or above 1,000 guineas. This, Ticknor says, could not have been done in 1846.

As to California, so many New Englanders are gone there, who, with emigrants from New York, &c., all understand each other, that they govern and legislate for all the rest, and have improvised a constitution far superior to any of the new ones planned and overthrown, or now on trial in the last two years in Europe. The determination not to have slavery bespeaks a wholesome prevalence of Northern men. In France I fear the new law of schools will place them far too much under the Catholic priesthood, and the same cause will prevent the Italians from becoming fit for representative government, and they will be voted incapable because those above them are determined they shall remain children. But periodical revolutions such as we have witnessed of late are the lamentable alternative of not acting honestly towards the millions, and having no faith in human progress.

I am glad you have planned your new book. Adolphe Brongniart's article on the genera of fossil plants [2] is a masterly production so far as I can judge. It is only by this double view, first botanical, then geological and chronological, that the subject can be thoroughly exhausted. There is some hope even for the *Old* World, when such essays can appear under an experimental republic with universal suffrage.

Agassiz, I am told, is about to marry a young lady of good Boston connections. If so, he will be a New Englander for the rest of his life, and will be the founder of a school of

[2] *Tableau des genres de végétaux fossiles*, 1840, *extrait du Dictionnaire Universelle d'Histoire Naturelle.*

zoology (for he has many pupils) of a high order, as he
teaches them to dissect and go into all the minutiæ. His
enthusiasm is catching, especially when he has a good soil
to work upon.

Believe me, my dear Bunbury,

Ever affectionately yours,

CHARLES LYELL.

To LEONARD HORNER, ESQ.

Berlin : August 8, 1850.

My dear Horner,—The Brocken Inn, where we slept four
days ago, is a curious sign of the length of time that the
Germans have been lovers of travelling for the sake of admiring natural scenery. It has been built fifty years, its
foundations standing 3,500 feet above the level of the sea, in
the latitude of London or thereabouts, and the spot being
buried some nine months of the year under deep snow. It
is solely for the pleasure of those who admire wild scenery,
who like to take two chances of an extensive view from a
summit usually enveloped in vapour and to see the sun rise
or set, if they cannot do both, as we did. I was very desirous
of testing by an examination of this mountain the grounds
of Von Buch's theory of the granite here having come up
' in a bubble,' and his singular notion that its dome-shaped
form and even outline support this hypothesis. Now close
to the said Brocken House is one of those tabular masses
which we have at the top of many a Grampian Hill, of
granite, and which corresponds to the Cornish tor, the rock
decomposing in tabular masses. It is called the Witches'
Altar, twenty feet high, and about two or three miles from
this I found the Schnärcher, which I understand are mentioned in Goethe's ' Faust,' to consist of two magnificent
tors of granite, the highest of the pair being 100 feet high,
and showing unequivocally to what an extent *denudation* has
gone, and how absurd it would be to imagine that the present surface is the original one of the mass of granite here
exposed. Lonsdale's remarks, unpublished I believe, of the
modern tors now forming in the granitic Scilly Islands, show
how the sea is producing similar outstanding peninsulas or

'drongs,' as they call them, in the Shetland Isles. I have
no doubt that the Devonian slates once enveloped the top,
as they do now the sides of the Brocken. So much for this
' bubble.' Its indefatigable author is on a pedestrian tour
in Switzerland, as Ritter tells me, so I have left my card
only at his door. The upper chalk of St. Peter's Mount,[3] to
go back to the beginning of this tour, is very beautifully displayed, and a young pharmacien of Maestricht, Bosquet,
author of a paper on fossil tertiary species of Cypris, was an
excellent guide, showing me where the change from our
chalk fossils began. He and his master had a fine collection
of shells, reptilians (monitor of Maestricht, huge turtles, &c.)
to show us. De Koninck, at Liége, is a first-rate paleontologist, and gave me a great insight into the carboniferous
and tertiary formations of the region. I was glad to observe
between Aix-la-Chapelle and Cologne, how the railway-cut-
ting exhibited the loess and its associated gravel and sand
at the watershed between the basin of the Meuse and Rhine.
But think of my surprise when I got north to the Porta West-
phalica, and found, 500 feet above the level of the Weser, on
the inclined oolitic beds, 700 (?) feet above the sea, a patch of
loess, just like our old friend on the Rhine, and full of the
three shells figured in my ' Elements' as the most common,
Succinea oblonga, half freshwater, *Helix plebeia*, and *Pupa mus-
corum*. Strombach of Brunswick, and the elder Roemer of
Clausthal in the Hartz, were much surprised to hear of this
find of mine. To explain it will, I suspect, require some
changes of level which have not entered into the calculation
of many here, who think the Scandinavian drift of the Northern plains the last geological phenomenon of importance to be
accounted for. Charles Bunbury will be interested to learn
that the newest examination of the shells of the six divisions
of the Devonian of the Hartz leads Roemer to think that
none of the older beds (slates, limestone, &c.) of this chain
are so new as the coal. Sedgwick and Murchison thought
the uppermost of the divisions might be carboniferous. Now
in this are numerous plants, Lepidodendrons, one Astrophyl-
lite, Calamites, but all save one Calamite perhaps, of distinct
species. This would help to confirm Goeppert's view as to

[3] Near Maestricht.

the specific distinctness of his Silesian flora of the Old Red. Goeppert has lately seen the Hartz flora, and says the *species* are not carboniferous. This would harmonise well with the close analogy in generic types, and the distinctness in *species* of the fauna of the Old Red when compared with that of the coal. An excursion with Strombach to an anticlinal axis called Asse, near Brunswick, was most instructive. It is clear that the keuper and chalk were formed in connected basins extending from England to North Germany. So was the Wealden, which supplied first-rate shining coal for the steam-engines at Minden and elsewhere, our locomotives, 'Blücher' and others, which carried us along twenty miles an hour over the vast drift-covered and sand-covered plains of the 'über-geschwemmtes land.' Yet ridicule was thrown by us for seekers after coal in the Wealden, where the limits of the same vast delta existed. The needles of a pine are as mani-fest in some of this Hanoverian Weald coal as are ferns, Sigillariæ, &c., in the older Stein coal. As to the botany of the Hartz, it seemed a tame affair after the White Moun-tains, and it was impossible not to compare them as we rode up from a woody region to a bare summit of granite covered with the same *Lichen geographicus.* The spruce fir, with scarce any intermixture even of birch, covers the hills, and up to the very top some dwarf trees are seen, or so near that I believe if sheltered they might grow much higher. At the summit a white anemone new to us was still sparingly in flower, and plenty of it in seed covered the ground. There are also *Trientalis Europœa* in fruit, *Galium saxatile, Gna-phalium (dioicum ?)* in flower, besides the commonest Lyco-podium, *Lichen rangiferinum* and *islandicus*, with several *vulgar* plants, such as *Ranunculus flammula* and *Calluna vul-jaris,* also the cranberry *Vaccinium.*

What most surprised me in the Hartz geologically, high and low, granite and limestone, marble or trap, whether at 700 or 3,500 above the sea, was the total absence of all glacial marking, whether polishing, or furrowing, or stria-tion, or *roches moutonnées,* or erratics. Such a contrast to the southernmost part of Sweden or corresponding latitudes in North America! Yet some of the Scandinavian drift reaches the southern base of the chain.

I find there is Eocene tertiary under the drift in the environs of Berlin, and brown coal with palm wood under the Eocene shelly clay. I have seen the fossil shells.

Mary is sending a letter about more modern affairs. Ritter has just called again. Farewell, and believe me, with love to all,

Ever affectionately yours,

Charles Lyell.

To Miss Carrick Moore.

Dresden : August 3, 1850.

My dear Miss Moore,—I find that Mary is sending off a letter to you, so I take one of the sheets which we got on the summit of the Hartz mountains as a memorial of our having performed what was once a great feat; ascending to the top of the Brocken, which is covered with snow seven months of the year, and though in the latitude of ' The Cedars,' [4] presented to us a set of mountain plants agreeing for the most part with those with which we were familiar in the Grampians. It is 3,500 feet high, which to Germans accustomed to that endless plain which stretches from Bremen to the Oural, and which on our way from Berlin to this place we traversed in a north and south direction, may well appear a magnificent height. I was well satisfied to be carried by railway to near the foot of the Hartz, and by a carriage which carried all our luggage to the inn at the top, for by going up one road and descending by another, I was able to examine the granite of which it is composed, and other points of geological interest for which one has little time and courage when all one's energy and patience are required to contend with physical obstacles.

Our visit to Potsdam turned out most propitious, Humboldt giving us a good deal of his company and conversation and the *entrée* to the finest view from ' Sans Souci,' where the King was residing, not open to the public. The three palaces are certainly fine, and an oasis in a desert must always have a peculiar charm, so that the gardens pleased us much, and the great fountains 120 feet high, which I believe I

---

[4] Where Miss Moore resided, near Richmond.

could stand admiring as long as any child. But I confess
that nothing pleased me so much as an exact restoration of
one of the buildings of Pompeii, an atrium or outer court of
a house, and a bath room, both of which form an integral
part of the palace called Charlottenhof. These were erected
by the present King [5] when Crown Prince, the bath being
used. The court has no roof in the central part, and as it
was a fine day, the blue sky, reminding me of Italy, had a
beautiful effect. Looking at the beauty of the proportions of
the room, and all the statues and paintings which adorned
it, one could not but feel how little we have outdone the
Greeks of a small provincial town, 2,000 years ago, in taste,
even in the palaces of our days. The restoration also by
Lepsius of one of the temples of Philæ on the Nile, in the
museum, all the columns and walls being painted as on the
original, a facsimile, only reduced to one fourth of the
original size, is an excellent way of carrying one back 3,000
years. On the whole, however, the Berlin collection of
Egyptian antiquities did not appear to me to be comparable
in extent to that of the British Museum.

We go up the Elbe to Saxon Switzerland to-morrow. The
weather both here and in the Hartz perfection. I sat long
before the Madonna di San Sisto to-day, and can feel its
beauty. The wealth of pictures is overwhelming to one who
is not able to give weeks to them.

Believe me, with kindest remembrance to your father and
mother and sister,

Very truly yours,
CHARLES LYELL.

To LEONARD HORNER, ESQ.

Liége : August 28, 1850.

My dear Horner,—I believe in my last letter I mentioned
that I was going with Credner a geological expedition to
try and determine the relative age of the loess of Thuringia
and the northern erratics. He took me to critical points,
and I returned with a quantity of new facts leading me to
believe the loess to be for the most part posterior. As near

[5] Frederick William IV.

Stuttgard the loess passes downwards in some places, as at
Weimar and north of Gotha, into calcareous tufa and
travertin, the latter being used as at Rome for a building-
stone.   This calcareous formation is from fifty to eighty feet
thick, and I searched in it for gyrogonites, seeing many
stems of Chara, and found them probably recent species, as
the shells, freshwater and land, are of living species.   But
a great prize was a mass of tuff with four eggs, which I
bought from the workmen near Gotha, perhaps of a tortoise.
They are fixed in the solid rock, and were entire when
found.   They were fifteen feet deep in the tufa deposit, and
in the same beds were bones of *Elephas primigenius, Rhinoceros tichorhinus, Ursus spelæus,* &c., and recent *Limnæa*
and *Planorbis.*

The beds I have been fossilising in between Dresden,
Weimar, Gotha, Eisenach, Cassel, Giessen, and Frankfort
are triassic.   The muschelkalk everywhere, as I formerly
knew it at Baireuth.   The keuper with Calamites, some growing like the East Virginian ones at right angles to planes
of stratification; the lower member, the Bunter, usually
without organic remains.

I was glad at Halle to be taken by Professor Giebel to
pay my first respects to the roth-todt-liegende, or Permian
conglomerate, and the contemporaneous porphyries out of the
ruins or rolled fragments of which it is chiefly made.   Again,
at Eisenach, where there is a kind of section of the Thürin-
genwald chain, I found the Wartburg, where Luther threw
the ink-bottle at the devil, is made of this same conglomerate
800 feet thick, and Charles Bunbury will be interested in
hearing that the Psarrolites occur in the lower beds of the
lowest member of the Permian.

With love to all, believe me ever affectionately yours,

CHARLES LYELL.

To CHARLES BUNBURY, ESQ.

11 Harley Street, London : September 14, 1850.

My dear Bunbury,—I am truly glad to hear not only of
your botanising in the Urwelt, but also that you do not mean
to let the plants remain for nine years more as unknown as
they have been for the last nine million years.

When I met Colonel Codrington and Sir J. Wilson in town since my return, they said they could see by my looks I had been holiday-keeping in the country. So much for that whirl over the Scandinavian drift and Thuringian trias which seemed to you, when you read an account of it, as anything but a repose after London exertions. But when a man has had to write and print a book and Presidential Address, and dine and preside at the Geological Society, and carry on society, and an 1851 Royal Commission, all at once, you may conceive what a rest our tour must have been, though the grass did not certainly grow under our feet anywhere. We went to bed and got up with the sun, usually an explanation of my returning better and Mary fatter, which perhaps you will scarcely sympathise with, being like your late uncle (and as Audubon says the mammalia generally, considered as a class, are) a nocturnal animal. To give you an idea how far from easy it is to get at the repositories of treasures in your way, I asked everywhere for the names of Dresden geologists, and most people said Geinitz is the only man. At last, after writing to him, I learnt the name of an officer, Gutbier, whose work on Permian fossils I afterwards bought. He was quartered at Dresden. We arrived and I set off for Geinitz. Not at home. Lieutenant-Colonel Gutbier ordered out to a grand review before two Austrian Archdukes. Mrs. Geinitz assured me that there was not a soul, young or old, caring for stones or fossils in the whole city. Not believing this, I inquired diligently for dealers, and found there were two, who sold stuffed birds and insects. The first of these whom I beat up said he had no minerals, but once sold some to a teacher of painting, Professor Zeula. Off I went to where Zeula once lived, he had changed his abode; went off again. At last found him, an artist selling pictures and antiquities, and just going to lecture on painting, all very unpromising. But at last showing him my Anniversary Address, and making him aware that I was *somebody*, he gave me the overhauling of 200 drawers, more rich in well-arranged fossils, and among others Permian and German coal plants, than any I saw in public or private cabinets. He would, I think, have sold the collection, though he got them from a real love of geology. The moral of this is, never to believe that in any great

German city—and the same is true in France and the United States—there are no paleontological resources.

Believe me, with love to Frances, ever affectionately yours,

CHARLES LYELL.

To GEORGE TICKNOR, ESQ.

London: 1850.

My dear Ticknor,—I believe I said that I thought Bunsen would lose his time in entering into the controversy about the Letters of Ignatius, but after a conversation I have had with him in which he gave me a sketch of the points in his new German pamphlet not yet translated, I have changed my mind.

He contends that the Syriac MS.[6] not only proves that the passages in the first three letters which were relied on as proving the Apostolic succession were spurious, but also that the four other letters which were an expansion of the inter-polated passages were altogether forgeries. In short, 'the three Letters' translated by Cureton are all that Ignatius ever wrote. Bunsen has remarked that if Mr. Keble, our great Puseyite, could say in his preface to Hooker that it was a special intervention of Providence which led to the recovery of the Letters of Ignatius at a time when Episcopacy was in danger, what shall we now say of that Providence, which just at a moment when a certain party of the English Church are pushing the pretensions of the clergy to the verge of Romanism, has brought to light this MS. in the cellar of an obscure convent in the Libyan desert beyond Cairo?

Bunsen says that when the eight verses are omitted, the three letters are such as he might be supposed to have penned in prison (A.D. 107); but not so the other four.

I believe the Syriac MS. is of the fourth century. How early did these frauds and forgeries begin, and who shall say that all the three letters are pure and undefiled, or what degree of certainty have we in respect to the full and literal genuineness of documents still more important of which no MSS. earlier than the fourth and fifth centuries exist? I learnt with much surprise from Bunsen, *à propos* to forgeries

---

[6] Purchased for the British Museum from an Egyptian convent, containing the original Epistles of Ignatius.

and pious frauds of the primitive and mediæval Church, that
Luther published the last edition of his Bible without the
passages in St. John on 'the three heavenly witnesses,' but
that soon after Luther's death they reprinted his Bible with
them inserted. As they are wanting in all the four (?) older
MSS. of the New Testament at Rome, as well as the Codex
Regius at Paris, I have always considered it a great want of
good faith on the part of the Protestants, that we distribute
millions of Bibles without any mark to show to the multitude
that these verses are spurious. Am I not right in saying
that a great many, nay most of the Bibles used by the Unitarians at Boston, contain these verses without any italics or
marginal note to show that they were interpolated? I presume that if this be the case, it would be defended on the
ground that the Unitarians do, what our ministers do not,
viz. explain that the passage is not genuine. It is precisely
the same plan adopted here in regard to Moses and his
geology. The vulgar hear the first chapter of Genesis read
out without comment or the smallest explanation from
ninety-nine out of a hundred pulpits, and they grow up
in the belief of the modern origin of the globe, and the unity
of the creation of man and the globe, and all the inhabitants
which have ever lived upon it since the beginning. Hence
they regard scientific men with suspicion and with prejudice,
and yet no educated clergyman could now-a-days enter the
field against the popular creed of the geologist or astronomer,
any more than he would stand up and attack Erasmus or
Porson about the 'three heavenly witnesses.' In short, in
spite of the Reformation, the Bible is nearly as much a
treasure in sacerdotal keeping as if we had all gone over
with Newman to Rome; and in spite of Cureton, and Bunsen,
and Blanco White, one is sometimes tempted to ask whether
good faith and a regard for the sacredness of truth is not a
rare exception to the rule in Anglo-Saxondom at least, whatever it may be in Germany. The sectarian spirit of the two
divisions of the small Episcopalian Church in Scotland, the
'Drummondites' and their opponents, the High Church or
Puseyites, is blazing out almost as violently as the old Kirk
and the free Kirk; and then there is Romanism, and the new
edict even of Pius IX.! against the excellent new Irish

Colleges intended by Peel to be open to all religious denominations ; and the Wesleyans, who are now much worse than
the Establishment in setting themselves against progress,
and admitting no interpretation which Wesley did not
sanction, although he, had he lived, would have moved on
with the age.   This and the narrow views of our dissenters,
and the power of the English Church to substitute a sham
national education for a real one, makes one almost despair.
Even some of the most liberal of our clergy assume in their
pamphlets that the labouring class will be made unhappy if
taught too much, so they take care to confine them to the
merest learning of their letters, and never allow them to
think, paying schoolmasters thirty pounds a year in some
places, wages which none of their men-servants, who are
found in food, would accept.   If the people, or the laity,
should succeed in taking the matter into their own hands,
as in your country, I should have some hopes, but I do not
expect it.   A few months ago the new King's College of
London began a professional system of Divinity on the professorial plan, and forty-eight young men intended to graduate
in London, and never to go to Oxford or Cambridge, immediately entered.   I attach considerable importance to this
move, as it is the first formidable opposition to or competition with the old Universities, and men brought up in the
metropolis will have larger views.   The clergy must continue
to be our real educational rulers in this country, which I
believe is more parson-ridden than any in Europe except
Spain, if we consider how the higher education as well as
that of the lower orders, and how the laity, as well as the
clergy, are under the influence of this ecclesiastical body.
Ten years ago, before Agassiz had been many months travelling here, he told me he thought the prospects of science in
England very poor because of the power of the English
Church, and I was surprised that as a foreigner he should
have seen so far.   The promotion of the new Bishop Hampden was well meant by Lord John, but he recanted so
ostentatiously his liberal views, called by some Socinian,
and he has done so little to distinguish himself, that few are
much pleased with the nomination.

You ask me when I hope to get out any account of the

results of my last tour. I have only been steadily at work three or four months at the non-geological results of my nine months' absence from England, but get on fast, owing to having talked and thought it over with so many clever minds on the subjects which shook me. I have to steer clear of two errors,—first appearing to be too discontented with the state of things in England, especially as regards progress and education; and secondly, being too much an optimist about the corresponding parts of your system north of the Potomac or of New York city.

With love to your family, ever most truly yours,

CHARLES LYELL.

To GEORGE TICKNOR, ESQ.

Mildenhall, Suffolk : December 29, 1850.

My dear Ticknor,—It is long since I have written to you, and I believe I owe you a letter. We are a large family party here of Bunburys, Horners, and Lyells, and the christening of my nephew [7] (also my wife's nephew) has been an event to mark our Christmas festivities. I am now writing my Anniversary Address as President of the Geological Society, and I then hope to be able to work at some original papers, the results of observations made on the geology of the United States in 1845-6, and last year in Germany, after which we hope to make another expedition of several months in Germany, that I may follow up a line of geological research broken off last autumn. You will see by the papers that there has been a great Church or ecclesiastical ferment here, I can hardly dignify it with the name of a religious excitement, for I fear the people are not, like the Scotch, enough in earnest. If they were, we should no doubt have an outburst of fanaticism and sectarian bitterness for a time, but it would lead to more religious liberty, such as you enjoy, and our statesmen and legislators would be more free to adopt a Government scheme of education, which must be very limited so long as we have 17,000 endowed clergymen, and 10,000 perhaps dissenting ministers, equally wishing to monopolise education. The Pope's rash move, or rather the

[7] Leonard Lyell.

insolent way of doing it, has roused the Protestant feeling of
the middle classes, and the churchman sees that he must
draw in his horns, and not lay claim to the supernatural
power which our ordination service confers upon him of
retairing and forgiving sins, than which no Catholic priest
ever claimed more. So far, therefore, as tends to the checking of Tractarianism and semi-Romanist pretensions within
the Church, this excitement, with all its intolerance, may do
some good, but I fear it will bring about no reform in the
Church creeds, liturgies, ordination services, and baptismal
mystic rites, which a large majority of laymen look upon
either as superstitious, or as forms of words which are to be
gulped, or laughed at, or which, with a most worldly spirit of
compromise, they contend for as a political badge.

Oxford and Eton continue to rear up men who pass
through Puseyism to Rome, and the opinions of the higher
and middle classes are getting more and more widely
separated in religious matters—an effect which may one day
undermine the Establishment, although its authors are aim-
ing to increase their domination and supremacy.

Ever most truly yours,

CHARLES LYELL.

To GEORGE TICKNOR, ESQ.

11 Harley Street, London : January 10, 1851.

My dear Ticknor,—I have lately been giving a lecture [8]
to a large audience (gratis, and in a missionary spirit) at
Ipswich, to more than 1,000 persons, on ' Geology,' one of
the new provincial institutions, badly off for funds rather, but
with a beautiful museum of natural history. Being on the
surplus-fund committee of the Great Exhibition, I am in
hopes of getting these provincial institutes connected with a
central London Board, and that we may have examiners to
grant certificates to be required of all who take places under
Government, great or small. The Prince Consort is a host
in himself in forwarding education, worth all the English
Whigs put together. Oxford, and Puseyism, and Evangelism
and a State Church, and the narrowness of excluded sectari-

---

[8] On White Chalk and Progressive Development.

anism, are fearful odds against us. If once the clergy see we are getting under weigh, they will try, and I fear with success, to get the mechanics' institutes into their own power, in which case the cause of science will suffer. But at all events, as yet, we are always moving onwards in the right direction here, while the rest of the Continent is retrograding, and the influence of Russian and Austrian semi-barbarism is felt as far west as Hamburgh, while France, to her shame, is strengthening sacerdotal despotism in Rome, as well as military power at home. At a large party of scientific men at our club the other day at dinner, it was said, ' We shall all have to migrate to America soon, as so many approve, even here, of Louis Napoleon and his Pretorian guards and Jesuits.' If Holland, Piedmont, Belgium, and the few constitutional states which remain free from Cossack influence, would join with us and the United States, it would be easy without a war, but at present it seems to us that the semi-barbarous nations are going to have it all their own way. You augured ill from the beginning of the French move in 1848, but you could hardly have anticipated such a result, and that, too, with a steady rise of the French funds. No one can now write a word on politics to a friend in Paris, and private letters from thence abstain from politics just as if it were St. Petersburg. Senior has returned from Rome even more shocked with the persecutions and dungeons there than Gladstone was with Naples. Never were so many of the best men imprisoned in all the more civilised countries as now.

As a set-off against occasional conversion of Puseyite Anglican clergy to Romanism, and people of rank and property, we have some thousands of Irish peasantry and farmers quitting Catholicism for Protestantism, and the Irish Romanist periodicals furious thereupon; also several instances of theological works of free inquiry in an earnest spirit, printed by men who are suffered to retain professorships in Universities, although so outrageously unorthodox that ten years ago they would have been sent to Coventry in society for entertaining or confessing such want of faith.

Remember us to the Prescotts and other friends.

Ever affectionately yours,   Charles Lyell.

To CHARLES BUNBURY, ESQ.

53 Harley Street, London : April 22, 1851.

My dear Bunbury,—Heer's paper is most interesting to me, although ' Pereant qui ante nos, nostra dixerunt,' came almost to my lips when reading some of the pages. The first question is, how much faith one ought to have in his determination of the S. Jorge plants. The Corylus, for example, is it a hazel? He draws important conclusions from it, and the *Asplenium marinum,* on which he reasons.

The discussion on the former existence of an Atlantis is one on which I have made many notes and written much. It is very suggestive as treated by Heer, and better than I anticipated from his old essay on ' Madeira.' He does not appear to feel enough the contrast of the shells and plants, the forms so much more endemic and less European (I allude to the land-shells). It is really a splendid essay of Heer— allowing for future modifications. As I think I can prove that the islands, as islands, go back to the Miocene period, I feel the more interest in his speculation as to the original source of some of the plants from tertiary ' protoplasts,' as Dr. Latham would say.

I hope you will get a paper out or ready before I finish mine, if that day ever comes, for unless I fix the time I see that a life-time would never bring it about in the natural course of events.

Believe me ever affectionately yours,

CHARLES LYELL.

To GEORGE TICKNOR, ESQ.

11 Harley Street, London : May 20, 1851.

My dear Ticknor,—Since I last wrote to you I have seen Mr. Lowell, who has come to see the World's Fair, with which he is as much struck and delighted as anyone. I arranged with him that I should give my lectures in Boston in the autumn of 1852, instead of the winter, as my scheme is to return in the winter for several reasons. . . . You will see by the papers which will reach you with the same steamer

as this letter, how wonderfully the ' Crystal Palace '[9] has
taken, and how the prejudices of those who were incapable
of taking in such a ' new idea ' have given way before the tide
of popularity and fashion. *We*, the commissioners, are now
doing our best to give it an educational turn, but my hope
is that the chief good will be the admission of the million
to see so much of the result of the highest civilisation, such
as even the aristocracy who have travelled to see palaces
and museums cannot help admiring. Foreign nations in
general have come forward handsomely, and have sent property to an extent which if estimated by the mere money
value is really amazing. Turkey, Tunis, Egypt, and such distant points. To us, who know what the United States might
easily have done, it is a matter of some concern that, if they
sent anything, they should have done so little, especially
after claiming so large a space. There are two or three good
carriages and sleighs, and fortunately an Englishman sent
Power's Greek Slave, which with your Bostonian Wounded
Indian, is, I fear, the only two pieces of sculpture—an article
which forms a fine part and feature of the Exhibition. But
any one of the Atlantic cities might have sent a better turn-
out, and ought to do so now, were it only to show John Bull
in the month of September, in the course of which several
hundred thousand will still be crowding to Hyde Park, how
much faster they have gone than the present contributions
will give an idea of. Even now the Lyons people have only
half their splendid silks and velvets unpacked, for they had
no faith in our opening before the 1st of June. We have
often been wishing that you and Mrs. Ticknor and your
daughters were here, as we are staying longer than usual in
town, to receive friends who are coming up to see the Exhibition. In the beginning of July we hope to get to Germany,
viâ Belgium, to return in September.

Our nephew is gone with his parents to Scotland for a
fortnight's visit to his aunts. We shall miss him, and my
brother is to sail for India the first week in July, but I hope
for two and a half years' absence only.

Believe me ever most truly yours,

CHARLES LYELL.

[9] First Exhibition at Hyde Park.

To His Sister-in-Law, Miss Horner.

My dear Susan,—I sat next but one at the luncheon to-day given by the President and Vice-President of the Ipswich Museum to Prince Albert. Airy, President of the Association, between me and H.R.H. One of the incidents during the dinner was the Prince saying to me, 'I will show you a geological illustration in your way; there is a glacier' (pointing to a huge block of Wenham Lake ice), 'and here is the stream proceeding from its melting, and you see where it is flowing to.' We all looked, and the stream was just pouring over the edge of the table-cloth into Henslow's lap, who, as President of the Museum, was sitting in the chairman's place. He had barely time to escape being wet through.

But now to my subject. After luncheon the Prince took me aside and said, 'Are you to be with us to-morrow at the Commission, when we want numbers, and have difficult jury business to decide? I told him I could not be sure. 'Then,' said he, 'I must ask you to get, if you can, from your sister, a copy of that statement of hers.[1] I am sorry I have mislaid it; I have hunted for it in vain. Tell her if she will only put the *facts* fully, it will do. Evans is now at Venice, and we shall have lost no time. But to ensure a report I must put the affair in hand before I leave town.'

You may therefore, if you have not a copy of the original (which I do not think was too long), make a shorter one of the facts of the case as to the danger of destruction, &c. That is the pressing part of the business. But still a word or two about their being so badly shown would be useful, I think, and the cold and discomfort to artists in winter.

I am sorry you should have to re-write it, but we have at least the satisfaction of thinking it is not forgotten.

Believe me, my dear Susan, ever your affectionate brother,

Charles Lyell.

---

[1] Miss Horner had drawn up a statement of the danger and injury to Raffael's cartoons at Hampton Court from damp and fire. They are now in Kensington Museum, and under glass.

CHAPTER XXVII.

FEBRUARY 1852–NOVEMBER 1854.

DEPARTURE FOR AMERICA—ELECTRIC CABLE—GEOLOGY IN NOVA SCOTIA AND NEW BRUNSWICK—VISIT TO SIR EDMUND AND LADY HEAD—EX-PEDITION TO VERMONT—NEW YORK EXHIBITION—VISIT TO OSBORNE—MADEIRA—GEOLOGY—WONDERFUL ORCHID.

[Sir Charles Lyell sailed for Boston in August 1852, to give a course of lectures at the Lowell Institute, and returned to England before Christmas.

In 1853 he was invited by the Government to accompany Lord Ellesmere as one of the Commissioners to the New York Inter-national Exhibition, and in consequence of the illness of his chief, much of the official business and representation of the Commission fell to his share. He returned in July, and in the winter went to Madeira with his wife and Mr. and Mrs. Bunbury.[1] Early in 1854 he visited Teneriffe, the Grand Canary, and Palma, returning to England in April. He attended the British Association at Liverpool the same year.]

CORRESPONDENCE.

To WILLIAM GROVE, ESQ.[2]

11 Harley Street: February 14, 1852.

Dear Grove,—I heard the whole of your lecture[3] from a distant part of the room well and distinctly, and with great delight. It could not have been better or clearer, so far as the making an abstruse question comprehensible was con-cerned, and the exordium and peroration were beautiful.

---

[1] Lady Lyell's sister.

[2] Sir William Grove, Judge of the Court of Common Pleas, author of the *Correlation of the Physical Forces.*

[3] At the Royal Institution, the previous evening, February 13, ' On the Heating Effects of Electricity and Magnetism.'

As to the latter, I cited a letter of Liebig once on ' English Utilitarianism,' and you cannot put down too often the *cui bono* objectors. But you did more by taking the very highest ground in doing homage to the martyrs for truth. The extent to which the concealment of nearly all the newly discovered truths in every branch, moral and physical, is defended, if opposed to the popular notion, is one of the worst vices of the times, against which the shafts of satire should be aimed.

Speaking of martyrs, do you remember the low tone of morality with which, in a letter cited in Sir S. Romilly's Life, Mirabeau, with all his characteristic force and eloquence, apologises for Fontenelle. ' He well knew,' he says, ' that philosophers do not multiply like fanatics under the axe of the executioner or in the dungeons of the Inquisition,' &c., and went on to show how slowly and cautiously ' truth should be unveiled.'

Ever yours truly,
Charles Lyell.

To William Grove, Esq.

11 Harley Street: May 30, 1852.

My dear Grove,—I was much obliged to you for your note. We worked very hard for three or four hours on Saturday, and greatly improved the Report.[4] Do not criticise too much what we do, but lean to the indulgent side, seeing that the great point in striving to get up a University for the people, is to approach it in the shape of industrial instruction, and if we can through certain influences get Lord Derby to join (and this I expect you will see done), much good must follow. The Church has of course an exclusive right to *educate* the people, but the humbler task of *instructing* them may, however reluctantly, be conceded to laymen and statesmen, provided the latter proceed with due caution and courage. The

---

[4] It was desirable to get a union of the principal Scientific Societies, and this movement ultimately led to the juxtaposition of the Societies at Burlington House. Mr. Grove gave evidence before the Oxford University Commission (to whose Report the above letter refers), recommending more instruction in physical science, and the making an elementary knowledge of it an integral part of University education.

individual independence of the Royal Society, &c., if they move to a new site, must of course be guaranteed, but I confess I regard this as of most importance (I mean the juxtaposition) to the getting some system and organisation into the schools for the people. We must not be too nice about the machinery for the first movement. The weak part of the Oxford Report is the not saying more about physical science ; but it is a really grand Report, and all would follow if what they recommended were adopted.

Ever truly yours,
CHARLES LYELL.

To LEONARD HORNER, ESQ.

Government House, Fredericton : September 12, 1852.

My dear Horner,—In the steamer I had much pleasant talk, not only with the Heads,[5] but with several other mess-mates. The Americans seem all to agree that the vast German emigration pouring now into the United States is giving them a much more respectable and orderly population than the Irish, besides the advantage of being chiefly Protestant. Mr. Gisborne was on board, who has already manufactured a rope for an electric telegraph, to be laid down this season from Nova Scotia, by Prince Edward's Island, to St. John's, Newfoundland, whence a steamer will go in five and a quarter days to Galway, so that it ought to be possible before the end of 1852 to send a message from New Orleans to Vienna in five and a half days. Some money has even been subscribed for a rope 1,600 miles long from Newfoundland to Galway. But even Gisborne said that though he had subscribed largely himself, he would recommend no friend to do so; though he says the whole line would cost no more than two miles of the Box Tunnel.

I hired a pleasant private carriage with a pair of steeds, to take me and Dawson[6] on the way to Truro, eighty miles from Halifax, changing horses about every twenty miles, on

---

[5] Sir Edmund Head, Bart., born 1805, died 1868 ; eminent classical scholar and devoted to art, on which he published several works. Lieutenant-Governor of New Brunswick in 1847, and in 1854 promoted to the Governor-Generalship of Canada.

[6] J. W. Dawson, Esq., now Principal of Montreal College.

an excellent road, through natural woods, with very rarely
any house in sight, and bright sunshine, my companion
knowing the names of all the plants, and showing me among
other things the great ridges of huge boulders, six feet and
upwards in diameter, which the ice heaps up round the
borders of the numerous lakes of clear water which we passed
in a region of quartz, slate, and granite. We got out for a
few minutes to visit five wigwams of Indians, beautifully
roofed with birch bark. There are 2,000 of these Indians of
pure breed roaming over this wild country, and after returning to the carriage I could not help wondering that this
scene was within eleven days of my having walked the streets
of Liverpool, after what was complained of as a longish
passage, with a fair share of head-wind, in the least fleet of
the new steamers of Cunard's line. I felt much fresher for
the fallow which the voyage had given to my mind after the
excitement of London. I was going to peep into the wig-
wam, when a curtain was hastily drawn, and Dawson warned
me against intruding on the Indians uninvited, as they are
very ceremonious among themselves. A Baptist missionary
who was master of their language and gained many converts,
was expounding lately the parable, that when you are asked
to a feast you should take the lowest place, &c., when an
old Indian remarked, ' It is strange that Christ should have
spoken so to the white people. When a stranger enters our
tent, he sits himself down by the door, and if we welcome
him, we entreat him to come to the other side of the fire,
farthest from the door. But the white men will come in
unasked, or, if invited, will seat themselves without ceremony
in the best place.'

Next day, *September* 2, I crossed between Truro and
Amherst, the Cobequid Hills, by a different route from that
which I followed with Mary in 1842, and more picturesque,
displaying a great quantity of hard wood under which the
fern and 'sweet fern,' *Pteris aquilina,* and *Comptonia aspleni-
folia,* were growing together, which they do not in the White
Mountains, having there two distinct zones. I was amused
at seeing good-sized schooners of 40 tons, building high up
on the hills, 200 and 300 feet. In the winter a strong team
of oxen will easily draw them down over the snow into the

Bay of Fundy. Imagine them buried by a landslip, a future geologist proving by them the change of sea level! It is marvellous to see the larch posts of the electric telegraph scaling the mountains, and serving all winter, though trees are blown down upon it every great storm. It is instantly repaired, and pays a very great interest for the capital expended, and in this ' sparsely ' peopled country! My horses were ordered by it, and whenever I went into an office I found a string of messages waiting to be sent on: *e.g.* 'The ship " Anna " sails from Truro at 12 A.M., Thursday.' They also effect insurances of vessels at New York in a few minutes, giving the tonnage, &c., and receiving an answer from the Insurance Company. Dawson, the first evening, told his wife ' all well,' and asked if a child who was not well was better. The answer could not be given at once from Pictou, because part of the news from Europe brought by the ' America ' with us, was still streaming through the line to various parts of Canada and the United States, but while we were at tea the reply came, ' Go to the Post Office and you will find a letter from me.'

On the Cobequid Hills the red maple had already begun to turn to a brilliant scarlet. The ground was covered with *Kalmia* and *Rhodora*, both out of flower, the latter a kind of Rhododendron, with bluish green leaves, and said to be beautifully purple in spring when in flower. The asters, golden rods, and everlastings, cover at this season all Nova Scotia and New Brunswick.

J. W. Dawson has for the last two years been a sort of school missionary paid by the Government. He showed me a room at Truro, where he met and lectured to eighty schoolmasters, convened to exchange ideas and for mutual instruction. No hindrance arises in these schools from sectarian differences, they are free to all denominations, but all attempts to make progress in the higher education at the college has failed, owing to the Legislature having divided the money among a number of sects, each unable to pay qualified professors. Luckily, when this system began, the schools for the people had not been started, so they determined to profit by experience. One of the last endowed colleges, that of Fredericton, New Brunswick, is rendered

useless and almost without scholars, owing to an old-fashioned
Oxonian of Corpus Christi, Oxford, having been made head,
and determining that lectures in Aristotle are all that the
youth in a new colony ought to study, or other subjects on
the strict plan which might get honours at Oxford. I trust
that Sir Edmund Head may succeed in his exertions to get
something taught which the pupils can afford to spend their
time in learning. At present they must go to the United
States.

My companion, J. W. Dawson, is continually referring to
the curious botanical points respecting calamites, endoge-
nites, and other coal plants, on which light is thrown by
certain specimens collected by him at Pictou, and sent to
Charles Bunbury. He told me that the root of their pond
lily, *Nymphœa odorata,* most resembled Stigmaria in the
regularity of its growth ; and Dr. Robb showed me a dried
specimen, a rhizoma, which being of a totally different
family and therefore not strictly like, still suggests the probability of the Stigmaria having grown in slush in like manner,
and sent out rootlets. I was much pleased to see two Sigillariæ dug out with their four dichotomous roots, each again
dividing. It requires more labour than most geologists have
time for, to have the whole disinterring and washing of
trunk and root, allowing for occasional failures owing to the
external marks being occasionally wanting, especially if
near the base of the trunk and first starting off of the
roots.

Dawson and I set to work and measured foot by foot
many hundred yards of the cliffs, where the forests of erect
trees and calamites most abound. It was hard work, as the
wind one day was stormy, and we had to look sharp lest the
rocking of living trees just ready to fall from the top of the
undermined cliff should cause some of the old fossil ones to
come down upon us by the run. But I never enjoyed the
reading of a marvellous chapter of the big volume more.
We missed a botanical aide-de-camp much when we came to
the tops and bottoms of calamites, and all sorts of strange
pranks which some of the compressed trees played. The
so-called flabellaria, which I believe C. Bunbury thinks
are the stipes of ferns, puzzled us much. They abound

most with Sigillaria. The underclays with Stigmariæ are
wonderfully like those clays with recent roots under peat,
and sunk stumps, in the marshes of the Bay of Fundy.

The names which the sailors have made out of the original
French settlers' names are amusing. Point Demoiselle is
Muzzle Point; Mont Chapeau Dieu, so called because usually
capped with cloud, is Mount Sheepoddy; the great river
Petit Codiac is Petticoat Jack. But I have met with none
to rival an old friend in Musquito Bay, on the St. Lawrence,
Anse de Cousins, which is 'Nancy Cozens.'

After the Joggins we returned to Amherst and thence to
Dorchester, where a friend of Head's, called by some his
prime minister, Mr. Chandler, received us in a handsome
house, and his son took us next day to see a wonderful bed
or vein of asphaltum, the subject of a hotly contested law-
suit—a vein from one to eleven feet thick, of pure pitch coal,
or something like it, traversing fractured coal measures.
It is too puzzling to attempt an account of it. As Gesner
has consulted me, and the other party Dawson, we are trying
to persuade them to compromise the suit. About 3,000*l*.
pocketed already by the lawyers! The chief point being,
whether it be a bed of coal which passes by a crown lease,
or of asphalte or some mineral which would not pass.
Many of the most eminent of the scientific chemists and
geologists of the United States examined. The distances
are enormous. After Dawson left me at Dorchester, I travelled about 150 miles at one stretch in a stage coach towards St. John's, New Brunswick. Chandler and two
friends, good company inside. The population returns just
made show that New Brunswick, which nearly numbers
200,000 souls, has increased in the last ten years faster than
the average of Maine, Massachusetts, New Hampshire, and
Vermont, the four adjoining States. Nova Scotia is about
280,000. Both big enough and fertile enough to hold as
many as Ireland—noble provinces. They lose many of their
most enterprising colonists by desertion to the States, and
say 'we must get railroads in order to keep them.' Many
are off to Australia and California. Still they grow 23
per cent. in ten years. Sir E. Head met me at St. John,
and we examined the Falls together. It is very strange to

see a great river or tidal current, rush first in for several hours, and then out of a narrow gorge formed of metamorphic rocks, vertical beds of limestone and slate, invaded by trap and syenite. They ought to be called the Rapids rather. The vessels, a little fleet laden with timber, wait till the tide flows neither way, and then sail in or out. The whole harbour is beautiful. The sail up the river eighty or ninety miles from St. John to Fredericton is fine—picturesque, till a country of nearly horizontal coal measures produces tame outlines and low hills. The power of ice is ever present to me in these regions, transporting huge boulders in winter. The sea-ice carried an enormous angular mass of coal grit, twelve or more feet in diameter, from an adjoining promontory to the place where the colliers loaded. They were obliged to blow it up with gunpowder at low-water this spring. Our stay at Government House was very agreeable, and made up in some degree for losing the Heads for the last four years in town. He has many difficult points to grapple with in his colonial management, but Lord Grey supported him well. He is evidently much liked as Governor. We saw the Puseyite Bishop who had preached to us on board the 'America' enter, carrying a crozier in his hand! No one at home has yet ventured to perform this play, for which he cites an ordinance of Edward VI., in which I am told there are other popish ceremonies ordered which are not yet ventured upon.

I believe I mentioned in my last that Dawson and I found the skeleton of an animal in the middle of one of the upright trees of the Joggins, Nova Scotia. I thought it would prove a Labyrinthodon, and not a sauroid fish. Agassiz has seen the jaw, and what I supposed the humerus. He says it is something quite new, but he inclines to believe it will turn out something ichthyic. Dr. Wyman had a long work with me at it yesterday, and after he had begun by thinking its reptilian characters predominated, he ended by inclining the other way. He and Agassiz are to confer to-day upon it. On the whole, I fear it will not prove reptilian, though a remarkable fossil, and certainly in a strange place Agassiz is looking well, and preparing to publish a great embryological work on 'Paleontology.' He says 'he has revelled in the sea here.' The French Academy have voted him the first prize bequeathed by Cuvier, for having done the most since Cuvier's death.

With my love to dearest mamma and her daughters, and to the Bunburys, believe me ever affectionately yours,

CHARLES LYELL.

To LEONARD HORNER, ESQ.

Boston : October 30, 1852.

My dear Horner,—Since I wrote to you last I have made a geological expedition into Vermont, as I believe Mary mentioned in her letter, where I found a tertiary brown coal or bed of lignite, nine feet thick and very pure, associated with white porcelain clay, just as in many parts of Germany, near Leipzig, &c. No fossil shells, but I hope the fossil fruits may enable Charles Bunbury or some one to decide its age. I never saw before in the United States any deposit quite like it. I am going to give a lecture on some remarkable trains of huge erratics which attest in the west of Massachusetts, in lat. 42, what wonderful power the ice, I believe coast-ice, has exerted. James Hall [7] and I worked five days at it, and I think we can explain the manner in which the wind and currents distributed them in certain linear directions. Even in the neighbourhood of Boston, I have been doing with Agassiz, Dr. Gould, and to-day with Strepson, a young engineer and conchologist, some good work, but I have not time to enter into it. The discoveries of Agassiz in the Florida reefs will delight Darwin, and I shall have much to tell him. Meanwhile, if you see Charles Darwin, tell him that in Florida the effective reef-building corals do not build lower than about sixteen fathoms; so he was much within the mark. Agassiz believes in subsidence and the atoll theory. He saw one spherical individual of Porites fourteen feet diameter ! and thinks its age, and that of some Mæandrinæ, as vast as Ehrenberg thought.

Agassiz has in MS. sixty monographs on embryology of marine creatures, invertebrata. He confirms in most things

[7] Mr. James Hall, Director of the Government Geological Survey of New York.

the observations of Sir John Dalzell, where he has been on the same ground. He has upset all Ehrenberg's Infusoria, some are Vermes, some Crustaceans, some plants, some spori of plants, &c. And what, I asked, are Infusoria? The reply was, 'There are no such beings!'

October 31.—Mr. Peabody, a Unitarian preacher in King's Chapel, gave us a fine discourse to-day on the death of Daniel Webster, very simple and original. He dwelt on the great importance to a nation of a commanding intellect, as even those parties who differed entirely in their political views had the advantage of seeing every great question placed in a clearer light and on broader grounds, so that they were forced in opposing them to do so less ignorantly. There was no hyperpanegyric—which of course there has been in the various speeches made here. The union of all the distant States, and sects, and political parties, shows how national this country is. 'Uncle Tom's Cabin' will, I hope, do more good than harm on the whole. It is a gross caricature, because the very great number of kind masters, and of families where the same negroes remain for generations, is carefully kept out of view. But all the evils described, or nearly all, do now and then occur in a population of nearly four millions. As to Congress, it can no more interfere constitutionally than our Government to reform the harems and other abominations in Turkey.

My class is a very satisfactory and steady one, and being twice as great as has been seen in the Lowell Institute for many years, gives pleasure to its patrons.

Ever affectionately yours,
CHARLES LYELL.

To J. W. DAWSON, ESQ.

Boston : November 6, 1852.

My dear Sir,—I have very good news to tell you. Agassiz only conjectures that hollow-tree Joggins animal is a cœlocanth fish, possibly allied to Holoptychius, and assuming it to be a fish, then the remarkable bone which first struck us, and of which there are two, must be a hyoid bone because it could be no other. Yet he knows of no hyoid bone like it.

You see how very loose a decision this is, when the materials
are so ample.   Wyman begins to suspect an ichthyic reptile
allied to Siren, *Proteus anguinus* (Nemobranchus?), &c., as he
says there is a bone in them more like it.   So much for our
principal skeleton.   But you will be delighted to hear that
in the same stone Wyman has worked out part of a vertebral
column, seven vertebræ in a series, and three other detached
ones of the same dorsal and lumbar region, belonging to a
distinct creature, and which he at once pronounced a sala-
mander from the articulating surface of the ball-and-socket
joints, &c.   Afterwards, when it was shown to Agassiz, he
exclaimed, ' This is more reptilian than anything I ever saw
in the coal ! '   I now begin to regret that we left a single
fragment of the stone on the beach.   For Wyman worked
this treasure out of a most unpromising stone, like many
which I threw away.

Latest intelligence.—Dr. Wyman has just been here with
great news.   The first bone which we found is clearly not
the hyoid bone of a fish, but the iliac bone of a reptile.   Do
not say anything about it, as every hour he is advancing.
The iliac bone is so precisely ours, and the hyoid of a gar-
pike (the nearest which Agassiz could find) is so very *unlike*,
that we may pronounce ours to be a true ichthyic reptile till
he can gainsay us.   So we have two reptiles according to
this, and as only four individuals were previously known
in the coal of the whole world, I hope we have added 33¼
per cent. at one stroke to the reptilian paleontology of
that era.

Believe me, my dear sir, ever truly yours,

CHARLES LYELL.

To GEORGE TICKNOR, ESQ.

11 Harley Street, London : February 17, 1853.

My dear Ticknor,—I must add a few words to my wife's
letter to Mrs. Ticknor, though my head is rather too much
occupied with geology to have anything else worth telling of.
My publisher is seriously thinking of an edition of 5,000
copies for this, the ninth, of the ' Principles of Geology,'
with which I am very busy.   Macaulay says that Longman

told him that in England more than a million copies of
' Uncle Tom ' have positively been sold, far exceeding in
popularity—what think you, the ' Pilgrim's Progress ' or
Walter Scott ?—not at all—but James' ' Anxious Enquirer ' !
of which I at least never heard before. This perhaps explains far more of the run of Mrs. Stowe's book, than either
its talent, or slavery, or the charm of reading a new picture
of life, after our own social state has been hackneyed in so
many novels, good and bad. It is, in fact, the first book
which ever hit precisely the taste of the religious world and
of the profane ; 100,000 they tell me have sold of the ' Wide
Wide World,' and half that number of ' Queechy.' One
small printer who bought of Clowes the paper intended for
the Exhibition Catalogue (the privileged one) which was a
failure, and got it cheap because of its unusual shape, when
paper was not easy to get, cleared 3,000*l.* by 60,000 copies of
' Uncle Tom.' Murray ascertained this.

So now everyone is speculating in American reprints,
good, bad, and indifferent. They are poured in upon us like
gold from Australia.

Ever most truly yours,
CHARLES LYELL.

To GEORGE TICKNOR, ESQ.

11 Harley Street, London : April 26, 1853.

My dear Ticknor,—I wish to make to you the first
announcement of a sudden change in our plans, by which
we hope soon to have the pleasure of again seeing you and
other friends on your side of the water. About a week ago,
Lord Granville (President of the Council and Vice-President
of the Royal Commission for the Exhibition of 1851, which
still subsists, and of which I am a permanent member),
called here to say, that the Government had determined to
send out commissioners to the New York Industrial Exhibition. That it was somewhat contrary to etiquette, but
that they were ready to waive that, inasmuch as the last
Government of the United States had partially recognised
the New York scheme as national. Lord Ellesmere had
offered his services to the ministry, and they had thought of

me as No. 2, both to represent science and 'as one who
would be acceptable to the American people.'

I told him I must take time, at least a day or two, to
consider ; for unless I could calculate by great exertion to get
out my ninth edition of 'Principles,' now eight months out
of print, I could not go. I must also, I said, see the instructions, and I stipulated that if I went, I would not
undertake any share, much less a superintendence, of the
Report. Secondly, that I would serve, like Lord Ellesmere,
without any salary or pecuniary remuneration. These terms
being agreed upon and put in writing by me in a letter
to Lord Granville, and shown no doubt by him to Lords
Aberdeen and Clarendon, from whom he brought the message, I was nominated, with leave if I could not be ready
for the Government steamer in which Lord Ellesmere is
to go, to take a mail packet, in which I might sail with
my wife.

I also bargained that I might return as soon as the New
York business is over, as my scientific work, to be cleared off
before sailing for the Canaries in the autumn, will be much
interrupted by this affair. In short, I give them two
months in all of my time, in return for what is considered
here a great compliment usually. Some of your countrymen
here protest that the whole New York affair is a humbug.
But Ingersoll has sent in a note which confirms the ministers
here in their friendly resolutions. The Italian sculpture,
consisting of a hundred statues, is, I hear, beautiful, and
stimulated no doubt by the sales they made here in 1851.

I had resolved that nothing should persuade me to interrupt my regular geological studies, but there are many
inducements to make us accept this on such terms, and I
shall be glad if you will tell Mr. Everett, Lowell, Winthrop,
Abbott Lawrence, Hillard, and many others whom you think
would be interested. My wife will tell Mr. Prescott, but not
enlarge on the subject. I thought I might read to Lord
Granville a pithy passage from your last letter about the
state of parties—Pierce, Douglas, Cass, &c.—which was admirably fitted to show them that, politically speaking, this
might be a good move, or could not fail to be, for it is done
in a good spirit. Lord Ellesmere is, as you know, a most

accomplished man, and has really great knowledge and taste in the fine arts, and has a glorious gallery of pictures.

Macaulay is in an unsatisfactory state. Whether India or overwork, or both, have damaged his strong constitution, I cannot learn; but he seems to have grown very weak and ailing. But I hope he will rally.

Ever most truly yours,

CHARLES LYELL.

To LEONARD HORNER, ESQ.

Boston: July 11, 1853.

My dear Horner,—Lord Ellesmere's illness, and the uncertainty as to what I may be called upon to do in the event of his absence, has kept me in some doubt, but I hope still he will rally, as he usually does rapidly, and not disappoint the meeting, and this I expect will be the end of it. I find educational matters more eagerly discussed than ever, and a growing conviction that the whole system must give way, simply because with the doubling of the population and trebling of the wealth, students on the old foundations do not augment in number. They have cheapened education, but it will not do, and finally they are coming to the reluctant confession, that as a large part of the people will do without Latin and Greek, but will have science and modern literature, they must provide the articles they want.

In New England, at New York, and in Philadelphia, I heard precisely the same declarations, and they have been made at public meetings since my arrival. It is evidently a crisis. Fortunately they have neither mediæval endowments to render them independent of modern public opinion, nor ecclesiastical collegiate governments to prevent them from adopting a radical change. They are going to work in good earnest.

There are so many sectarian colleges, that it will be some time before any one grows big enough to subdivide sciences as they require.

Agassiz is quite recovered and at work. Dr. Leidy made professor at Philadelphia, aged only thirty-five, and a first-rate osteologist and comparative anatomist.

Ever affectionately yours,

CHARLES LYELL.

To HIS WIFE.

Osborne : August 23, 1853.

My dearest Mary,—I made out my journey and voyage very successfully. The parterre of flowers was in great beauty, and the views of the sea in spite of cloudy weather very pleasing.

I got here between six and seven, and Lord Clarendon arrived alone just after. I was very glad to find that he was to be the Minister in attendance. As usual, he has made himself very agreeable. He has been reading Lord Ellesmere's speech in my paper, and likes it much ; the Prince is to have it by and by. At dinner we had, besides the household, Count Mensdorf, the Duchess of Kent's nephew, Austrian Ambassador to St. Petersburg. After dinner, when four of the household played whist, the Prince had a long talk with Lord Clarendon and me about the United States, foreign politics, and University reform. Very frank, and both going as far as I do in most things, and never clashing in sentiment.

The Prince then invited us to join the ladies and sit down at their table, and I was asked by the Queen news of New York doings, and made them merry with Soft-shell, Old Hunkers, &c., and gave an account of the Exhibition prospects, United States prosperity, &c.

I shall try and drive over to Ryde this afternoon, if the Queen and Prince drive out. Lord Clarendon has proposed returning with me to town, which will be very pleasant. How they are overworked ! We parted at a quarter-past eleven o'clock, and Lord Clarendon found, to his dismay, five red boxes with State papers from London in his bedroom ! He told me he had hoped the interval of salt water would have stopped them. I am enjoying myself much, and only wish you were here. It is a very pleasant residence, like a small German Principality palace.

Ever affectionately yours,

CHARLES LYELL.

To Leonard Horner, Esq.

Madeira: January 1854.

My dear Horner,—I have been wishing to send you some account of my geological results in Madeira, but have been so actively employed in the field, that I have had barely time to write my notes each evening. Besides, I am continually modifying, and improving as I hope, my views of the structure of the island, which far exceeds in interest anything I had anticipated, as it does in picturesque beauty any region of equal extent I ever saw; Switzerland not excepted.

The sea-cliffs and innumerable ravines, both of great depth, display the rocks very finely, but the ravines and deep valleys make it a long affair to get from one spot to another, which looks an easy morning's walk. The horses can climb admirably, and even leap up a succession of rocky steps, to which I have now got accustomed, and feel safer than on my own legs.

My companion, Mr. Hartung,[8] is very zealous, and his agricultural, entomological, and botanical pursuits had made him in the last three winters get up the physical geography and language well, and some beginning of geology; so he helps me very much, and is an apt scholar. He also draws tolerably, and improves in this daily.

I have satisfied myself that Smith of Jordanhill was right in attributing a sub-aerial origin to the volcanic rocks of this island generally, although Vernon Harcourt has since disputed that opinion. I have now seen more than either of them did, and have visited the only part where marine formations are known above the level of the sea. They reach a height, not of 1,600 feet, as reported, but of 1,200, as determined by my barometer. I have found waterworn pebbles of the usual Madeira volcanic rocks in these same beds, which contain corals and sea-shells; such pebbles are entirely wanting in the volcanic breccias which all over the island contain angular fragments thrown out by explosions. There is one example of an impure lignite, and

[8] Mr. George Hartung, of Königsberg, author of *Travels in Norway*, &c.

a leaf-bed under basalt, in which I have been the first to find leaves ; a greater mixture of ferns with dicotyledonous leaves than Charles Bunbury has seen in any other tertiary formation. He has found about five species of ferns, Pecopteris, Sphenopteris, Adiantites (?), &c., and has not quite examined all my specimens, most of which have been obtained for me by a peasant, whom I have kept at work ever since I found out the spot, which like most of the best geological localities here, is rather inaccessible. One really ought to have a tent, so much time is spent in going and coming back, but there is much to see upon the way, which makes up somewhat for the exertion. On the great question whether the Curral is or is not a crater, on which Darwin will be curious, I have made up my mind that it is not. In general this island confirms his doctrine, that if all valleys were cut by rivers alone, they would be very narrow, though they might be of any depth, and that the sea is the great widening power. The question then arises, whether the two principal central valleys of Madeira, the Curral and that of the Serra d'Agoa, could be as wide as they are if due to aqueous and fluviatile erosion alone, aided by the gradual upheaval of the rocks to the height of 1,200 and perhaps more feet, which the marine strata have reached, during which upheaval the original flattened dome has I presume acquired a more convex form, in accordance so far with the ' Erhebung's ' theory, provided time enough be allowed for the uplifting. But I find the beds very much nearer horizontality in the central region, where they are highest, and the dip which they have from 3° to 7° is by no means always away from the central valleys, as stated to be the case in the Caldera of Palma. Besides, these central valleys radiate outwards to the sea, both to the north and south. unlike the case of Palma ; the dome was of an elliptical form, as there was not one habitual volcano, but a chain of rents, like a miniature Andes.

In the middle or axis of the chain, sub-aerial volcanic matter, 4,000 feet deep, is seen to be piled up. The basalts accumulated between numerous cones of eruption, and then flowed away in all directions, occasionally encountering and burying lateral cones, some of which are exposed in the lofty sea-cliffs. What surprises me is, that the sheets of basalt are scarcely, if at all, more inclined than at first, where nearest the great focus of eruptions, whereas they have a slope from 9° to 13°, and sometimes rather more, as they recede from it. Dikes are innumerable in the central region, and are fewer nearer the sea, unless where a transverse or north and south volcanic axis branches off, and presents on a smaller scale an epitome of the whole island. One of these, at Cape Giram, has its anatomy finely laid open by a nearly vertical sea-cliff 1,600 feet high, where I have counted 120 dikes, some of them running from the base of the cliff to the top. Even in this section the basalts are most horizontal where the dikes are most numerous. The modern rivers have left beds of pebbles 100 and 130 feet above the present channels; precipices which bound them are so steep that higher ones cannot be preserved, for they get undermined and fall down.

Believe me ever yours affectionately,

CHARLES LYELL.

To LEONARD HORNER, ESQ.

Lazzaretto, Santa Cruz, Teneriffe: February 21, 1854.

My dear Horner,—Although I have only time to give you a brief outline of the results of my last two or three weeks' work in Madeira, I must endeavour to do this before I am absorbed in this new field cf interest in the Canaries. At last I found beds of pebbles, which I believe to have been rolled by rivers under currents of lava in Madeira, and like the fossil plants alluded to in my last letter, and more of which were discovered up to the last, these pebbles prove that there was here a volcanic island built up by a series of *sub-aerial* eruptions. I had also the pleasure of finding a set of trachytic lavas and tuffs 900 feet thick, at Porto da Cruz, all newer than an older and more inclined basaltic series, showing that one set of movements was over, and the erosion of one set of valleys completed, before some important portions of the volcanic operations had begun. Such conclusions you will see, respecting the north-east of the island, are in harmony with the different ages which I ascribed to the Cape Giram beds in the south. I was able to prove by repeated visits to that grand sea-cliff section, that the juxtaposition of the highly inclined and the horizontal beds was owing to a great dislocation, but I must not dwell on that now.

Besides some beds of waterworn alluvial gravel covered by trachytic lava at Porto da Cruz, I found at Camera de Lobos, near Funchal, some tufaceous alluviums, with slightly abraded blocks overspread by layers of columnar basalt. But after a nine weeks' search, I never met with a single river bed with lava resting upon it, no old fluviatile gravel conforming to the present valleys, or even approaching to a coincidence of position with what are now the lowest levels upon which a stream of lava had flowed. I therefore supposed that Madeira during its growth was not like Auvergne or the Olot district in Spain, but like Etna, where there are no torrents or rivers or springs, not even in the Val del Bove, where the cattle have to be supplied with the water of melted snow, or like Mouna Loa in the Sandwich Isles, where a dome of much larger dimensions is, I think, described by Dana as without running water, though not without rain.

Von Buch saw but little of Madeira, or it would certainly have been regarded by him as a 'crater of elevation.' If the Curral, and the Serra d'Agoa, two great central valleys, be given up as craters, and regarded as valleys of erosion, still Madeira would resemble the Mont d'Or and the Cantal all the more for not having any great central crater. The existence of a marine deposit at one point north of the east and west axis, rising to about 1,200 feet above the sea-level, shows that here, as in the case of Etna and Somma, there has been upheaval in some shape and to a certain amount, and in Madeira, as in the case of the Neapolitan and Sicilian volcanos, the earlier eruptions were submarine, for the shells and corals are associated with rolled volcanic fragments and tuff. But besides this, there is such a tilting of some masses of solid lava away from the central axis as to imply that much of the principal movement was of a kind to cause the beds to dip away from the force of eruption, although, as I said before, there are large spaces at great elevations, from 2,200 to 5,000 (?) feet high, where there are plateaus at or near the centre of the island, where the beds of basalt have a slight dip, not more than I conceive they may have had originally. In this respect Madeira differs from Palma and Teneriffe, so far as I know the latter by Von Buch's descriptions.

A great proportion of the dikes are older than the inclined basalts, and others as old, and it would seem something of a paradox if we found that the grand explosion and upheaval was reserved till all the ordinary volcanic energy was spent, although we might perhaps imagine that when the safety-valve was closed, the subterranean force might expend itself in uplifting and injecting the rocks below. Let us see then whether we can reconcile the different theories of the eruption and elevation schools, so that by the aid of both we may account for what I have regarded as facts of a conflicting character. In the first place we have abundance of evidence that had there never been any elevation whatever, Madeira would have been a lofty island about 4,500 feet high, formed by a long series of supramarine eruptions which gave rise to mountains more than 4,000 feet high, consisting from top to bottom of volcanic tuff, and breccias and scoriæ, lapilli, cinders, and lavas, and slightly inclined basalts flowing away in all directions, east, west, north, and south, for miles, perhaps far beyond the present limits of the isle, and not below the sea. I have fossils, leaves of ferns and dicotyledons (more than 150 specimens) showing that before the island was formed, it was clad with vegetation. In the Curral we see very deep into the interior, to within 1,700 feet of the sea-level, and that in the middle of an island above 6,000 feet high. No signs there of any corals or shells, or such wearing or rubbing off of the angles of the stones included in the breccias as I invariably found in the Porto Santo and S. Vincente marine volcanic conglomerates.

This hypothesis, therefore, of the upheaval of the island, so far from explaining its structure, would involve us in many perplexities. In the first place there are no raised shingle beaches, no tuffs containing sea-shells, a leaf bed only 900 feet above the sea-level covered by tuffs and lavas 1,200 feet thick, &c. Mr. Smith's [9] suggestion that some parts of the coast of the isle have sunk since the volcanic period would

[9] Of Orotava.

o 2

perhaps explain much more, but I will not at present discuss that view.

The conclusion to which I am coming is this, that there has been a prodigious succession of periods of volcanic eruption both along the central east and west axis of eruption, and along certain north and south lines, and finally, and at a later season, along the coasts. In this small chain, thirty or thirty-five miles long, as in the greater one of the Andes, all the volcanos were not active at the same time. Some were extinct ages before others broke out; yet there was such a similarity in the products of all, and in the mode of rending, injecting, and dislocating the rocks already formed, that the general result throughout the whole chain was very uniform, the general structure and composition nearly the same, and the numerous local exceptions only discoverable after much study.

But I must conclude, for the arrival of the Cadiz steamer has made us determine to avail ourselves of its aid to visit the Grand Canary, and Mary, Mr. Hartung, and I go there to-morrow morning; but I shall leave others to send you all the news not strictly geological.

The climate of Madeira was like that of a good English summer, never too hot for work in the field, and never cold, even when one was wet with a shower. In the small grounds of the Lazzaretto here, over which we are allowed to walk, Messrs. Bunbury and Hartung gathered forty-eight wild plants in flower. With love to all, believe me ever yours affectionately,   CHARLES LYELL.

To MRS. BUNBURY.

11 Harley Street, London : June 10, 1854.

Dearest Frances,—I was very glad to learn from your last letter to Mary that you had not forgotten the barometrical calculations.[1]

I received the other day a letter from Joseph Hooker from Hadleigh, in which he gives all his reasons for thinking Coniferæ as highly developed dicotyledons as any others, and above monocotyleds. In this he differs from Ad. Brongniart. Henslow sends word that he concurs. I should

---

[1] Made by Mrs. Bunbury in Madeira.

like to show the letter to your husband and get his criticisms. But it is a precious document, as I have to allude to the question when treating of the carboniferous flora.

Please send all possible objections to the views of Hooker and Henslow. It is curious that in a number of ' Silliman ' just received, Agassiz, in a paper on ancient faunæ and floræ, asserts that they were at every ancient period as *varied* as now, but that the grade of the highest was not so high as now. The supposed fact of *all* the Stonesfield mammalia being marsupial made him despise the pouched animals. Afterwards Owen found the Thylacotherium (or Amplitherium) to be placental or monodelphous. Now a grand discovery has been made last Wednesday. Owen read a paper on a new Purbeck (Upper Oolite) mammifer, a new genus of the mole family, allied to *Talpa aurea* or the Chrysochlore of the Cape, allied also to Thylacotherium. Several perfect lower jaws with teeth in good condition, a true *non-marsupial* mammal.

Truly the oolite is looking up, and it is the more striking since the chalk (Wealden and all) still refuses to yield up any of its mighty dead, if it has any to disgorge, as I expect it has. But if not, it only shows that their absence is no argument that in older beds more advanced forms were not developed.

What does Charles think of Hooker's notion that there were some grasses in the coal? No doubt it would better suit the progressive theory if they came in long afterwards. It would be more orthodox that Nature should first try her hand at some easier feat. If I might paraphrase the lines of Burns,

> On ferns she tried her 'prentice hand,
> And then she made the grasses, O;

but I must really have done. The Duke of Argyll, at the *levée* yesterday, had a long talk with me about my speech at the Geological Society on Owen's paper, and he said, ' You will never get over the absence of mammalia in the coal which has been so well worked, and contains more dirt beds than the Purbeck.' This was a good hit, and, as I confessed to him, the strongest point on that side; but on negative evidence the coal was without land quadrupeds, birds, saurians, chelonians, ophidians, land crabs, air-breathing mollusca (save two individuals which I found), insects, save three or four individuals, no helices, no lymneas or planorbis, &c. &c. In short, it proves too much when the marine fauna was so perfect.

Ever affectionately yours,

CHARLES LYELL.

To LEONARD HORNER, ESQ.

11 Harley Street, London : August 12, 1854.

My dear Horner,—Edward Forbes is come to town for a short stay, and I must get out my Grand Canary and Madeira fossil shells to show him. Murchison has come out with a very useful book on ' Siluria,' well done, and much new matter on Coal and Old Red, as well as older things. I shall have to controvert one chapter on ' a long period when the invertebrate animals alone existed.'

I asked M'Andrew yesterday how many vertebræ or any remains of dead fish he dredged up on the Dogger Bank or the great ' Ling Banks ' or cod-fishing grounds off Shetland. To my astonishment he replied, ' I never drew up one bone or tooth, though sometimes live fish came up in the mud. Once I got what I thought was a tooth of some vertebrate animal, but it proved to be of an echinus.' He imagines that future geologists will find no signs of the existence of fish in the sand and mud of the great fishing-banks of Newfoundland ! Yet you remember in the ' Principles ' I had a ' bone-bed ' in the sea of the Hebrides, so we have an Upper Silurian bone-bed. Huxley is established at the Museum of Practical Geology ; he is coming on fast in Paleontology. Elie de Beaumont was here for two days, and made himself very pleasant. He is looking well.

Ever affectionately yours,

CHARLES LYELL.

To C. J. F. BUNBURY, ESQ.

53 Harley Street, London : November 13, 1854.

My dear Bunbury,—I have been intending from day to day to write and thank you for your former letter, in which you gave me an account of the conversion of the Ægilops into wheat, which I shall keep for next edition of the 'Principles.' Hooker seems to believe it, though in general not so prone as Lindley to entertain such ideas.

When we were at Charles Darwin's we talked over this and other like matters, and Hooker astonished me by an account of an orchideous plant [2] sent, I think, by Schomburgk from Brazil, and which on different branches of the same individual specimen contained flowers previously referred to three distinct orchideous genera, some of the said genera having had several species referred to them. You probably know about this, which will figure in C. Darwin's book on 'Species,' with many other 'ugly facts,' as Hooker, clinging like me to the orthodox faith, calls these and other abnormal vagaries. I think Joseph Hooker said that each of the three so-called genera, now resolved into one and the same species by Sir Robert Schomburgk's specimen, were morphologically distinct on many different points. I have learnt all I could from the Bury paper of your lecture, which was, I think, very well imagined, and not dry, as it might have been if too botanical. There are no end of desirable illustrations of plants, if one could set Henslow to work in his Michael Angelo style ; a name rarely conveys an idea to a class without a picture.

Of my four species of Bryozoa from the Grand Canary, one is recent and three unknown, so says the first-rate authority, Mr. Busk. I imagine the age may be Miocene or falunian; but this is a mere guess as yet. The San Vincente sub-marine beds were volcanic, or contemporary with eruptions, whether upraised before Madeira existed as a volcano I cannot say. But as the mass of Madeira is of supramarine origin, this San Vincente bed must have been its foundation, I suppose. It is a good question. Porto Santo is strictly an analogous case, an upraised volcano of submarine origin, and a superstructure of sub-aerial lava.

Sebastopol is an anxious business. The lateness of the season is, I suspect, chiefly owing to Lord Aberdeen never having intended anything but a demonstration in the Mediterranean, and never in earnest. This I think I know

[2] *Catasetum*, described in the *Fertilisation of Orchids*, by Charles Darwin, and referred to in the *Origin of Species*.

through private sources. The Czar knew it, and did not budge, of course.

I hope to comply with your friend Lord Arthur Hervey's wishes [3] some day, not a distant one. Meanwhile I am very glad your lecture went off, as I fully expected it would, so well, and that you are going to repeat it. I am sure that what Dr. Franklin said of Wesley's field-preaching and extemporaneous discourses, that the most effective were those he had often given, so is it with lectures. Each time you will improve the same subject; you know better the value of the hour, and measure the relative importance and popularity of the subjects better, and waste no time in saying how little time you have, though I do not hear that you did so, but most men do.

With love to Frances, believe me, ever affectionately yours,

CHARLES LYELL.

---

[3] The present Bishop of Bath and Wells, who requested him to give a lecture at Bury St. Edmunds.

CHAPTER XXVIII.

JANUARY 1855—AUGUST 1856.

DEATH OF EDWARD FORBES—BERLIN—DR. HOOKER'S INTRODUCTION TO
'FLORA INDICA'—HUGH MILLER—ENTOMOLOGY—MACAULAY AND PRES-
COTT'S HISTORIES—FOSSIL BOTANY—ON THE DOCTRINE OF SPECIES—
SILESIA—RIESENGEBIRGE—BOTANY—GEOLOGY OF SAXON SWITZERLAND
—BARRANDE'S 'COLONY' OF UPPER SILESIAN FOSSILS.

[In 1855 and the two following years, Sir Charles and Lady Lyell
spent several months of each year travelling on the Continent, always
with the view of geological inquiry. The University of Oxford, his
Alma Mater, conferred upon him the degree of D.C.L. in 1854, and he
attended the British Association at Glasgow that year.]

CORRESPONDENCE.

To His Sister.

53 Harley Street, London : January 2, 1855.

My dear Caroline,—I enclose a letter from Sir Walter C.
Trevelyan of Wallington, supposing you must be the individual alluded to, who having found a rare insect, wished
to keep the *habitat* to yourself. I wrote to say that we
never kept it a secret, but that as I had given away all the
duplicates I had here, I would ask you if you could send one
or two to Sir W. Trevelyan, by post in a small box with cork,
or as you think best, and I asked him to acknowledge
direct to you their safe arrival, if safe. He alludes to poor
Edward Forbes's death, which is indeed the greatest loss of
an active scientific friend I have ever sustained, and as he
was but thirty-nine, so unexpected. We are getting up two
London memorials to him, to which I subscribe. His death
and Mr. M'Ilvaine's of New Jersey, and Lord Cockburn's,
are a great many in one year, and among geologists much

younger than myself, whom I often saw, Mr. Strickland, run over by a railway.

It makes one feel as one grows older much as the officers in the Crimea must feel, that they are surrounded by companions whom one may never see again, a few weeks or months hence, and it seems strange how little one is checked in one's own proceedings, and how one coolly purchases a lease of sixteen and a half years of this house.

With my love to all, and a happy new year, believe me your affectionate brother,   CHARLES LYELL.

To GEORGE HARTUNG, ESQ.

53 Harley Street, London : January 21, 1855.

Dear Hartung,—Your letter, dated January 4, reached me several days ago. You make steady progress as a geological artist, and I think the view of Funchal and the S. Martinho range with Cape Giram and Pico Bodes range in the distance, is a particularly truthful sketch.

I chipped off with a small hammer, some three weeks ago, nine small fragments of rock, from as many specimens which I collected when with you, and sent them duly numbered in a letter by post to M. Delesse at Paris. No mineralogist and analytical chemist is perhaps so good an authority as he is. In a few days he sent back an answer, and had found, by aid of the microscope, minute crystals of all the minerals, larger crystals of which were in the big pieces I kept here. First I sent the ' yellow concentric ' from Sitio do Poizo and another place, calling it basalt. He replies, we in France should call it basalt, but I did not expect that you would; as it is made exclusively of greenish felspar and olivine without any augite, ' therefore would you not in England call it trap?' Certainly I had imagined that the darkish grey colour was derived from augite, though I was aware that it consisted in great part of felspar. Other rocks, which I called basalt, he also named the same.

I hope when you return to London to know much more about the Madeira and Canary Island rocks. I shall send you a copy of my book with the report I have given of the result of our joint observations. It would be better for you and me to see a country of recent or active volcanos next. But you are well employed in Madeira. Pray go to the Pico of Camera de Lobos and see the lava (?) or scoriaceous mass on the flanks of the hill which I often wished to go up to.

The Bunburys and Lady Lyell desire to be remembered to you. I will try and report in my next letter on the Grand Canary organic remains, some of which are curious.

> Ever most truly yours,
> CHARLES LYELL.

To LEONARD HORNER, ESQ.

Berlin : March 24, 1855.

My dear Horner,—We have spent our time very pleasantly and happily here.[1] I returned this morning from an expedition to the Kreutzberg to see the boulder sand and clay, which I think has undergone a good deal of rearrangement since it first began its travels from Scandinavia. My companions were Beyrich, Ewald, and Dr. Roth. One or two mornings' work with Mitscherlich have been very profitable. He, Gustave Rose, and Dr. Roth went to Vesuvius and Stromboli and Vulcano. At Stromboli those three men established the important fact, that from the modern crater of Stromboli have proceeded *continuous* streams of lava twelve feet deep and 100 broad, of dolerite as compact as the ordinary lavas of Etna and of exactly similar composition, and that the same lavas are now and have been from the first on slopes of 15°, 20°, and 29°. What shall we think now of Elie de Beaumont's and Dufresnoy's assertion that no *continuous* stream could form on the least of these inclinations.

Scacchi's account of the eruption of Vesuvius in 1850, and all that happened there between 1845 and 1850, a copy of which Mitscherlich lent me to read, is highly interesting, and quite unfavourable to Von Buch's theory and to Dufresnoy's facts and views. In one paper Scacchi is so astonished at Dufresnoy's assertions, who never walked once over one whole side of Somma, that he says in so many words, 'that if the French *savant* was ever at Naples, he could never have gone out of his house, or only by night.'

It is strange what influence Von Buch exerted, for he

---

[1] With Chevalier and Madame Pertz, Lady Lyell's sister.

has made both Ewald and Beyrich entirely disbelieve all the glacial hypothesis. The other day I told Mitscherlich I would convert them both, and he said (both of them being present, and laughing at the joke), 'No, you will never do that, for the one (pointing to Beyrich) is like a stone, and the other like india-rubber; you think you are making a great impression, and then find next day that up he comes again just in his former shape.'

Ever affectionately yours,

CHARLES LYELL.

To DR. JOSEPH HOOKER.

53 Harley Street: July 17, 1855.

My dear Hooker,—I have to thank you for your valuable volume on the 'Flora Indica.' Your introductory essay is a great treat, and I must study it carefully with reference to the 'Principles of Geology,' as I may some day have an opportunity of correcting and enlarging my views. I have also to observe that I was gratified at your note at p. 40, because Bunbury once told me that Henfrey, in some general work of his, had said that Edward Forbes was the first to strike out that line. I never thought it worth while to allude to this, as I have tried not to spend any time or words in making 'reclamations,' though, I suppose, that if my books had not had so large a sale, I should have given way to that weakness, as I see so many others do.

I am glad you have two full indexes. I would have put the whole into one. I am sure that, as a rule, it is best, however heterogeneous the subjects, to have one only at the end, for one finds a word easier in a large than in a short index, and one knows exactly where to find it. But this is a trifling matter in a volume which one sees at once is the result of such a vast body of new facts. The maps are most useful.

De Candolle has just sent me his two volumes, but I have not had time to look at them.

I have had a week's steady work in the Wealden district, studying the problem of 'denudation,' but I must conclude.

Ever affectionately yours,

CHARLES LYELL.

To Leonard Horner, Esq.

Shielhill : September 5, 1855.

My dear Horner,—You mention in your letter to Mary
how much you were interested in Hugh Miller's ' Schools and
Schoolmasters,' a biography every page of which we read
with delight, though I quite agree with you in wishing he
had kept to his prose. I found the man quite equal to my
expectations from his books. I had always missed seeing
him before, and narrowly did so this time, as he arrived in
Edinburgh the day after us, after a three weeks' ramble.
He welcomed me most heartily, and was entirely at my disposal for two days. and would have accompanied me longer,
whether in the field or the museum, had I had time. His
remarks on other geologists were very just, and his criticism
also, with a generous appreciation of their merits. He is
quite as willing to learn as to teach, and I got a great many
new ideas, both in going over his beautiful collection, and in
walking over Arthur's Seat and Salisbury Craigs. I had
a walk to the top of Blackford Hill with Charles Maclaren,
who was very well, and full of geology. Dr. Fleming has a
notion of giving us something on Edinburgh, and I wish he
would. I had two good talks with him, and a walk over the
Calton Hill, and over the coast of Fife from Kinghorn to
Kirkcaldy. I view the Edinburgh and Fife rock with very
different eyes now, since Madeira and the Canaries. They
are in part intrusive, in part *sub-aerial* in my opinion, in part
estuarian, like the associated Burdie-house coal measures,
which last are the lower portion of the coal series.

Ever affectionately yours,

Charles Lyell.

To Charles J. F. Bunbury, Esq.

53 Harley Street, London : October 6, 1855.

My dear Bunbury,—I am afraid, from a letter which
Symonds has written to me this morning, that you have had
rather more than an ordinary share of bad weather, but I
hope you will make out a visit to him.

In Phillips' new edition of his ' Geology,' just out, he makes

the Lingula beds Cambrian, just as I do, which I am glad
of, as, however Murchison may complain, it is really we that
are adhering to the original divisions and names adopted by
Murchison and Sedgwick.   It would be wrong to give up
the term Cambrian just when we are beginning to have a
distinct fauna for it, as Salter was the first to show here,
and Barrande in Bohemia.   Sedgwick's attempt to take the
Lower Silurian into his Cambrian is even worse than Murchison claiming all that is older than the Devonian as apper-
taining to his Silurian.

Tell Symonds that I yesterday had a talk with Murchison about the Dumfries beds.   He says, when he first saw
them he felt, as he does now, that no such sandstones,
speaking lithologically, were ever seen by him in England or
elsewhere in the true trias.

I am glad that Frances has made Mr. Symonds' acquaintance, and that she is enjoying herself.

Believe me ever affectionately yours,

CHARLES LYELL.

To HIS SISTER.

53 Harley Street, London : January 5, 1856.

My dear Caroline,—I am actually going to dine at Lady
Coltman's alone, for Mary is not well enough, having been
in bed instead of breakfasting with me, for the second time
only since our marriage.   There was something like an epi-
demic at Mildenhall, though certainly Leonard and Frank[2]
were merry enough, and some others, but I presume that
Harry and Katharine gave you the news of our merry-
making there.

I was shocked to find poor Curtis[3] so nearly deprived of
sight, and expecting to be soon quite blind; the night I met
him there was a discussion at the Linnæan Society as to
the cause of the very unusual number of moths, especially
of the Noctua family, in the summer of 1855.   President
Bell, who is now proprietor of the house which White owned
at Selborne, said that there had only been one-fourth the
usual number of birds' nests in the spring of 1855 ; others
attributed the abundance of the *Lepidoptera nocturna* gener-

[2] His nephews.   [3] The entomologist,

ally to the scarcity of birds killed by severe winters. Curtis
said this could scarcely explain the case of the Noctua
tribe, since they get through their first change the autumn
or summer before they come out in wing. The result was
that many rare moths were common and many new ones were
taken. I suppose the ' Entomologists' Annual ' will give you
some report of this.

I had some good geological walks with Charles Bunbury,
most of them dry enough, although on the immediate
borders of that fenny region which Macaulay describes
early in his third volume ' as saturated with all the moisture
of fourteen counties.' Macaulay's is certainly a work to
read slowly, and to buy, not borrow. It is full of all kinds
of interest, and even the criticisms one hears on all sides
show how much it occupies the public mind. I hear that
his health is worse than ever. He has a bad cough, and few
expect we shall ever see two more volumes. How often he
must have waded through a shelf of MS. to be able to say in
a single sentence that there is no evidence in such and such
memoirs of certain popular stories. Some single chapters
imply the perusal of a room full of books and records.

Prescott's ' Philip II.' is very interesting. Not a few
readers complain that there is a want of expression of virtuous indignation on the part of the historian at the crimes
of Philip II. But he makes him out to be a villain of the
deepest dye, and all from his own royal letters. In regard
to the *auto-da-fés*, or murders perpetrated against heretics,
I think Prescott naturally makes great allowance, as in
those times every well brought up young lady in Spain who
was not a dissenter, would have sat and seen a Protestant
burnt alive with no small satisfaction, and when the Inquisition were torturing thousands to death, with the sympathy
of all or nine-tenths of the Spaniards on the same side,
one must not judge the monarch as severely as we should
Isabella II. if she should try now to imitate her Most
Catholic ancestor.

With my love to all, believe me your affectionate brother,
CHARLES LYELL.

To DR. FLEMING.

53 Harley Street, London: February 6, 1

My dear Dr. Fleming,—Lord Wrottesley, at the insta ᴵ ᶜᵉ presume, of the Committee of the British Association, left out two or three passages, in which in a tone complimentary to members of the ecclesiastical profession who are in earnest as churchmen, I had explained my reasons for objecting to ecclesiastics holding chairs of Natural History, &c. You know my reasons because they are in print, and will, I hope, refer any objector to my chapter on ' Oxford and University Reform ' in my first ' American Travels.' You will also point out as a verification of my arguments founded on past experience, three remarkable illustrations of the soundness of my objection to the clergy, especially if they be eminent; that Buckland, from the time Le was preferred to a Deanery, held and neglected his Geology Chair at Oxford; that Sedgwick did the same at Cambridge, or was less efficient when resident Canon at Norwich; and worst of all, that Henslow, whose influence when resident in the University was so great over the young men, no sooner became the conscientious pastor of an ignorant Suffolk flock, than three-fourths of his usefulness as Professor of Botany was at an end, and still he retains his chair! Say that I was in some degree a prophet rather than a bigot. I hope I am not to infer that the Natural History Chair is inefficiently filled in Auld Reekie. I never saw the Professor, and God knows gave him no help.

Believe me ever most truly yours,

CHARLES LYELL.

To CHARLES BUNBURY, ESQ.

53 Harley Street, London: February 19, 1856.

My dear Bunbury,—I was very glad to hear an account of the *Hypnum fluitans.*

Hartung has written to Professor Heer to invite him to send his S. Jorge fossil plants to me, that I may submit them to your inspection. I have made acquaintance with Mr. Wollaston, the nephew and representative of Dr. Wollaston, who has spent several years in the Madeiras, camping out on the Deserta and in Porto Santo and on all manner of so-called inaccessible heights in Madeira, pitching his tent on the summit of Pico Ruivo among other spots.

I have learned much conchologically and entomologically from him. The beetles are wonderful in their distribution, each island and almost every rock having its own. So many peculiar to the Madeiras, several hundred, and apterous although of genera not wingless in Europe or elsewhere, &c. Query, was it not foreseen that wings would only cause them to be blown out to sea and drowned? The winged kinds may date back to the period of the old Atlantis.

The shells common to the Madeiras and southern Europe, about a dozen of them, can be almost all proved to have been introduced by man, and the date of their radiation from Funchal is often known. A *Lymnea truncatella* got in thirty years ago, and has gone all over the island. A friend of Wollaston received a flower from Europe in a pot very lately, and found five species of European helices alive, buried in the mould. But although the fossil species of Madeira and Porto Santo are seventy-five in number, not one of them belong to any of these modern immigrants, and in the whole seventy-five, only one British, *Helix lapicida*, is a species living out of Madeira.

The evidence of a connection between the Atlantic group of islands is not strong unless it was at a very remote and almost Miocene epoch. But Wollaston thinks the insects can hardly be explained without it, though there was no union with Africa. He thinks there must have been a land of passage to enable some Canary Island species to travel northwards to the Salvage and Madeiras, that the migration was from south to north.

I hope when you are next in town to make you known to Wollaston. He is of independent fortune and no profession, has published on the Coleoptera of Madeira a splendid 4to, which Hooker and I have got the Athenæum to purchase, and you will do well to read the introductory remarks.

Ever affectionately yours,

CHARLES LYELL.

To Professor Oswald Heer.

53 Harley Street, London : April 23, 1856.

My dear Sir,—I received a few days ago both copies of your most valuable essay on the 'Plants of S. Jorge,' and am very much obliged to you for them. I sent off one to Mr. Bunbury, who is reading it with great interest, but slowly, as he does not read German as fluently as many other modern languages. I will, however, take the liberty of observing, that your style is so clear, and the sentences of such moderate length, that it convinces me that what we complain of in regard to the involved sentences in most German authors, is not the fault of the language, but of the writers.

Mr. Bunbury sees by the figures you have given that four of your ferns are wanting in our collection, namely : *Trichomanes radicans, Osmunda regalis, Asplenium Bunburianum,* and *A. Marinum.*

On the other hand we have three very well-marked ferns, and perhaps a fourth, which you have not.

One of the most remarkable of our dicotyledonous leaves is not in your set. Mr. Bunbury agrees in the identification of *Pteris aquilina* and *Oreodaphne fœtens.*

I am extremely pleased with your discussion of the old 'Atlantis,' a subject on which I have been preparing some observations, especially in regard to the fossil and recent shells. MM. Lowe and Wollaston have of late been seven months under tents in Madeira, Porto Santo. They have satisfied themselves that *Helix pisana* is not truly fossil, but lies bleached and dead on the top of the Caniçal and Porto Santo shell deposit. They believe it came into the island with man. Indeed they admit no European shell to be truly fossil except *Helix lapicida* in Porto Santo, now lost in the Madeiras as a living species.

They have found *Helix tiarella* living in the north of Madeira. But these and other discoveries do not alter the data on which you have reasoned materially.

I have told Wollaston of your Laparocerus. I am in hopes that Mr. Bunbury will still find some novelties to communicate in the way of species, though he and Dr. Hooker are

very timid in identifying species in comparison with yourself. The specimens of the fossil *Myrica Faya* which you
have figured do not agree closely with any of the varieties
of the same plant which Mr. Bunbury brought from Madeira
and Teneriffe.

We have more numerous and finer specimens of *Wood-
wardia* than yours, and hope you will remember that Mr.
Bunbury recognised this plant in my paper in 1854 to which
you refer. He thinks it is *W. radicans*, as he allowed me to
say at that time.

When I asked Mr. Wollaston why there have been no
coprophagous or dung-eating beetles created in Madeira, he
answered because such already existed in the island, and
they are now playing their part. I asked, ' What did they
feed on before the era of man ? ' and he said, ' Probably on
decayed wood.' Believe me, my dear sir,

Most truly yours,
CHARLES LYELL.

To C. J. F. BUNBURY, ESQ.

53 Harley Street, London: April 30, 1856.

My dear Bunbury,—I rather think that Heer stayed later
in the year than we did in Madeira. Wollaston says that a
botanist had almost better never go to Funchal at all, where
the nine years' fire burnt all the native flora, and left a clear
field for the foreign invaders. He speaks with rapture of
getting into the interior and camping out, at the height of
4,000 feet and upwards, on the Fanal and other native
woods.

After what he tells me, I feel how fortunate I was in regard to geology, for certainly Funchal is a first-rate station
for that science.

I certainly believe that the Madeiras were islands in the
Miocene sea, and that most of the indigenous shells came
there after the separation, and the reason of the flora being
so much less peculiar than the land shells is partly that the
latter have comparatively small powers of migration, especially across salt water.

Are Carices among the swamp plants which have a wider

P 2

range than plants loving drier grounds? I have got some
of their seeds (Carex) from Dr. Boott, and am making experiments as to their power of vegetating after long submergence in salt water. The seeds of Carices I find float well
in salt water, and a marine current would carry them far
in a short time. Many seeds will not float at all. Allowing,
what you say is admitted by botanists, that the temperature
of pools of water is more uniform than that of the dry land,
the seed of a Carex thrown ashore and getting into a freshwater swamp behind a line of sea-beach, or cast on a delta,
would be likely to thrive, while other dry-land plants would
not find a fitting climate.

According to Maury's last chart of the Atlantic, and his
deep-sea soundings, I think the only way Madeira could be
united with Europe and the Azores is by land stretching
round and crossing the Atlantic north of the latitude of the
Azores, and then coming south again.

What you say of the difficulty of supposing a connection
with America is a good hint. Wollaston rather exclaimed
against it as new and doubtful, entomologically speaking.
Agassiz always maintained that the European flora was
like that of the United States. If so (*Smilax, Comptonia,*
and other genera were cited in common) some Miocene
forms coming down to our times might create a generic
resemblance between the Madeira and American floras.

The *Ancylus fluviatilis* is said to have been indigenous in
the Madeira rivers, and lately the Portuguese introduced un-
intentionally the *Lymnea truncatella,* and it has run over the
island in thirty years, and is said to have appeared even in
pools and ruts in the roads. If this be so, they have powers
of spreading which require investigation. I suspect that
their spawn adheres to water-birds and water-insects?

When Huxley, Hooker, and Wollaston were at Darwin's
last week, they (all four of them) ran a tilt against species
farther I believe than they are deliberately prepared to go.
Wollaston least unorthodox. I cannot easily see how they
can go so far, and not embrace the whole Lamarckian
doctrine. Huxley held forth in a lecture last week about
the oxlip, which he says is unknown on the Continent. If
we had met with it in Madeira and nowhere else, or the

cowslip, should we not have voted them true species? Darwin finds, among his fifteen varieties of the common pigeon, three good genera and about fifteen good species according to the received mode of species and genus-making of the best ornithologists, and the bony skeleton varying with the rest !

After all, did we not come from an Ourang, seeing that man is of the Old World, and not from the American type of anthropomorphous mammalia? Pray write soon.

Ever affectionately yours,

CHARLES LYELL.

P.S.—M. de la Harpe of Lausanne called here one morning, and when I expressed my misgivings as to Heer's confidence about species founded on mere fossil leaves and fragments of leaves, he said that I ought to see Heer's collection and mode of work, and I should have more faith. I told him that Wollaston could not have felt secure without a leg or one of the antennæ of the beetle, not even that it was of the division of the Curculionidæ to which Laparocerus belongs, much less as to its not agreeing with any of the numerous known species of Laparocerus, a genus peculiar to Madeira.

De la Harpe replied, 'I am not at all surprised. Mr. Heer once, from a fragment of an insect's wing from Œninghen, said what it was, and the whole insect has since turned up fossil, and is exactly as Heer drew it prophetically. No one,' he continues, 'who does not make a study of elytra or wings as a separate branch, or of the leaves of plants, can imagine the resources of that branch when isolated, and when the entomologist or botanist feels he must rely on that and that only.'

All this convinced me at least that Heer is a man of genius, who impresses those who come into personal contact with him with a complete faith in his powers.

De la Harpe himself has converted Salter into a far greater belief in 'foliology' than he had before, and I saw the table in Jermyn Street covered with the leaves of numerous species of figs (Ficus) to prove that many of the Isle of Wight Eocene leaves are true figs by their venation. C. L.

To DR. JOSEPH HOOKER.

Hamburgh : July 25, 1856.

My dear Hooker,—A few minutes after you left me on Sunday, I fell in with the two copies of your Essay,[4] and have had one as my companion for two days, during a calm voyage to this place. I had read it before, but now I have digested it, and so the spirit moves me to thank you for it. Had I published when I returned from New Brunswick an Essay introductory to the Geology of that British colony, I should not have sold more than half a dozen copies. But had you gone to a publisher who knew his business, and who would have suggested or insisted on some general title, as 'On Botanical Classification, with Considerations on the Limit of Species in the Vegetable Kingdom' (and then in small type, 'In Special Reference to the Flora of New Zealand'), 'being an Introduction,' &c., you would then have sold a great many for the benefit of science, yourself, and all of us, zoologists included. Some day or other I would have you get out the whole of it almost verbatim, with selections from 'Flora Indica,' to serve as canons in natural history. This kind of work will be very indispensable from some one of authority, seeing where we are drifting to; for whether Darwin persuades you and me to renounce our faith in species (when geological epochs are considered) or not, I foresee that many will go over to the indefinite modifiability doctrine.

If so, it will not, or ought not, to make the slightest difference in regard to the rules you lay down, and in doing which you have with prophetic caution anticipated the possibility of many of your readers embracing the transmutation theory. But the species-multipliers will be delighted with a theory which sanctions to a great extent the conclusion that the boundaries of species are in the nature of things artificial, or mere human inventions, and therefore gives them a kind of right to affix their own arbitrary bounds. So long as they feared that a species might turn out to be a separate and independent creation, they might feel checked;

[4] Introductory Essay to the *Flora Antarctica.*

but once abandon this article of faith, and every man becomes his own infallible Pope. In truth it is quite im-
material to you or me which creed proves true, for it is like
the astronomical question still controverted, whether our
sun and our whole system is on its way towards the constellation Hercules. If so, the place of all the stars and
the form of many a constellation, will millions of ages hence
be altered, but it is certain that we may ignore the movement *now*, and yet astronomy remains still a mathematically
exact science for many a thousand year.

You must go into the doctrine laid down at p. 13 with
fuller explanations, and it must come to be understood that
they who make species must be consistent with themselves,
and make them all of co-ordinate value. If they will not
admit your two cedars to be one species, they must then
deny the authority of Linnæus and other great naturalists
who have made the primrose and cowslip one and the same.
I cite this at random, but some such inconsistency should be
exposed, and they should be set down as totally wanting in
philosophical power and their synonyms never cited.   I
could furnish good examples in conchology of such utter
neglect of relative value in the species invented by the
ignorant, who at the same time adopt the names of their
predecessors, which comprehend a much wider range of
varieties in one species of the same genus.

The speculations as to the quondam connection of the
antarctic lands now separated by ocean are very interesting.
I think Darwin does not enough allow for a suggestion
which you advance, that the land from which species now
common to A and C migrated, may have been in the space

B now occupied by the ocean. I always incline to the idea
that existing continental areas are in great part of post-
eocene date, as Europe, north of Africa, and a large part of
Asia best known, certainly are. For similar reasons spaces
like B may have been the compensating areas of subsidence.

Your idea of the antarctic species going north by the
Andes, and of the Panama Andes having once been loftier, is

very grand and probable. The marine (living) shells each side
of the narrow isthmus of Panama are very distinct, and it is
most probable that in the older Pliocene period that isthmus
was higher and broader.

Your occasional criticisms of Edward Forbes, to whom you
have done ample justice, interested me much, and some day
I should like to talk them over, as well as a multitude of
other topics. I expected you would have used ice-rafts more
freely in the antarctic ; coupling them with oscillations of
level in the post-miocene ages, one might get a most compli-
cated succession of states of the same area, geographical and
glacial.

I fear much that if Darwin argues that species are
phantoms, he will also have to admit that single centres of
dispersion are phantoms also, and that would deprive me of
much of the value which I ascribe to the present provinces of
animals and plants, as illustrating modern and tertiary
changes in physical geography.

I am also wholly at a loss to account for such facts as you
told me, of the *Lysimachia vulgaris* in the Australian Alps,
if we embrace the indefinite variability speculation, doing
away with all creation, and substituting a power of individuals to produce offspring unlike themselves, or a pos-
terity referable to distinct species. For if such were the
influence of external causes, one would think that the *Lysimachia*, or *Capsella bursa-pastoris*, could not retain its
character after wandering over the globe, and that it would
be contrary to the doctrine of chances that the individuals
descending from common parents of a different species
should be found in the southern and northern hemispheres
with identical characters, each having of necessity got into
this new state or permanent variety (or species) by quite a
different set of changes, both as to climate and co-existing
animals and plants.

If such results were possible, I should expect the re-
currence of the same species in distinct geological periods,
after they had become extinct, or in abeyance for an intermediate period or two. I have seen the Zamias and Cycads
of the Botanic Garden here, but have not yet seen Lehman ; however Otto, the inspector, showed Victoria Regia in

flower, and all sorts of Nymphææs, which are killed, root and all, every winter, and yet are got up in seven weeks in great beauty.

I am just starting to see Wiebel, the geologist.

Believe me ever sincerely yours,

CHARLES LYELL.

To LEONARD HORNER, ESQ.

Warmbrunn, at foot of Schneekoppe, Riesengebirge : August 8, 1856.

My dear Horner,—We have just been taking a walk in a fine evening's sunshine, to see the warm springs and the scenery of this beautiful watering-place; the top of the Schneekoppe, 4,954 feet high, quite clear, but the Schnee-grube, only 300 feet lower, capped with clouds and looking grander for the moment and loftier than its conical neighbour. We came here to-day from Breslau, thirty-five miles by rail, and about thirty more by extra post, over a good road, and through a very fertile and thriving country. The railway took us through a brown coal country, partly covered by Scandinavian drift.

When we left the plain, about 600 feet high, and got about 1,000 feet above the sea, we were in carboniferous and trap rocks chiefly, and now in granitic. Ferdinand Roemer and Goeppert were very attentive during our stay at Breslau. We had not seen Roemer since he came to us in Hart Street ten years ago, before he went to Texas; he has seen much, and is well read in geology. Goeppert showed me his coal plants, and tried to persuade me that his Stigmariæ were not roots, but a perfect plant of some aquatic species. His specimens in amber are very curious, and worth seeing, and especially under the microscope. Some large ones having the whole surface filled with circular convex indentations explain the specimens of copal which you may have seen on my mantelpiece in London, and which puzzled Charles Bunbury, and if I remember right, Robert Brown. The amber which has these markings is always found buried in the earth, and with a coating of decomposed amber, which comes off; and Goeppert says that much copal is dug up out of the soil at the foot of trees, &c., which produce it. He has a process by which he can give this kind of surface to amber.

August 9.—Yesterday Mary and I reached the highest summit of the Riesengebirge, an expedition well worth making from Warmbrunn. The latter place is about, I presume, 1,000 feet high, but I missed my barometer much, after my Madeira experience, having got accustomed to read off at once from the instrument, without calculation, an approximate height. The first seven miles was over a level region, where the granite peeps up in small knolls, and where there are numerous ponds, with oak, lime, birch, and fir. This region contrasts well with the fir-clad hilly region, where we left the carriage and began the ascent, I on horse-back, and Mary carried on a chair by two bearers. We had seven miles to go to the top of the Schneekoppe; first a mile through a straggling village, Seydorf, with orchards and gardens, then into a wild country of spruce fir, with very vigorous undergrowth of the whortleberry, which only allows the heath, *Calluna vulgaris*, to grow here and there in patches. At every two or three miles, a small farmhouse, fitted up as an inn, serves as a resting-place, and as we had occasional showers, enabled us to dry our cloaks and umbrellas and to take refreshment while the men smoked, &c. After passing some clear trout streams gushing down their granite channels, we climbed at last to the region of Knie-holtz, or brushwood, where the spruce firs which had exclusively formed the woods below become stunted, only three or four feet high, and nearly all the ground is occupied with a short fir, looking like dwarf Scotch fir, but I have gathered same branches with cones to ascertain. I believe in the guidebooks it is named *Pinus pumilis*. Here a species of *Persicaria* in flower almost equalled the dwarf pines in height. Iceland moss, gathered here for lung complaints, was very abundant. The snow prevents any tree from getting up above two or three feet, but there were no ' stag's horns ' or bare boughs such as I saw on the ' White Mountains '—in the same zone of stunted trees. We then came to the bare region where there are no trees, and much rock covered with *Lichen geographicus*, and at the top some 300 or 400 feet, where the slope of the granite is inclined at an angle of 35°;

but a zigzag road enabled us to ascend without difficulty. I was disappointed at not finding the vegetation at the summit more alpine. The same *Persicaria* we had seen far below was still conspicuous, a species of *Hieracium*, a few plants of heath (*Calluna*), a *Luzula* very abundant; these and a few others we have preserved, and a red lichen which looked like paint on the rocks, or the colour of the French soldiers' trousers, and a black lichen, with many others; but so many of the plants looked like those far below, that I wondered that 5,000 feet, even in the latitude of London, did not seem so arctic as the Clova hills, or hardly as the Clune, near Kinnordy. A traveller showed me the common Monkshood, and said he gathered it near the top; we could not see it, but it was conspicuous below in the woody region. Patches of snow had remained only eight days before we were on the summit.

After having been so many days between Hamburgh, Berlin, and Breslau, in the great plain of Scandinavian gravel, it seemed quite strange, as it will do to you in the Hartz, to be in a region of purely local drift. Nothing but the wreck of true Riesengebirge granite and gneiss, which comes up to meet the granite on the south side of the highest ridge near the top, not one far-drifted pebble.

For part of the day, the granite cone of the Schneeberg had a banner of vapour streaming from it like Etna, when the rest of the range was free from it. Joseph Hooker describes this as very singular in reference to the isolated and loftiest peaks of the Himalaya, and tries to explain it meteorologically.

In the stunted fir-region, a dandelion-looking flower, which we have kept, was pointed out by our guides as 'Arnica.'

At Breslau (the easternmost point I have ever reached in my travels) Goeppert showed us his splendid brown coal tree, formerly called *Pinus proto-laryx*, 26 feet in circumference, but another 33 feet has since been found. The state of preservation of the wood, which divides like deal in a peat bog, is to me very strange; for no matter whether the formation be called the lowest 'oligocene,' according to Beyrich's nomenclature, or by any other term, it is certainly about the age of the Isle of Wight series, below the Hampstead beds, about the top of my Middle Eocene.

I must send this off, trusting to Mary to say how pleasant a time we passed at Berlin, and to tell unscientific news. Humboldt was very courteous. It was a great pleasure to hear all were well. Mary came down quite fresh from the mountain, but as I had not been on horseback for an age, ten hours' riding with only two off the saddle, made me tired, till a good ten hours of bed restored me.

Ever affectionately yours,   CHARLES LYELL.

To LEONARD HORNER, ESQ.

Prague : August 23, 1856.

My dear Horner,—Geinitz showed me much at Dresden. Not many miles from that city, to the north-west, there is a region where numerous deep coal shafts are sunk 1,200 feet deep, through chalk and Permian into coal, so that they are turning the country into a manufacturing region, and getting it covered with railways, and threatening to make the capital so smoky that the Madonna di San Sisto and the other wonders of art may ere long run the same risk as the contents of our National Gallery.

Geology has gained much by these speculations to win the coal at such depths. They go through chalk marl, under which is the lower Quader, a sandstone representing our upper greensand, and sometimes greenish. The roth-liegendes beneath this is of various thickness, associated with porphyries and syenites, which come in in unexpected places, and interfere with speculation and contracts. The coal is often excellent. The specimens of rothliegendes conglomerate showed that brecciated character which made Ramsay conclude a glacial origin for some of ours, but these Saxon cases would not help him much, as the angles are often worn off. Yet as contrasted with Old Red and New Red, the subangular character of the fragments in these Permian agglomerations holds good, and is very singular.

Our exploration of Saxon Switzerland with Gutbier as a guide was thorough. The scenery of that beautiful region derives its chief beauty and peculiarity from the upper

Quader, a mass of quartzose grit, like some millstone grit when coarse, and when fine just like much of our tertiary ' Druid Sandstone,' being often 800 feet thick, and rent in two opposite directions, and the joints often open, and these joints facing the rivers, and brooks, and valleys quite perpendicularly, with here and there intervals of two, four, or eight feet between two joints removed, you have ravines and separate columns innumerable. As this mass is above the level of the Pläner, which is lower white chalk, or grey chalk (the clunch clay of Strata Smith), it follows that it represents our white chalk. In other words the Quader of Saxony, a grit wholly deficient in calcareous matter, corresponds to the most purely calcareous rock of Great Britain, and yet contains here and there the same shells.

I was glad we made an excursion to the country on the left bank on the south-west side of the Elbe in Saxon Switzerland, for there I saw the same Quader unbroken, and forming large platforms, instead of such isolated masses as at Königstein, Lilienstein, and others which occur chiefly on the north-east side, though by the way, Königstein is an exception on account of the winding of the river. From the Schneeberg, 2,300 feet high, where the cretaceous or Quader reaches its highest point, we had a fine view into Bohemia, the Toeplitz region, and of the Mittelgebirge, where the Milleschauer, a phonolitic mountain, very conical, is 2,500 feet high, and in another direction saw the beginning of the Erzgebirge. Returning to the Elbe at Tetschen, Gutbier resumed the railway to return to his fortress of Königstein, of which he is sous-commandant; a great place, which the late King of Saxony, a lover of science, gave him, because he was a naturalist as well as a good soldier and military engineer, though not in the engineer corps. He regretted missing Charles Bunbury when he was at Dresden, as he had published on the fossil plants of the Saxon Permians. We learnt the names of many wild plants while with him, but mostly English ones or Scotch. I don't know whether *Prenanthes purpurea*, which was new to me, is British. The extreme scarcity of land shells, owing doubtless to the absence of calcareous matter in the Quader, is quite striking. Yet on the walls of the fortress of Königstein, where artificially lime had come with the mortar, we found two *clausilias* and a very singular planorbiform variety of *Helix lapicida*, which may throw light on one of my Madeira fossils.

My excursion to Toeplitz and Bilin gave me a good idea of the chalk, and brown coal, and basalt, and syenite, and the infusoria beds of Bilin.

The number of shafts sunk into the tertiary to obtain brown coal is so great between Toeplitz and Aussig, and the tall chimneys so numerous to pump the water out of the mines, that you would suppose yourself in an old coal country. Yet it is only Upper Eocene.

The gorge through which the Elbe flows south of Aussig displays fine sections of this brown coal in the form of sandstone with much basalt, and afterwards cretaceous beds, Pläner and Quader. We then got into the valley of the Moldau through the old coal and overlying cretaceous, till near Prague we entered Barrande's Silurian of Bohemia, and here I found this great workman and discoverer himself, very much disposed to show me the principal localities of the 1500 ! ! species of Silurian and Cambrian fossils which he has brought to light, nine-tenths of them distinct from the *species* of Scandinavia and England. Sternberg had only worked in the old coal.

How little did I think when La Place once pointed out to me, from the top of the Observatory at Paris, the number of new convents rising on all sides, and new Jesuit establishments, and expressed his indignation at it, that among other results of this bigoted course which Charles X. was then pursuing, would be the banishment of the Bourbons, first to Edinburgh and then to Prague, where, twenty-four years ago, Barrande, the tutor of the Duc de Bordeaux, began geologising! He took me to the quarry where he found the first orthoceras, and where we picked up many. He went with it to the Museum at Prague, where they assured him that Count Sternberg had left nothing for anyone to do in this region. Barrande has now 500 species of Cephalopoda from these beds, proving among other things, as he quite admits, that the mollusca, at any rate, were high enough in the scale in those very early days, and the very class of whose history from first to last we know the most. I have been in two immense quarries, one in limestone and another in trap alternating with Silurian schist, exclusively worked by Barrande for fossils, and there are others of the same kind in all directions. A perfect knowledge of the Bohemian language was indispensable for Barrande, and I have heard him talk with the peasants most fluently. For twenty-four years he has gone on steadily investigating this region, which is twenty-four French leagues in its longest diameter. I mean the fossiliferous part of the old rocks. It is very difficult to remember Barrande's nomenclature, as he only uses letters. I have already been twice on the ground of this famous *colony*, as he terms it, and, explain it as we may, it is the most singular and, at first sight at least, anomalous fact I ever remember to have verified in paleontological geology.

As he puts it in his work, it is a sort of anachronism, not by any means equalling the Petit Cœur case of supposed coal plants of the Alps occurring in lias, but a small approach to the same kind of entanglement of two distinct faunas or assemblages of species, which but for these few exceptional localities would have been supposed to have had nothing whatever to do with each other, or to have belonged to two distinct (though consecutive) epochs.

Yours affectionately,   CHARLES LYELL.

To HIS SISTER.

Vienna: August 28, 1856.

My dearest Caroline,—I found your letter among many others in the Post Office, and was much interested with the entomological news. I had quite made up my mind that Wollaston would be very glad of the contents of the box, for knowing you have a good eye for insects generally, and a practised one for *Elaphrus lapponicus*, your remark to me that the batch was undersized struck me much, and I thought the Wilsons had gone up to some higher bog and got a smaller variety of *E. lapponicus*, which would, I suspect, have been more curious in Wollaston's eyes than this *E. uliginosus*. I have told my bookseller to send you a copy of Wollaston's book on the variation of species as illustrated by insects, that you may see how he is at work. I have had no time to

look even after *Lepidoptera diurna* on this tour, but have
been struck with their perfect identity, such as came in my
way, with those of England. At Dresden I saw the ' Cam-
berwell Beauty'—I forget the Linnæan name—but at any
rate the English name is, if I mistake not, much older
than Linnæus,—a splendid insect on the wing. Although
I have missed Professor Reuss, the geologist of Prague,
who was travelling in Moravia, I have been very lucky in
finding the men I wanted at most of the principal places I
have visited. The Prussians have a University (equal in
the number of students to Bonn, though not equal to
Berlin) at Breslau, the capital of Silesia. Herr Goeppert
and F. Roemer, who have published good works on geology,
were very attentive. At Dresden Professor Geinitz, whom
I missed when last there, accompanied me on an excursion.
But I was particularly fortunate at Prague to find Barrande,
a Frenchman whom I had before known at Paris and
London, who was tutor to the Duc de Bordeaux formerly, and
is now chief manager of his estates. When Charles X. went
into exile to Prague in 1832, Barrande settled there, and
finding the Germans were neglecting the older (Silurian and
Cambrian) rocks, and that they were rich in fossils, he set to
work, and spent all his own private fortune in opening and
working quarries expressly for the fossils. How many thousand pounds sterling he expended in twenty-four years I cannot say ; but he told me that in order to explain the eighteen
metamorphoses of a trilobite called *Sao hirsuta,* he collected
20,000 specimens, and they cost 5,000 francs, or 200*l.* He has
at length got 1,500 species, and nearly all of them are peculiar
to Bohemia, or new. He showed me these in his museum, some
of which, however, I had seen at Paris, and I went four days
on an excursion to understand the quarries and position of
the beds, and some very extraordinary results at which he has
arrived. He is a very remarkable man, with a good deal of
simplicity of character, so rare among Frenchmen, and much
commoner in Germany. Our friend Colonel Gutbier, with
whom we spent three agreeable days in Saxon Switzerland,
complained that Humboldt had lost this simplicity by living
so much in Paris. I found him, however, at eighty-seven, just
what I knew him more than thirty years ago, quite up to

all that is going on in many departments. Think of his eyes, when in these days of cheap postage, his letters cost him 100*l.* a year, and he answers and reads himself nearly all!

Ever your affectionate brother,
CHARLES LYELL.

To DR. FLEMING.

Vienna: August 31, 1856.

My dear Dr. Fleming,—I was very glad to receive your letter at Prague. I have been much pleased with what I have seen of those parts of Saxon Switzerland, as it is commonly called, which I had not explored when there some years before. At Prague I made four excursions with M. Barrande, who has perhaps done more work than anyone in our time in paleontology and the field united. He explained to me on the spot his remarkable discovery of a 'colony' of Upper Silurian fossils 3,400 feet deep, in the midst of the Lower Silurian group. This has made a great noise, but I think I can explain away the supposed anomaly by adopting Barrande's printed explanation about the removal of a barrier, and by allowing for ten per cent. of 'peculiar' species in the so-called colony—a point, I believe, often overlooked. But I will not go on, lest you should not have read about this famous puzzle. What between Sweden and Bohemia we have now 120 species of the Cambrian (or Primordial) fauna all distinct, and all the *genera* of the trilobites save one, from the Silurian. As Barrande names his primordial fauna C, it is easy to remember that this stands for Cambrian. In all, he has 500 species of Cephalopoda, scarce one of them English or Swedish. They show that the mollusca in early days were not of simple form, but of high degree. About 130 gasteropods, but not one of Lamarck's canalicu'ata, so I suppose the carnivorous cephalopods did their work, as Dillwyn used to say, long ago.

The Silurian fauna of the United States agrees considerably with that of England and Sweden, which may have formed one great province becoming gradually different in its remote parts, while Bohemia, Spain, and Portugal constituted a second province, having all the species except a few brachiopods distinct. Besides these, there must have

existed contemporaneously another province, namely, the
mother-country of the famous 'colony.' All the facts stated
by Barrande respecting the latter are perfectly true, as I am
now convinced. Murchison has been over the ground, and,
as Barrande tells me, has printed his adhesion to Barrande's
view so far as the section is concerned. Fortunately the
Moldau, or rather the great natural rent through which this
river passes, lays open clear and continuous sections displaying most beautifully the succession *en masse* of the rocks
composing the groups D, E, F, G, including the 'colony,' so
that he who runs may read, provided he has Barrande with
him to cross-examine. I never saw Silurian fossils in such
abundance except in a few strata in Sweden; but here they
pass through many thousands of feet. Yet the whole fossiliferous area is only equal to one-sixtieth part of the
Adriatic. As Barrande himself has calculated this, I wonder
he remains such a finality man. I remember at the Geological Society when Sedgwick and Murchison used to argue
with me exactly on the grounds now taken up by Barrande
in proof of a beginning of life on this globe, founded on the
notion that no fossils would ever be found below the stiper
stones. Now that a totally distinct fauna has turned up,
and that the transformations of some are traced from the
egg to the adult, the discoverer is just as sure that here at
least we have the true beginning. We expect to be home
about the end of October.

Believe me ever truly yours,

CHARLES LYELL.

CHAPTER XXIX.

SEPTEMBER 1856–AUGUST 1857.

GEOLOGY OF THE NEIGHBOURHOOD OF VIENNA—STYRIA—SALZBURG—DIS-
CUSSION WITH HEER AT ZURICH—FOSSILIFEROUS QUARRY AT SWANAGE
—VISIT TO SWITZERLAND — FOSSIL PLANTS — ELIE DE BEAUMONT'S
FRANK OPPOSITION.

CORRESPONDENCE.

To LEONARD HORNER, ESQ.

Middendorf, Styria : September 15, 1856.

My dear Horner,—We are here in a small inn in the
Styrian Alps some 2,500 or more feet above the sea, and
where they have six months of snow from December to May.
On our way from Gratz to Ischl, it is a beautiful country, of
green meadows in the foreground, and cultivated fields on
steep slopes, with usually a background of limestone mountains, with perpendicular precipitous sides often of bare rock,
but with larch and spruce on every available ledge.

I left off at Vienna with my geological news. Before
quitting it I contrived to make excursions with different
geologists to the chief points of interest both in the Miocene
tertiary basin, where the beds are nearly horizontal, and the
secondary and Eocene tertiary rocks, where the strata are
at high angles or vertical, as usual in the Alps, of which
we have here the eastern extremity, although they are in
fact continued in the Carpathians, with the same strike and
character. The commencement of these last mountains
is well seen from some of the heights near the northern
suburbs of Vienna, one of which called the Himmel, near
Sievering, I visited in company with Mr. Cevarowitch, one of
the officers of the Government Geological Survey. In that
hill are splendid quarries of highly inclined Vienna sand-

stone, corresponding to much of the Alpnie Flysch, and deserving the name of Azoic as much as certain rocks so called in Scandinavia, in which fucoids have been found, as they have been in this Viennese sandstone. I could not help looking on it with great interest, even from its nega- tive characteristic of containing no fossils, whether in the Viennese or Swiss Alps. There seems little doubt that part of it is of Jurassic and part of Eocene origin, for some few nummulites have been lately detected in one region and specimens of Aptychus in another, but they must be wonder- fully rare, since so many magnificent quarries have been opened in it for building and paving stones, and the whole Viennese corps, who have proved themselves very sharp- sighted collectors, pronounce the formation to be hopelessly barren. I wrote to Barrande to commend to him these ' Azoic' beds, as he argued with me at Prague that no living beings could have inhabited the globe before his ' primordial fauna,' because some of the sandstones below his primordial were precisely like in mineral character to others lying above it, and in which fossils have been met with. Edward Forbes used to say that no one who has dredged much can ever wonder at rocks being wholly devoid of organic remains.

I was fortunate, just as I was on the wing for Styria, to see a young geologist who has been kept always at field work in the Alps by their Survey, of the name of Stur, very much in earnest, as indeed they all are, but who is by far the most ardent generaliser of them all. He knows no language but German, but Studer arrived in time to help me in one of my conversations with him. Stur is, I believe, the first who has attempted to give a series of diagrams showing the state of the Alps from the earliest periods. He makes out eight or nine periods of elevation, or of convulsion rather, or movement.

Upon the whole the quantity of work done and doing in Austria, and the number of young rising men of education and some of them of family, interested in geology, is very striking, and as they are all connected with mining establish- ments, and therefore professionally occupied, or with Im- perial Museums, they will continue to pursue the science. It is a great pleasure when one stops at a coal-mine, as I did at Leoben, a small place between this and Bruck in the Styrian Alps, to find several gentlemen of education managing the works. It was a tertiary coal, forming a bed of coaly lignite actually some forty feet thick, of Miocene age, with fossil plants. There are many basin-shaped masses of such lignite formations in the Styrian Alps, in some of which freshwater shells (a few of them of living species) accompanied by *Mastodon angustidens, Anthracotherium,* &c. occur, a good deal disturbed and yet horizontal as compared to the distorted, coiled, and overturned older stratified masses of this wonderful chain.

Boué made himself very agreeable and useful to me in two excursions. He told me that at one time, having a fright about French political affairs, and expecting that Thiers would make war with England, he sold all his money out of the French funds, and having a great opinion of the United States, bought into theirs, just when they happened to be most depressed at the time of the panic after the 'repudiation' affair. In a few years he found his fortune exactly doubled by the rise of the United States state and other securities, upon which he sold out and invested in land and houses in Austria, to which his wife belonged. Here he felt rich for many years, and has still a house in Vienna, and a delightful country house near Baden, an hour and a half by railway from Vienna, the environs of which city are most agreeable and picturesque and geologically varied, as you would expect on the flanks of the Alps, or where they join the tertiary of the plains. But the increase of prices and enormous pressure of taxation has of late made Boué feel poor, and unable to travel and geologise as he would wish.

Hallstadt : September 20.—It is raining here, and the snow has actually fallen last night on the higher part of the hills not more than 4,000 or 5,000 feet above the sea, which rise abruptly 2,000 or 3,000 feet from this beautiful lake which is under our windows, separated from us only by a garden full of Dahlias and China asters. We made out our expedition to the valley of Gosau very successfully yesterday, and saw enough, though the clouds came down too fast, to convince us that it has not been overpraised by lovers scenery. The fossils of the chalk, consisting chiefly of univalves, might well lead a geologist not familiar with the

Blackdown beds, to imagine that the formation was tertiary, although the species are different, and the hippurites and other cretaceous forms ought soon to rectify the idea.

I have fortunately obtained here from the Bergmeister two specimens of moderate size of the Hallstadt (upper triassic) limestone in which an ammonite is associated with an orthoceras in the very same fragment. They say that Von Buch came here to ascertain the fact, and confessed that it overturned a favourite theory of his. I spent an agreeable day and a half with Unger, and I gave him a copy of Charles Bunbury's paper on the 'Carboniferous Plants of Cape Breton,' which he (Unger) was glad to have. In this limestone region we have been finding many land shells, especially in our excursion to the three beautiful lakes through which the Traun runs, or rather from the highest of which it rises, so justly praised by Sir Humphrey Davy.

Ischl: September 20, 1856.—We have just arrived here, the clouds covering the mountain-tops and leaking a little at intervals. We found letters from Berlin to Mary, and yours to me. I am glad you saw Ewald, and that he approved of my attempt to explain Barrande, who has half, but only half, given in. I accused him of a downright anachronism, which of course he did not quite like; but it was really a solecism, and he ought to thank me if I can help him out of it. No doubt Ewald is right in saying that the word 'colony' is wrong in the exact way in which Barrande uses it. But if we consider it as a colony of an antecedent state of the fauna of E, the term will do. The first European settlers in Virginia were a colony from England, though not of England when she had more than doubled her population in the days of Queen Victoria.

With love to all, ever affectionately yours,

CHARLES LYELL.

To C. J. F. BUNBURY, ESQ.

Salzburg: September 27, 1856.

My dear Bunbury,—Your letter was forwarded to me from Innspruck to this picturesque and beautiful town. There is so much to see here, and in the snowy Alps immediately to the south of it, both in geology and scenery, that I have determined not to go so far as the centre of the Tyrol, as it would be too hurried a journey to do any justice to it, and the shortening of the days and the snow on mountains only 6,500 feet high, which fell last week and is very slowly melting, has warned me that the best season has gone by. Yet we have had to-day a clear blue sky, and a sun often too warm to expose ourselves in, without shelter, and the tops of the Great and Little Watzmann as steep as the Aiguilles of Mont Blanc, 8,000 feet high, have formed a grand feature in our landscape in an excursion in the neighbourhood. I am trying to understand what members of the 'Alpine Limestone,' as it was formerly called, are Jurassic, which Triassic, and which Cretaceous. It is strange to think that rocks so devoid of fossils to a cursory observer, should have yielded of late, by dint of searching some particular spots and beds, such a multitude of species, 1,200 invertebrate for the Upper Trias alone.

The sandstone with *Fucoides intricatus*, a part of the 'Vienna sandstone,' seems to be made out pretty clearly to be a member of the chalk above the gault. I collected some of the fucoids yesterday, which were so fine that I shall send you specimens when I return.

Ever since we have been in the Alps, we have not ceased to admire the profusion of autumnal Gentians, consisting of three kinds, which we are told are *G. ciliata*, a fringed sky-blue one with a fringe on the corolla, *G. autumnalis* (or *Germanica*), a purplish one with many flowers on a stem, and *G. asclepioidea*, with several rich blue flowers on a long stem. The latter abounds most at higher levels than the other two. The *G. Germanica* is often as thick as some purple orchises in boggy ground in England. We observe that on the same stem, some flowers have four and some five divisions of the corolla, and a corresponding number of stamens respectively. Perhaps this is common in many other common genera. They regard *Colchicum autumnale* as a pest here. It adorns the meadows very much, and is as thick as if in a garden plot at very various elevations. This town is more than 1,309 feet high, though commanding a view of the great plain of Bavaria, and looking low with the Alps on one side.

I have been observing the land and freshwater shells
with some attention. After collecting some thirty species,
I have never seen that large and conspicuous English shell,
Helix aspersa. As it is not fossil in the loess of the Rhine, I
suppose Woodward may be right in saying 'that it came in
with man and the dog.' *Helix arbustorum* swarms above all;
H. nemoralis (or *hortensis*) not very common. On the whole,
perhaps six in twenty-five not British. I am glad you
alluded to A. De Candolle's book, which he sent me, and
which I have always been meaning, and still mean to read.
Hooker's doubts as to the *Pinus pumilis* being the Scotch fir
interested me much, for though the possibility struck me, I
gave up the idea when I saw how positively it ranks as a
true species in the books. Its cones seemed to me as large
as those of a tall Scotch fir, which has the effect of making
it look peculiar, and more unlike than its mere dwarfishness
would do. It never gets so low as the (Spruce?) on the
White Mountains, but I saw the *Persicaria* striving to top it,
as did *Lycopodium dendroides* presume to do in the case of
the dwarfed American fir trees. I saw no gradual passage,
but the change from a region where the snow does not lie so
long, to one where it endures for a great many months, is
rather sudden on the Riesengeberge.

With love to Frances, ever affectionately yours,

CHARLES LYELL.

To PROFESSOR GEORGE HARTUNG.

Munich : October 5, 1856.

My dear Hartung,—I was very glad to learn by your
letter which I found here, that we may look forward to the
pleasure of seeing you in London on your way to Madeira.
I have made progress, since I saw you, in German, having
had to talk some whole days with good geologists who
know no other language, so I hope to make out what you
write in your own tongue, and can always get the aid of my
wife when in doubt. I am well pleased that you have found
in conversing with your friends, how much (and very
deservedly) the connection of the history of the Madeira and
Canary volcanos is dependent for one of its greatest interests
on the various groups of organic remains.

To give you some idea of the difficulty of bringing all the evidence to a focus, I may mention, that when I applied at the British Museum to get the corals, Zoantharia of Porto Santo, &c., named, they told me there was no one in England who could do it, and they had just sent over their own to Paris to be named by Milne-Edwards.

So I have got my old friend the former secretary of the Geological Society, Mr. Lonsdale, our best authority, to rndertake them, and his report is, I hear, ready. The Bryozoa have been examined by a competent English geologist, Mr. Busk, as a favour to me. As to the shells, I have already worked hard at them, and hope to have a list of sixty species from the Great Canary. I spent a day in comparing the two species of Cypræa with Mr. Gascoin's shells, who has the finest collection in Europe of Cypreadæ, unequalled in any public collection. The result was, that they are both new species. You remember them perhaps, and that I thought they would turn out living species, to which they have, until closely compared, a very marked resemblance.

As to the land shells, I have given much time to Wollaston's unrivalled collection, and hope to have some new and original results to work into the paper, but then the detailed lists of species, recent and fossil, which must be compared, constituting the *pièces justificatives* of what is said in the body of the paper, must be given separately.

Ever most truly yours,
CHARLES LYELL.

To LEONARD HORNER, ESQ.

Strasburg : October 22, 1856.

My dear Horner,—I believe that my last geological news was written to you from Salzburg, after which we visited a most beautiful part of the Alps, immediately south of Salzburg, Berchtesgaden and the Königsee, the latter a beautiful lake, surrounded by limestone precipices 4,000 feet high, and some of which look as if they were horizontally stratified, but which appearance is found after much study to be a delusion. I was greatly perplexed by it, for I knew that no single uniform calcareous deposit in the Swabian or Jura chain could boast a thickness of 1,000 feet, where

it was clearly to be made out in reference to age and relative
position, and I was unwilling to believe in the sudden
development and augmentation of each calcareous formation
on their entering the chain of the Alps. Fortunately I fell
in at Berchtesgaden with Mr. Gümbel, who has been employed
for the last four years with the construction of a geological
map of Bavaria by the government. He had just given
several months to the very region which was puzzling me,
and assured me that he had ascertained that the limestone
which appeared 4,000 feet thick or more, was only 350 feet
thick, and was the same member of the lias which I had elsewhere seen under the name of ' Dachstein,' so called from a
snow-covered alp of that name near the valley of Gosau.
He declares that the splendid cliffs of limestone bounding
the lake of Königsee, and seen far beyond, forming the surface of the snow-covered ' Great and Little Watzmann Alps,'
are a mere facing of stone, cleft by horizontal joints in
imitation of the bedding. These alps are between 7,000 and
9,000 feet high. According to this explanation the name of
Dachstein is not a bad one, for a deposit that covers or roofs
such lofty mountains. I was surprised, before I fell in with
Gümbel, at meeting with a species of Orthoceras in a quarry of
limestone in the Untersberg mountain near Salzburg, which
the Vienna geologists have set down as Dachstein. It is a
quarry which has supplied marble for the Bavarian Walhalla.
For I had thought it marvellous enough that I had seen several,
of species of that ' paleozoic ' form, in the Hallstadt or upper
triassic group. But Gümbel assured me that the marble in
question was even still younger than the Dachstein, being
what he terms a member of the Oolite called the ' Adneter
Kalk.' I afterwards learnt that very lately a *belemnite* has
been detected in the Hallstadt beds, so that they are travelling downwards, while the orthoceras is ascending upwards
very fast.

    At Munich I called on Prof. Wagner, who very politely
put off a journey he was just starting on, for a day, in order
to show me over the collection of the late Count Münster,
which they bought for the University of Munich. The specimens in the Solenhofen lithographic stone are splendid, as I
daresay you remember. But what attracted me most was a

newly arrived set of fo sil Miocene mammalia, from a calcareous breccia near Athens, already in part described by
Wagner. Not only Dinotherium, *Mastodon angustidens*,
and a new large Machairodus, but a large Miocene ape, and
a great creature of the sloth or megatherium family, called
Macrotherium. It may perhaps be a great sloth which
Cuvier determined from a single phalangeal bone in France,
which I always thought a grand *tour de force,* and which some
called in question.

Liebig showed us his newly built laboratory, and a friend
of Gümbel's, Mr. Hessling, took me over a new Physiological
Government Institution, and a separate one for Anatomy.
The King seems favourable to science. In the first of the
two last-mentioned buildings was an aquarium for experiments on living freshwater animals. For several years the
pearl-bearing *Unios* (*U. margaritifera*) have been bred in
numbers. It takes eight or ten years for a pearl to form.
There are three kinds of external deformities which clearly
show whether there are pearls or not inside the shell. By
knowing these you may, without the labour of opening them,
cast them back into the water if they are barren, as a
vast majority are. It is a singular paradox which Liebig
discussed with me, that the thick-shelled pearl-producing
Unios live in the lakes of the granite and 'Greywacke'
regions, where there is scarce a trace of lime in the water,
whereas in those lakes and ponds in the limestone districts
where there is a very large proportion of carbonate of lime,
the Anodons with their very thin shells abound. It seems
to depend on vegetable food.

At Ulm I made an excursion to see some freshwater beds
about the age of the 'Mayence basin,' which, everywhere
almost on the Continent, geologists call 'Lower Miocene,'
and I have seen so many signs of *freshwater* and *land* species
passing from strata of this age to the supposed equivalents
of the Falunian, that I may perhaps have eventually to
adopt this same classification, and call my 'Upper Eocene' a
'Lower Miocene.'

For if it is once established that the line cannot be
drawn anywhere without being somewhat arbitrary, it may
be better not to let too much be absorbed (as Hamilton

said) ' in the Eocene vortex.' My argument of the perfect
distinctness of the French and English and Belgian *marine*
shells of my ' Upper Eocene' and those of the Faluns still
holds; but perhaps Beyrich's investigation, and Sandberger's,
may eventually make out a passage even there, and Hébert
thinks he has many links by aid of the Bordeaux basin.

At Tübingen I found Quenstedt, as I anticipated, a
hard-working, enthusiastic, and original man. Like Barrande, he has opened large quarries in the lias and oolite
exclusively for fossils, and has obtained an unrivalled series
of ammonites, brachiopods, belemnites, &c., besides some fine
reptiles and fish. He finds the beds admirably defined by
their characteristic fossils over an area of thirty or forty
miles of the Swabian Jura. He showed me the various beds
in the field, and we collected the commonest species *in situ.*
Gryphea arcuata (or *incurva*) though so abundant only extends vertically through eight feet. The number of species
of ammonites, which he proves to be varieties of one, is very
great.

Oppel studied for three years as Quenstedt's pupil,
then travelled in France and England to compare the
Keuper and Jurassic beds of Swabia with those of England
and France, and out of this came the joint work of Oppel
and Suess on the ' Koessen' beds, now proved to be of the
age of the ' bone-bed' of Würtemburg and England.

Last year Oppel thought the bone-bed near Stuttgardt
contained liassic marine shells, but they now prove to be all
distinct. You may remember that for several editions back,
(if not from the beginning,) I have always put the bone-bed
in the trias, and not, as other English geologists did, in the
lias. This opinion is now made out by the marine shells to
be true. It would have been strange if the fish and reptiles
had misled us.

At Zürich I was glad of an opportunity of discussing
with Escher v. der Linth and Heer, both fresh from Vienna,
the great question of the St. Cassian beds, or the discovery
of the long sought for marine fauna of the upper trias.
Also the real position of the Flysch ' Vienna sandstone;'
whether, as Schafhäutt assured me, it was under the nummulitic and of cretaceous age, or, as Gümbel said, above it,

and as he almost proved to me by sections near Salzburg which I saw, or, as they told me at Vienna, partly Eocene and partly *Jurassic*! Escher is clearly of opinion that it is Eocene—an azoic Middle Eocene—perhaps 2,000 feet thick, but some think that all the Glaris fish belong to it. They showed me the celebrated skeleton of a bird in the black slate of Glaris. The rock may well pass, as it did once, for a 'transition schist.'

Bunbury will be interested to hear that Heer, who was with Escher in Italy after the Vienna meeting, found all the Monte Bolca plants without exception to be different from his ' Lower Miocene ' (my Upper Eocene) flora. Heer showed me his plants of Swiss Miocene and Pliocene localities. The number of American forms is what he most dwells on. How abundant the plane trees must have been, and yet there is no indigenous European Platanus. But the insects of Œninghen struck me most. The colours of a *cimex* so well preserved, and of several species of *Buprestis*, and so many Brazilian forms as he makes out, especially a large hydrophilus, of which he had at first only the elytra, and yet he ventured to say it was a Brazilian form of the *Hydrophili*, and now the discovery of a perfect individual has proved it. But this was not guessing. They have a magnificent collection of living species of insects at Zürich, which it took Heer seven years to arrange. Of beetles alone, or Coleoptera, there are no less than 30,000 species!

The fruits and even flowers of some of the plants of Œninghen are wonderfully preserved.

A coprophagous beetle, ' *Aphodius*,' in the Swiss lias, foretells the future finding of a liassic mammifer.

At Paris I learnt from a letter of Wyman, that he has at last an *Ophidian* from the old coal of Ohio—a great step—besides new batrachian bones. I also found Hébert hard at work. No less than five individuals of the large ostrich-sized Gastornis from the conglomerate under the Plastic-clay are now found, and with them a large lophiodon and crocodiles and a coryphodon.

Since I wrote what I said above about the controverted point where to draw the line between Eocene and Miocene, I have called on Deshayes, and he was reading the French

translation of my manual. He said he entirely agreed with
me as to classification, hoped I should never give up, and
would show me a multitude of new proofs which he had recently obtained by work in the field in the Paris Basin in
support of my old plan (and his) of considering the ' Grés de
Fontainebleau ' as Upper Eocene, and as quite distinct from
the Falunian type—an opinion which Edward Forbes thought
demonstrated by his latest discoveries in the Isle of Wight.

Believe me, my dear Horner, ever most affectionately
yours,   CHARLES LYELL.

Paris: October 26.

To CHARLES BUNBURY, ESQ.

53 Harley Street, London: January 13, 1857.

My dear Bunbury,—Mr. Beckles called on me about a
month ago, to show me the bones of the huge Iguanodon's
foot (tridactyle, like a bird!) which he had found. I showed
him *Stereognathus ooliticus,* then in my charge, and urged
him to go and open a quarry at Swanage in the dirt-bed,
and go after higher game.

He took me at my word, and soon wrote to say that the
first day produced two reptiles, and the second a jaw (query
if mammalian) which he sent me by post.

It was a new genus of insectivorous mammal! Mr.
Bristow, of the Survey, then told me that Brodie at Swanage
had been collecting for two years without sending to Owen.
I wrote and begged for a sight of the new *Mammalia,* said to
be found by him in the Purbecks. Up came a box from
Brodie, and with three new species (of two new genera) of
mammals, one as big as a hedgehog, and with the *skull*
which Dr. Falconer interpreted to me. Then came box after
box from Beckles, every day a new form of reptile or mammal,
till I have got up to indications of twelve species of mammalia (including the original Spalæotherium) and nine
genera!! The biggest is one-third bigger than a hedgehog.
The last was made out by Falconer to have the dentition of
the kangaroo rat, or of that family, with the back molars
like *Microlestes* of the Trias! I have shown all to Owen,
and he confirms Falconer's determinations. I am figur-
ing this *Hypsiprymnodon microlestoides*—a pure vegetable

feeder ! All these twelve are got from an area not larger than my drawing-room, and from a bed only three inches thick.

I have casts and beautiful drawings made at my expense when last at Stuttgardt of the Microlestes, and am much pleased at knowing what that oldest of yet found mammals was.

So the ' Noch-nicht-gefunden-seyn ' (a capital specimen of a German substantive) of the Angiospermous plants in rocks older than the chalk, offer no reason to anticipate the rarity of warm-blooded quadrupeds. I asked Hooker whether he did not infer a rich mammalian fauna from the genera being almost as numerous as the species; he said judging from plants, a *poor* one ! for in tropical islands the genera are very numerous proportionally, and the flora poor, but he added, they afforded no argument as to contemporary continents. But then, surely the number of species being almost as great as that of the individuals, argues a rich fauna.

Falconer reckons on large beasts as we go on. I am very glad of the cranium, a well-developed one, some evidence higher than lower jaws. Had these mammals turned up just above the Wealden, they would not have been so useful in warning men not to speculate on negative evidence in regard to no land creatures being found in aqueous deposits.

With my love to Frances, ever affectionately yours,

CHARLES LYELL.

To LEONARD HORNER, ESQ.

Bienne, or Biel : August 7, 1857.

My dear Horner,—You will have learnt from Mary's letters that we have been having a prosperous journey through Belgium and the old Rheinthal to this delightful country, which every time that I revisit it always seems more full of charms and wonders than at first. My visit at Liége was very useful, as besides talking over with Koninck the present state of Belgian geology, I got a full day in the Devonian rocks with poor Dumont's *locum tenens* (and I hope his future successor), young De Walque, who has published a good paper on the lias of Luxembourg. At

Aix-la-Chapelle I found Dr. Debey most ready to show me his splendid and unique collection of cretaceous fossil plants. I went with him to the hills in the neighbourhood to satisfy myself of the real position of the beds in which this new and peculiar flora has been detected, and the first hour convinced me of the principal fact, namely, that the sands and clays in which this vegetation is preserved are decidedly inferior to the white chalk with flints, which with the chalk without flints, and chalk marl, are of moderate thickness and rest upon a cretaceous greensand like some of our upper greensand, and immediately beneath this are the plant-beds. I found belemnites and numbers of our common chalk fossils. On the second day I made an excursion in a new direction in company with a younger geologist, Ignaz Beissel, son of a late merchant of Aix, who has entirely devoted himself to the marine beds above the Aachenian. I also paid two visits to a schoolmaster of the name of Joseph Müller, whose collection of the cretaceous beds of the neighbourhood is large, and who has published on them.

Dr. Debey thinks he has a hundred species of the Australian family, the Proteacea—of course I cannot judge whether he exaggerates the number, but certainly he allows many forms to belong to one and the same species. As Robert Brown, when I last talked with him about these plants being referred to *Banksia, Protea, Grevillea, Dryandra,* &c., suggested some difficulties and doubts, I was glad to find that Debey had been at Paris, and had shown Adolphe Brongniart some of his specimens, and he had agreed with Debey and remarked that the *Dryandra* was undistinguishable so far as the leaves go. But Debey has some fruits also, and he showed me the epidermis of some of the leaves so well preserved in the fine clay, that the cellular structure and the stomata can be seen under the microscope quite as clearly as in a living plant, and these stomata in the Proteacea are differently arranged from those in other families.

He has about fifteen species of Coniferæ; one of these agrees very closely in its leaves and the form and shape of the cones with *Wellingtonia.*

The ferns are most beautiful, some forty in number, many in fructification, and the spores bearing microscopic examination. But the numerous forms of ordinary dicotyledonous leaves are doubtless what will most astonish botanists, who thought that such plants made their first appearance on the earth in the tertiary period. Ettingshausen of Vienna is to publish jointly with Debey an account of the flora. It will be as remarkable a lifting up of the curtain which concealed from us the botany of the Chalk period, as the Purbeck discoveries are in reference to the land animals of the Upper Oolitic era. After I had come to a very strong opinion that these Aix beds and their plants were at least above the Gault, and below the White Chalk, I was told at Bonn that Ferdinand Roemer has lately published a paper expressing the same opinion, which I am glad to hear. I was much confirmed by seeing at Bosquet's, when I went over to spend a day with him at Maestricht, four genera of land plants found in the Maestricht chalk (two of them dicotyledonous angiosperms) *common* to the Aix beds.

It seems that the mineral character of the European cretaceous deposits changes rapidly after one crosses the zone of paleozoic rocks which run through Belgium and the adjoining parts of France and Germany, and the plants found by Debey grew, I presume, on some of the land formed by the dividing ridge of paleozoic rock (Carboniferous and Devonian) which parted the two cretaceous seas.

I picked up a few in the quarries. Among the monocotyledonous plants there are no palms, which I wondered at considering that they play a part even in the Miocene flora. I found much silicified coniferous wood in the sands. Araucaria has been made out by leaves and fruit. Some thin layers of regular coal occur here and there.

At Bonn I found Dr. Otto Weber, who has published on the Brown Coal Plants of the Siegberg district—a fine Miocene flora older than the faluns. Here again I was shown fossils of the genus *Protea*, and of *Banksia, Dryandra*, and *Hakea*, all of that Australian family which seems to have predominated in the Cretaceous, to have flourished largely in the Eocene, and to have figured still, though less largely, in the Lower Miocene. But I have since learnt from Heer that this family of Australia continued to exist in Europe in the Upper Miocene, and Gaudin of Lausanne has

even found one species allied to Dryandra in the Older
Pliocene near Florence. As I had become interested in these
plants I was glad to take a good look at them to refresh my
memory in the garden at Popplesdorf, which were taken out
of the conservatory. The gardener gathered leaves and
flowers of *Protea* for me, and leaves of *Hakea, Banksia,* and
others. You remember the large fossil frogs from the
Brown Coal ; they were contemporaneous with these plants,
which lived when a great many American forms, *Liquidambar,
Smilax, Taxodium, Comptonia,* &c., were abundant in Europe,
and when the eruptions which formed the older volcanic
rocks of the Siebengebirge took place. Von Decken went
an excursion with me to Kessenich, where we saw the
Devonian beds and the gravel of the higher platform, and
where I told him how much light your Egyptian researches
had thrown on the origin of the loess, which, as my views
were rather different from those which Von Decken had proposed, led to some amicable discussions. He is a most
agreeable companion and full of information, and I think
very sound in his theoretical views.

The quantity of work which Hermann Von Meyer had been
doing at Frankfort since I had seen him some twenty months
before, surprised me as much as ever. He has a fair
quantity of daily official business, after which he devotes his
time most systematically to paleontology. He is his own
artist, and a first-rate one. The economy of time and temper
arising from his being able at once to draw what a scientific
eye can alone see correctly, must be great, and will alone explain the prodigious amount of good work he is able to get
through. When it is announced that he is going to write a
monograph, every collector in Germany and many provincials
in France send him their specimens. It is very rare that
any are detained more than three weeks.

Thus in three years specimens belonging to 271 individuals of the Carboniferous reptile Archegosaurus were sent
to Frankfort, and drawings made of all. It has produced a
splendid monograph of the genus, comprising two species
already in part published in the ' Paleontographica.'

Another on the Pterodactyles of Solenhofen has also
been brought out by Von Meyer since I was last with him.

Yet he has now scarcely a fossil reptile in his house, after having described eighty species from the three members of the trias, and figured all. I found his table covered with some twenty new species of an Oolitic Crustacean of his new genus *Prosopon*, sent to him from various quarters. He draws on transparent paper, so that the lithographer turns it and sees it through, and therefore has not to reverse it. This greatly increases the accuracy of the copy.

Our windows here (Lausanne) overlook Gibbon's garden, in part of which the hotel is built. The view of the Lake of Geneva and the Alps very beautiful, and the weather not too hot. Mary is quite well.

Believe me ever yours affectionately,

CHARLES LYELL.

To LEONARD HORNER, ESQ.

Lausanne : August 10, 1857.

My dear Horner,— I am glad you have been thinking of the future as well as of the present of the Geological Society. . . .

. . . . My taking the office a third time is out of the question. I have done a fair share of that duty, and hope to continue for years travelling, making original observations, and above all going to school to the younger, but not, for all that, young geologists whom I meet everywhere, so far ahead of us old stagers, that they are familiar with branches of the science fast rising into importance which were not thought of when I began. Such is the case with young Gaudin here, a fossil botanist who in age might almost be my grandson.

When a vacancy occurred in the Institute by Buckland's death, Elie De Beaumont sent me word that there was a party for me, but that he should use all his influence in another direction. I presume that his motives for this extraordinary message were various. He had lately received hospitality at my house in London, and had been as usual on most courteous terms; in Paris always ready to furnish me with unpublished Government maps of France, &c., and he may have wished to act openly and frankly ; also to let me know in return for my enmity to his opinions (or as they always say in Paris) to himself, '*mes ennemis*,' &c., meaning ' my theoretical opponents,' that he had the will and power to thwart me in what he really imagines is the great object of everyone's ambition. His message did not open my eyes to his course in the election, for I knew that before, but was a gratifying testimony to the existence of a party in my favour.

It would be the height of the inconsistency of human wishes or expectations, to receive the compliments paid me by the younger geologists since I left town, and at the same time to expect the opposite, older, and more influential men of science to confer honours on the leader in a new school, and as they think heresy.

Prof. Bunsen at Heidelberg avowed at the dinner he gave us that all his taste for geology had been derived from my writings. Here is a man whom many in Germany rank above Liebig. He is aware how I am opposed to many prevailing opinions. Marcou, in writing to me about my ' Supplement,' tells me he is gradually coming round to all my opinions, even my theory of climate. Morlot at Lausanne writes to say that he has been giving lectures in that town, using my ' Principles ' as his textbook, ' for he is one of my school.' Cotta, my translator, writes in the same strain, all regarding me as the head of an opposition party, opposed to those who still on the Continent, and especially in France, have the power in their hands.

In one of my last conversations with De Beaumont he used the word *étranger* in reference to some eminent French geologist—I forget whom. I exclaimed that I always thought he was a Frenchman. He explained that he merely meant not a member of the Academy. I was so much diverted and surprised at this classification of all scientific men into those who did, and those who did not belong to his Academy, that I betrayed my amusement. He went on to assure me that it was the way in France to speak of those who were not in the Institute as foreigners, *étrangers* (outside barbarians). He imagines that to get into it is the great object of every man's ambition.

Believe me ever very affectionately yours,

CHARLES LYELL.

To LEONARD HORNER, ESQ.

Zürich : August 15, 1857.

My dear Horner,—We have returned to our beautiful room here, or to one which has the same views of the Lake and the Alps, having made a most successful tour through no small part of Switzerland, by Soleure, Neuchatel, Lausanne, Vevay, Freiburg, Berne, and Aarau. But I must try and resume my sketch of my geological proceedings. I said something of the day which Hermann v. Meyer, with his usual liberality, gave up to me. He lives very isolated at Frankfort, though in communication with all Germany and part of the French provinces, and when anyone calls who appreciates his labours, it must do him good.

Wishing to make an excursion, I got Professor Kaup of Darmstadt to introduce me to a Major Becker, of the Darmstadt Engineers, a geologist, who drove with me into the Odenwald, and showed me Rothliegendes, various trap rocks, and some tertiary limestone of the Mayence basin (Lower Miocene). I was not a little surprised, just when taking leave, to find that my companion was a brother of Becker, the Prince's secretary.

At Heidelberg I had some good talk with Professor Bunsen on the theory of glaciers ; Tyndal versus James Forbes, &c. ; and Professor Blom showed me many fossils illustrative of the triassic rocks of the neighbourhood and the Odenwald, and also of the Permian. They have made out the age of their sandstone, Bunter, &c., better than I had imagined. The Rothliegendes is certainly very angular in its pebbles in that region, though not so much so as you found at Eisenach.

Professor P. Mérian at Basle invited us to his country villa, and told me who was, and who was not at home. I missed Fridolin Sandberger at Sächingen on the Rhine, which he had just left, but was lucky enough to find Heer at Zürich, disposed to devote as much time as I could give to show me his collection and talk over Swiss and German tertiary geology. He had lately returned from Italy, or at least had been there since I was here last year, and had seen De Zigno's oolitic plants at Padua. Charles Bunbury

will be curious to know that after great accession from all
countries this flora remains exclusively confined to Cryptogams, Cycads, Conifers, and without any dicotyledonous
angiosperms.

I perceive that Heer is trying to frame a progressive
theory for plants, though he is a good deal put out by finding
a Paleoxyris in the Coal, one of the Bromeliaceæ. In fact the
monocotyledons do not seem as yet to keep their place in
the chronological system as they should do if they knew
their real rank in the order of development. Some of them
appear before their time. It is, however, striking to observe
that the tendency of geological facts (or opinions) carries a
man who is working in a new field, and an independent
thinker, into the speculation that nature began with cellular,
and went on to vascular cryptogams, from lichens and seaweeds to ferns, and slowly got up to Coniferæ and Cycads,
then to different divisions of dicotyledonous, apetalous,
polypetalous, and gamopetalous in the order of their perfection. Although Heer is too well aware of the exceptions to
his rules, and even of the impossibility of classing the dicotyledons correctly according to relation, dignity, or perfection,
yet the attempt shows how seductive such a generalisation is.
So long as it is admitted that man came last, and the idea
of progress is cherished as the only way of uniting that
fact with paleontological data, I suppose these views will
find favour. It seems the only prospect of a complete
system, of uniting all into one grand whole, the supposed
absence of fish in the oldest rocks, with the coming in of the
Mammalia last of all, and with a parallel series of progressive steps from the algæ to the lilies and the roses. But it
might be better if we were rather less ambitious. This
eager desire to solve the whole problem may mislead zoologi-
cally, botanically, and geologically. I suppose most men
prefer a doubtful system which enables them to group
together a great many facts, than to have none. I spent
three days in Heer's collection. He is continually finding
fruits which bear out the generic determinations previously
obtained from leaves alone. The fruit of *Cinnamomea
camphora* among others, the leaf of which was so well known,
also certain maples common to the lower and upper molasse,

as are a great many plants, helping to link together what I now call Lower with Upper Miocene, and justifying my substituting the former term for Upper Eocene in reference to the Mayence basin, Hampstead, Isle of Wight, &c. What is singular is this, that Heer finds all the insects and plants, 1,000 species of the former and 700 of the latter, from the freshwater molasse, to belong to extinct species, whereas the marine beds which separate the Lower from the Upper molasse contain *shells* like those of Touraine or the faluns of the Loire, of which a third are recent. Possibly some of the plants may be identical with the living, though Heer has apparently good grounds for distinguishing them, often by the fruit, as distinct. But the more one examines his specimens and hears his reasons, the more faith one has in him and in fossil botany. I now have little doubt that the whole of the Œninghen beds are Upper Miocene.

Heer convinced me that the Madeira plant sent to him from S. Jorge was the *Oreodaphne fœtum,* and if Charles Bunbury found *Laurus Canariensis* among mine, it must be because both are there, which is not improbable. The glacial phenomena appear to me more than ever wonderful. There is much lecturing here in small places, very like the United States. I met with a M. Zollikofer, who has done good original work in the Italian glaciers, starting to lecture to the watchmakers of Chaud de Fonds and ten other places.

Ever affectionately yours,
Charles Lyell.

CHAPTER XXX.

AUGUST 1857–SEPTEMBER 1857.

LEAVES ZURICH TO SEE THE LIMESTONE QUARRIES OF SOLEURE—GLACIAL
PHENOMENA IN SWITZERLAND — THE RHONE GLACIER — ZERMATT—
VIESCH—THE MATTERHORN.

CORRESPONDENCE.

To LEONARD LYELL (*six years old*).

53 Harley Street, London : April 13, 1857.

My dearest Leonard,—Mrs. Nisbet[1] brought me the
Actinia on Saturday evening, and the next morning Aunt
Joanna took it to the Zoological Gardens, where I have since
seen it, looking quite well, and likely to live till you come
back.

There are a good many of the same species (*Actinia
Mesembryanthemum* is the name of it) in the Fish House, and
almost every one of them differs a little from the other, and
I could only see one which resembled yours in colour, and
that one was not exactly the same. It is what naturalists
call a variable species.

I am not sure that zoologists understand the meaning of
that beautiful row of blue tubercles which surround the base
of the feelers or tentaculæ. Some individuals seem not to
have them.

If I can manage to run down with Aunt Mary to Brighton
some day before you return to town, we will see what we can
find on the beach when the tide goes down.

With my love to Frank and Arthur, believe me your
affectionate uncle,

CHARLES LYELL.

Give Rosamond a kiss for me.

---

[1] The faithful housekeeper, who was forty-three years in his service.

To LEONARD HORNER, ESQ.

Zürich, St. Gallen : August 16, 1857.

My dear Horner,—From Zürich we went partly by rail.
and in part by voiturier to Soleure, where I went to see the
quarries of ' Portlandian ' white limestone, almost a marble,
in one bed of which all the turtles (of several genera) have
been found—all it is said with the carapace upwards and the
plastron down, a position in which they might be expected
to settle. The most striking phenomenon in these large
quarries at Soleure was an extensive shelf of the limestone,
from which a covering of about eight feet of solid unstratified
mud, full of angular and subangular boulders, had just been
removed. These boulders, with now and then a block large
enough to be called an erratic among them, were many of
them striated, polished, and scratched on one or on all sides.
But the ledge of limestone below was smoothed in the most
beautiful style, as well as traversed with parallel furrows.
The sections of innumerable large *Nerinœa* on the surface of
the polished ledge made a beautiful show. There is no rock
which receives and retains glacial markings so readily and
faithfully as a compact limestone, provided it be covered with
mud by the glacier, for it loses all its striæ by a few months'
exposure in the open air. It is singular to observe what a
heavy and solid thing this fine mud, called ' glacier mud,' is.
Perhaps it has often been pressed under a great weight of
ice. They have discovered lately in the railway excavations
that they can advantageously blast them with gunpowder.
The erratics here are part of the supposed doings of Charpentier's great glacier, which walked across the great valley
of Switzerland from Monts Blanc and Rosa to the Jura, with
a thickness of ice of some 4,000 feet, then abutting against
the limestone chain, and rising nearly to its culminating
ridge spread itself on each side, after leaving Pierre à bot
(the *toad* stone) above Neuchatel, and some still more enormous angular masses on the cretaceous and oolitic rocks of
the Secondary Chain.

In order to escape from the necessity of appealing to such
a gigantic mound of ice, I ventured, you may remember, to

suggest that the sea may have floated the Alpine erratics to the Jura, as Darwin has shown that the ocean now carries on ice rafts the rocks of the Andes to Chiloe, arranging them there with no small regularity. But the entire absence of marine remains in the associated gravel, mud, and moraine, whether here or anywhere in Switzerland, the conformity of the distribution of the travelled blocks here with the shape of so many valleys, and above all, the sight of the Alpine snows at Berne and elsewhere, has made me strongly incline, with Charpentier, Agassiz, and others, to embrace (as James Forbes did) the theory of a terrestrial glacier. With Desor at Neuchatel, I revisited Pierre à bot, and saw the zone of blocks of the Monte Rosa talcose granites and gneiss at one level on the Jura, and the Protogine blocks of Mont Blanc at another. Afterwards Morlot accompanied me from Lausanne to a spot in the middle of the great valley between the Lakes of Neuchatel and Geneva, where the glacial appearances are splendidly displayed; and lastly, when I had got out of the domain of the colossal glacier of the Rhone (or the Vallais), and at Berne was in the region of the supposed ancient glacier of the Aar, I had a grand day with Escher von der Linth, and went over all the arguments for and against the land- and the sea-ice theory, examining the old moraines around Berne, and 'the gravels' (as Blackadder used to call them on Strathmore) or the stratified old alluvium consisting of rearranged (or *remanié*) boulder stuff, which have been thrown down by rivers since the retreat of the ice. At Berne Escher pointed out to me that on the right side of the moraine of the Aar, you have fragments of the rocks which, far above the Lakes of Thun and Brientz, belong to the right side of the valley of the Aar, and on the left side those derived from very different formations occurring on the left side. It is evident therefore that the two lakes were then full of ice. Another interesting point is this. It is clear that a greater glacier like that of the Rhone coming down the Vallais—filling the Lake of Geneva—rising as they assume 3,000 or more feet above its level, and crossing to the Jura, must have blocked up the mouths of the minor or lateral valleys. Places were also pointed out to me, one of them near Vevay, where the old colossal masses of ice of the Rhone-glacier blocked up tributaries which in summer brought down pebbles at points far above the level of the Lake of Geneva, 1,000 feet or more. At these points of junction a mixture of stratified alluvium proper to the said tributary torrent, and of unstratified mud and blocks from the Vallais, are observed in spots on which it seems most unnatural for any such accumulations to have taken place, unless one admits Charpentier's theory or some modification of it. Now, just such old moraines of mixed character, called here 'diluvium glaciare,' are observable in Forfarshire at the openings of lateral valleys into the larger one of Clova, or Water Esk, only explicable by imagining the deeper glen to have been once filled with ice. Indeed, if the hypothesis now generally adopted here to account for the drift and erratics of Switzerland, the Jura, and the Alps be not all a dream, we must apply the same to Scotland, or to the parts of it I know best. All that I said in May 1841 on the old glaciers of Forfarshire (see ' Proceedings of the Geological Society ' for that year) I must reaffirm, and the glacier of the Rhone is comparable to our principal one of the Tay, which came down by Dunkeld and by Coupar to the south of Blairgowrie, and then through the lowest part of Strathmore, characterised occasionally by masses of actinolite schist which could only have come down from the valley of the Tay. This glacier went across the strath, and in a straight line by Forfar to Lunan bay, where it reached the sea, being traceable by its *débris* for a distance of thirty-four miles, and therefore by no means unworthy of comparison with some of the old Alpine ice rivers. And we have also in Angus our ' Pierres à bot,' as I stated in my paper on the Sidlaw range, which is our Jura. The block of mica schist which I measured on the hill of Turin, a hill 800 feet high, resting on the Old Red Sandstone within forty feet of the summit of the ridge, was thirteen feet long, and must have come from the Grampians, probably very far from the northwest. It is a fine monument of the transporting power of ice, as displayed in Scotland, but Pierre à bot is really more wonderful when seen again, than when I first beheld it, so vast and angular, so clearly resting on a limestone chain, with the great tertiary valley between it and the Alps. There can in the first place be no doubt that ice was the carrying

power, and the distance travelled by such blocks and others
is the same whether our hypothesis employs floating ice or a
land glacier. Escher has pointed out to me that in several
cases where the valleys bend at sharp angles, as in the case
of the Rhine above the Lake of Constance, the floating ice is
out of the question. He has established this by the aid of
a very peculiar rock called the granite of Pontelyas, near
Trons. Fortunately this granite is exceedingly unlike any
other in Switzerland or the Alps, large regular crystals of
common felspar in a base of green felspar, with another dark
mineral (I forget the name) dispersed like mica. As fragments are traced from its starting point we have positive
proof of its origin. I told Studer, Escher, and Morlot that I
was disposed to embrace fully the land transport theory, but
I thought they took too little account of the probability of
considerable and unequal upheavals. When we consider
that in Scotland, on the Clyde side, we have post-glacial
upheaval to the amount of between 200 and 300 feet, and
1,500 feet in Wales (Moel Tryfane), and 700 or more in
Canada at Montreal, how much more may we presume that
in and near the Alps (which I take to be the greatest centre
of movement in the later geological periods of all Europe),
there will have been vast and unequal changes of level after
the dispersion of the Alpine blocks in all directions, northward, southward, &c. !

I am happy to find that Escher, who knows the geology
of the Alps more minutely than anyone, has quite realised
the idea of the probability of its great foldings having been
effected very slowly, and that the lateral pressure exhibited
on both sides may have been brought about at the same slow
rate at which the central nucleus crystallised, and in the act
of crystallising expanded. Even the Eocene strata turned
slowly by metamorphic action into granite and gneiss in some
places.

I ought before this to have told you that after diligent
search the geologists have been unable in any part of
Switzerland to find a single marine shell in any moraine,
or any part of the boulder clay. In this respect the glacial
formation is precisely what I know it to be in Forfarshire ;
and as our Scotch gravels, with the exception of a few

patches near the sea, and not many feet above its level, are also
without such remains, so here the stratified gravels formed
since the glaciers retreated, and often 200 feet thick or
more, are without them.   This to be sure is mere negative
evidence, but lately Morlot has detected some wood very
well preserved in an old moraine (700 feet thick) opposite
Lausanne on the Lake of Geneva, and at Lausanne itself in a
similar position the bones of the marmot, the living species
(*Arctomys*) in unstratified boulder mud, full of striated and
polished pebbles.

In regard to my belief in the greater height of the Alps
and other modern changes of level, I cannot say more than
that this important mode of facilitating the explanation of
the phenomena has been far too much neglected.   My idea
is that all the mountain chains, lakes, and valleys have remained as before, and I quite agree that none of those
astonishing lateral-pressure folds can be of post-glacial date.
But when all northern Europe was colder, an addition of
2,000 or 3,000 feet to the altitude of the greatest chain
would have had a prodigious effect in the local augmentation of cold and ice, and if the same causes which at several
distinct and successive geological epochs, have produced
more movement in the Alps than almost anywhere else, did
also in the glacial and post-glacial period contribute to
intensify the Alpine as contrasted with the Jurassic oscillations of level, the slope down which the streams of ice crept
along may easily have been greater than now.   A generation must die off before geologists will know how to make
use of an ample allowance of time,

Ever affectionately yours,

CHARLES LYELL.

To LEONARD HORNER, ESQ.

Altorf: August 29, 1857.

My dear Horner,—My last letter on the geology of my
tour related chiefly to glacial phenomena, and before I get
among the glaciers themselves, which we are fast approaching, I must try and make up an arrear of some ten days or

more during which I have seen much of the tertiary, and
something of the secondary rocks of Switzerland, and a good
deal of their geologists. Mary will have told you of my
having attended the meeting of the Swiss naturalists. I
was tempted to do so by hearing that their doings only
lasted three days, and that it was very easy to see and converse, instead of listening to long and often dull memoirs.
There is also much brotherhood among these Swiss. They
give me very much the idea of men who meet together for
mutual instruction, and as real lovers of natural history
and science. I saw most of Heer and Escher, Mousson also
of Zürich; Mérian of Basle, Langen of Soleure, Gaudin of
Lausanne, De la Harpe of do., all men whom I had seen in
my tour, were present; also Ziegler and Desor. Trogen in
Appenzell, where they met, is near the boundaries of the
tertiary (molasse) and Alpine region.

When you see Falconer, pray tell him that I visited all
the localities in which the Zürich proboscidians named by
him in their museum had been found. The *Mastodon
angustidens* of Winterthur has proved a splendid fossil.
The story has gone through all Switzerland how Falconer
threw up his cap to the ceiling when he saw it. According
to his suggestion it was sent to Kaup, and it turned out, as
Falconer thought, a nearly entire specimen; upper and lower
jaws, in one of the latter the second tooth seen below the
milk tooth. Also the incisor. Moreover a young Mastodon
of the same species. It is from the Upper molasse. I
visited the quarry,

I also went to Dürnten, and saw the beds from which
Elephas antiquus was extracted, and Falconer will be glad to
learn that I obtained a tooth, which though not perfect, I
feel almost certain is *Rhinoceros leptorhinus*. The shells
Valvata and *Cyclas* of living species. The plants also (Heer
was with me), *Pinus abies, P. sylvestris, Betula alba, Phrag-
mites vulgaris*, and a few more. No *Cyrena* as yet. The dip
of the beds Escher suggests may be due to subsidence when
the valleys were cut through the Pliocene, and when the
water was withdrawn from this peat-like deposit. He says
the molasse is quite horizontal below. At first I hoped I
had got a decided case of post-pliocene upheaval or change

of position in Switzerland, to favour my post-glacial oscillations mentioned in my last. But I fear that the subjacent undisturbed molasse precludes this inference.

By the way, talking of Escher and upheaval, I was not a little pleased to find how thoroughly he goes with me in doing things slowly. No one in Europe is so well acquainted with the stupendous folds and inversions of strata in the Alps, and yet he believes it all took place without any interruption of the habitable state of these mountains. Had man been there, he thinks he would not have known what was going on.

I end this at Visp, where the walls of most of the houses are rent with the earthquake which shook them two years ago. Perhaps after all there *is* something still in progress. The shocks were felt for 300 miles by 200.

Yours very affectionately,

CHARLES LYELL.

To LEONARD HORNER, ESQ.

Domo d'Ossola: September 10, 1857.

My dear Horner,—We received your letter of September 5th at Visp, and shall leave Mary to answer it, and continue my sketch of the geology of the country we have lately visited, where among other things I have seen the glaciers of the Rhone, of Zermatt, and two smaller ones adjoining it, those of Zmutt and Findeln, besides that of Viesch which Escher recommended me to see, and which comes down from the Bernese Alps towards the Vallais, whereas the Zermatt group descends from the region of Monte Rosa. I think I have learnt a good deal, for I have been surprised at finding the signs of glacial action diminish so rapidly in proportion as I got nearer and nearer to the glaciers, until it became least evident when I stood on and touched the glaciers themselves. You will remember that when I was at Soleure, 150 miles from the source of the supposed ancient Rhone-glacier, I was struck with the quantity of polishing and grooving and scratching of the limestone rocks, and of the pebbles in the old moraine or unstratified mud. The erratics also on the Jura there, as well as near Neuchatel, and the moraines of gigantic size 800 feet deep at Lausanne or opposite it on the south side of the Lake of Geneva, struck me much, and near Trogen and Winterthur and round Zürich, it is wonderful how abundant are those peculiar memorials of ice-action which neither the landslip, nor the avalanche, nor the vibrations of the earthquake, nor any known agent but ice can produce, and which the torrent whether of mud or water cannot cause, but on the contrary immediately effaces.

I had imagined that as the Reuss has strewn the country below the Lake of Lucerne with erratics and glaciated (or polished, furrowed, and scratched) blocks and pebbles, that when I followed for a day and a half's journey the valley of this same Reuss up to its source from Altorf to Andermatt and beyond, I shall find the evidence of the old glacier still more clear. But I positively could not find one proof, except some talus-like moraines and two or three *roches moutonnées*, of ice-action. Had I searched longer I should no doubt have met with them, but I had expected them to obtrude themselves on my notice as they do in the low country of Switzerland far from the Alps. Crossing the Furka I came down upon the glacier of the Rhone, examined its lateral moraines (of small size), the blocks on its surface, and afterwards its terminal moraine, with a view of ascertaining what proportion of the stones, whether angular or rounded, had been so dealt with by the ice, that if I brought any of them away, they would be recognisable as *glacial* by a practised eye. I began to count, but not finding one in a hundred, no, nor in two hundred, was soon tired out, and I repeated the same experiment afterwards on other glaciers and with the like result. Several times when I and three or four companions (guides and volunteers) looked sharply for hours in the moraines on or around the glacier, we observed a few good examples of smoothed and scratched boulders which were unmistakable, but they must each have made one only in many thousands—a singular contrast to what we witness when we go far away from the spots where alone Nature is now manufacturing specimens of this peculiar kind.

As to the medial moraines which travel down on the sur-

face of the ice, I do not see how we are to expect any fragments of rock in such a situation to differ from the talus of
the foot of a precipice, except occasionally that such
moraines have had some friction between the ice and the
rocks which bounded the tributary glacier. And in regard
to stones which have been ground against the bottom, I cannot help suspecting that after glaciation they most of them
get rolled by the powerful torrent which runs beneath the
glacier for many miles, and bursts out usually from a large
arch at the end. The triturating power of these Alpine
torrents is enormous, and in a short distance they would
obliterate every groove and scratch.

In the glacial period, when the weight of ice was enormously greater, when in the region of the Alps there was so
little melting, when glaciers at present only ten, fifteen, and
twenty miles long and from 300 to 1,000 feet deep, were 50 to
100 or even 150 miles long, and 4,000 feet deep (and if there
is any truth at all in the generally received theory of the old
Swiss glaciers such must have been their gigantic dimensions),
one may readily grant that the pressure and friction were so
much in excess of what we now see, as to explain the contrast between the ice work done in the olden times and that
accomplished in our own days, to say nothing of the probably disproportionate length of the periods compared.

But you will naturally say that I am going too fast in assuming that the glaciers are a *vera causa*, and adequate to do
all, if only enough of time and of cold be allowed. This is
the great question, to answer which I came this year to the
higher Alps, being desirous to see the proofs with my own
eyes. I began with what I *did not* see, because taking me
by surprise it made the greatest impression on me, and because it removes many difficulties which I had in Forfarshire,
where in passing through Clova, and over the granitic portion of the Grampians as far as Deeside, I was disappointed
at missing good evidence of glacial striæ, &c., although I believe there were glaciers there in the glacial period. Our
Scotch erratics, our *till* or unstratified moraines, and in some
places our dome-shaped rocks or *roches moutonnées*, are just
what we find here even in those places where it is excessively
difficult to meet with smoothed pebbles or blocks with parallel

grooving. It seems that most granites and nearly all the crystalline schists will not readily take or retain such markings. Limestone, and still more Serpentine, are the best, but the limestones must not be exposed to the atmosphere, or they soon decompose superficially and lose their distinctive markings, and I have examined blocks even of these calcareous and serpentinous stones by hundreds on the glacier and in its moraine, without meeting with any indubitable signs of glacial action. Such negative facts have surely not been dwelt upon enough by Agassiz and many other writers on this subject. But now for what *I did see.* First, the small size of the lateral moraines of the Rhone Glacier some way above its termination, and the equally small scale of the terminal or frontal moraine of the same, astonished me. I saw afterwards some twice as big (thirty-five feet high perhaps, and twenty or thirty broad), and I have read of some much larger. The old moraines of Switzerland are 150 and 200 feet high, to say nothing of that one opposite Vevay and Lausanne, said to be 800 feet. I suspect that Zollikofer, in saying that the lateral moraine of the glacier of Macugnaga S. of Monte Rosa is 150 feet high, means the old moraine of the extinct glacier, for he speaks of vegetation on it, whereas I never saw even a lichen on any recent moraine. Hooker I think saw recent moraines in the Himalaya 200 or 300 feet high! The steepness of the modern terminal moraine of the Rhone on both sides, and of others since seen, at angles of between 35° and 45°, and so sharp at the top as scarce to afford room for one's foot, is singular. There is an old terminal moraine 120 yards in advance of the new one of the Rhone Glacier, covered with wild plants, some in full flower, and cut through in two places by the river, showing entire absence of stratification, its height only fifteen feet, width 90, slope in parts 37°. The next I saw was the Viesch Glacier above Brieg, or in the Upper Vallais.

Although it is annually advancing, yet the melting back of its termination during the heat of this unusually hot summer, gave me what I wanted, an opportunity of seeing under it, or of observing the floor of granite on which it had pressed eight months before. My guide took me under an arch of ice, and I saw a rounded and domed surface of granite, smooth and with straight furrows, a quarter of an inch deep, exactly in the direction of the onward movement of the glacier. Although there is usually a dripping of water from the ice in such places, it was so dry here that I had to blow out of the rectilinear grooves the fine polishing powder as it has been called. Not only were there furrows on the top, but also on the sloping sides of the granite boss, and they were also nearly parallel to the others, because I suppose the plastic (or viscous?) mass of ice, with stones frozen into it, enveloped the granite, and moving the whole of it forward in the same direction, cut channels in the sloping sides as well as on the tops. All this was what I had read about, but I was pleased at seeing another sign of the freshness of the action of the glacier. Some dark-bluish slate rock had left streaks of colour which I could wash off—it could not grind the granite, but had lost a portion of itself and left its mark on the granite. A short distance away from the new moraine are abundant signs of old moraines and *roches moutonnées* and rocks *in situ*, grooved in the direction of the valley, but I had not time to look for striated pebbles. I could find none in the existing moraine, only one imperfectly furrowed piece of mica-schist.

Next I explored the Zermatt Glacier, two days at its termination, and another day high up above its middle portion at the foot of the Riffelhorn, crossing it and seeing all the medial moraines. A few years ago this last examination would have required one to camp out. Now an hotel more than 7,000 feet above the sea, renders it easy to one who is not fatigued by a mountain ride, and who can stand half a day's walk over the ice. One of my guides had acquired a keen eye for glacial markings. I was assisted by an intelligent Swiss, the young landlord of the Riffel Hotel, also by Mr. Sclater of Oxford, whom I had met at Sir W. Jardine's, and by another of my guides. After hours of search, I began to despair of ever seeing amongst thousands, one truly glaciated transported or travelling block. At last, under the Riffelhorn, on the right lateral moraine, we saw a splendid angular mass of granite which measured in length 59 feet, width 49, height 42 *Paris* feet (I had borrowed a tape measure of 45 feet on the French scale); nearly all of one of its sides was beautifully polished and furrowed, while the other sides were perfectly rough. In some places a very distinct set of small scratches parallel to each other, but not to the prevailing grooves, were visible. By confining our attention to the green or serpentine stones, we obtained here and elsewhere in the modern (and in the ancient moraines which appear in the same valleys at heights of several hundred feet) a very few well-scored pebbles, and now and then in the old moraines a fine surface of serpentine, with two or even three sets of striæ.

The rapid advance of all the glaciers I have visited is manifest. In the chaotic mass which the terminal moraine of Zermatt (or Görner Glacier) presented, I saw beams of wooden chalets and roofing tiles. In the moraine of the Zmutt Glacier I remarked a large rock with lichens on it. This was so unusual that I walked round it, and found on the other side of it a fine larch tree with its leaves still green. I then discovered that the glacier had closed round a small island of rock covered with large larches, and had detached several huge fragments of the rock, and pushed them into the moraine. Farther on I found on the outer slope of the moraine, which was fifteen feet high, and sloped at 35° to 46°, no less than eight large larch trees with their roots buried in the mass of mud and stones, and their boughs and trunks protruding in horizontal and oblique directions. Both in the Zermatt and Findeln glacier-moraines I found fresh green turf of meadows rolled up and mixed confusedly with the mud and stones. I was shown some meadows where the Zermatt Glacier had penetrated some eighty feet last year, although in other places it had moved forward much less. Wishing to determine its average rate of progress, and learning from the curé of Zermatt that it had reached a certain brook which descends from the Furg Glacier sixty years ago, I mounted the glacier with two guides, and measured the distance, which gave a mean annual advance of forty-three French feet. In the last sixty years it has walked over the site of forty-three chalets of men or beasts; three very lately. They usually remove most of the woodwork before the ice comes and shoves the moraines against them.

To one in search of the recent doings of the ice, or wanting to study the geological monuments which an extinct glacier would leave to after times, it would be more advantageous to visit the Alps in a period of recession. One wishes to see what is going on at the very bottom, where the pressure and friction are greatest. I often congratulated myself that I was not trying to convert a sceptic by showing him what a glacier in full activity could do. It is like taking a Wernerian who denied the igneous origin of trap to a very active volcano, such as Vesuvius. He would see nothing but cinders and fireworks, and, for lack of power to observe what was doing far below, would go away a worse sceptic than he came. But, thanks to the summer heat, I saw enough to make me believe the more abundant evidence which others have seen in those glaciers where the rocks are more favourable to a display of polished and scored surfaces. Even the few I found imply the existence of thousands, just as if you find one or two marine shells in a deposit after searching a day or two, you expect that time and patience may bring to light a whole marine fauna.

On visiting the Findeln Glacier I ascertained that it is in full march forwards, ploughing up the meadows and felling fir trees, and has been so for two years at least, and has gained, they say, still more remarkably in height, above sixty feet in advance, and more in altitude, so say the neighbours. Before that time they say it was stationary.

Its aspect is to the south of west, that of Zermatt NNE.; and if, as Darwin says in reference to the Andes, the amount of summer heat has more to do with the growth or decrease of a glacier than the quantity of winter's snow, such a difference of aspect or exposure to the sun and certain winds may go for much, at least for a few years, though when there is a continued augmentation of snow on this part of the Alps, and of glacier ice, all the glaciers of the same group must eventually share in the same movement. At present the fear of the Findeln (or Findelen) people is that their enemy is now making up for lost time. I was able to ascertain by means of a water-lead constructed a month before my visit at the end of the Viesch Glacier, that the ice had melted down, or lost in height, exactly five feet at its termination in the preceding four weeks.

Nothing can be more satisfactory than the identity of the fresh-made moraine to that older one which is often found at a short distance in advance of it, and of larger size, as in the case of the Findeln one. The same sharp sand and fine impalpable mud, making a mortar-like cement when baked in the sun, in which large and small angular stones, some few of enormous dimensions, are buried. When you trace the same formation almost continuously to the lower country, or find what may have been old lateral, and occasionally terminal moraines here and there with transported blocks, belonging to, or derived from mountains in the higher part of the same system of valleys, and accompanied with scored pebbles increasing in number as you descend, and with erratics and *roches moutonnées*; the whole leaves you no alternative but to infer that one and the same cause or *modus operandi* has prevailed throughout the whole area.

Even the marked multiplication and intensifying of the glacial agency, the farther one gets from the modern pigmy glaciers, seems to me now to militate against my former idea of employing floating ice to carry the far transported erratics to the Jura and other distant points. I want the grinding force of the denser and broader glacier which most of the Swiss naturalists have appealed to. The ice-rafts will not supply it (the more extended land-glaciers would). But here is enough in all conscience for one bout, so I must give you a respite. I have said not a word of the splendid scenery of the Monte Rosa Alps. The Mont Cervin or Matterhorn is as wonderful as J. Forbes represents it. Every day, as I saw Monte Rosa and some of its gigantic neighbours from new points of view they gained upon me, and had I stayed longer I might have doubted the justness of my first impression, that the Zermatt scenery, though grand, was not comparable to Chamounix. At any rate, it has the merit of being quite different. No one peak, not even the Matterhorn, can compare with the forms of the Aiguilles of Mont Blanc. We were greatly favoured with fine weather.

Ever most affectionately,

CHARLES LYELL.

CHAPTER XXXI.

SEPTEMBER 1857–DECEMBER 1857.

TURIN—GEOLOGY OF THE NEIGHBOURHOOD—EVIDENCE OF GLACIAL ACTION —VEGETATION OF THE RIVIERA—ZWERGLETHURM, NEAR VIESCH—VISIT TO LEGHORN—TOMB OF FRANCIS HORNER—MRS. SOMERVILLE AT FLO-RENCE- NAPLES—ROME—THE CAMPAGNA.

CORRESPONDENCE.

To LEONARD HORNER, ESQ.

Genoa la Superba : September 21, 1857.

My dear Horner,—It is more than a week since I last wrote to you after my arrival in Turin.   In that city I found several able geologists, especially Gastaldi, now em-ployed in the administration or chief management of a Government Polytechnic Institution.   Sella, a mineralogist and mathematician, and others aided me.   Last year, when at Vienna, I heard the *savants* of Italy much run down by the Austrians, but I am sure they only require fair play, and they are quite equal to those of Germany.   I did not see Plana the astronomer, nor Babbage's friend Menabrea, for it was vacation time, and the heat had driven all away before we arrived.   You remember Playfair, when he visited this country, said it was more likely to produce a Newton than we in Great Britain to produce a Raphael.   I believe them to have naturally an aptitude for science.

I made many excursions in splendid weather, accom-panied by Gastaldi, and in one case by him and Michelotti, who has published a work on the tertiary Lower Miocene shells of the Superga.   I have now had an opportunity of comparing the strata containing these shells and others im-mediately below them with the 'molasse' of Switzerland,

which they resemble somewhat mineralogically as well as in
their fossils ; but a good deal has to be done before an exact
parallelism of the successive sets of Miocene beds on each
side of the Alps can be made out. I have, I think, made a
beginning, and as there are some plants in the Turin or
Bormida series which I have urged Gastaldi to send to Heer
of Zürich, I expect we shall soon obtain much valuable
botanical aid from his identifications.

A comparison also of the extinct glaciers of the Italian
and Swiss sides of the Alps can better be made from Turin
than from any other place. Before my arrival I had seen on
the banks of the Lago Maggiore some good examples of
erratics and of moraines which had come from the Simplon,
but these, as you might suppose *à priori*, are far inferior to
those which have descended from the Val d'Aosta or which
belong to the ancient mighty glacier derived from the combined snows both of the Mont Blanc and the Monte Rosa
group of Alpine heights. This glacier, although perhaps of
less gigantic dimensions than that of the Rhone, has certainly left, as Gastaldi first pointed out in a memoir on the
subject in the French Bulletin, a far more imposing monument of itself on the plains of the Po than have the extinct
glaciers of the Rhone or the Rhine in the lower country of
Switzerland. You may well imagine that having fresh in
my memory the proofs enumerated in my last letter to you
of the recent advance of the Viesch and the Zermatt Glaciers
for a series of years, I was prepared to accept willingly (and
after my geological experience north of the Alps, to expect)
evidence of former extensions of the icy masses.

J. D. Forbes has well shown in his book on the Alps that
a glacier is a peculiarly sensitive instrument for measuring
the annual average of heat and cold, and that every slight
difference of temperature causes it to increase or lessen in
height and length. If therefore we had come to the conclusion paleontologically by the greater southern range of
arctic shells and quadrupeds (e.g. the musk buffalo and reindeer) that the cold once reached to lower latitudes than it
does now, and this at a period geologically speaking very
modern, we ought to look for signs of glaciers belonging to
the existing mountain chains on a scale corresponding to the

wintry climate implied by the southward migration of species the habits of which are well known to us. Not that I believe that geologists of this or the next generation would have got so far as to miss the phenomena which we are now only beginning to interpret correctly. I have often thought, would the rainbow have been missed by the most profound philosophers in optics? and yet it is as necessary a result of certain meteorological conditions as the glacier. It would, however, be inexcusable not to welcome the monuments which now stare us in the face when they accord so perfectly with conclusions derived from evidence so independent in kind as are the organic remains. As to those who feel at liberty, as Von Buch and Elie de Beaumont have done, to call in catastrophes when they are at fault, it is very unlikely that the followers of that school will in our time admit the agency, whether of floating ice or of land glaciers, to the extent to which we ought now to admit them. Even they who have most faith in the adequacy of actual causes, will require to dwell on the testimony which I have just alluded to, the arctic fossils ranging south, and the ready growth of glaciers when favoured by slight changes of temperature, in order not to be staggered when they stand amidst the vines and the maize and the mulberry trees of the plains of the Po, and are called upon to believe that a lofty mound or ridge 2,000 feet high, called the Serra, running out into the great alluvial flat, is nothing but the left lateral moraine of an ancient glacier. That it is so, I am now fully convinced.

I had asked Escher whether he knew any sections in Switzerland in which either the ancient or recent moraines exhibited that singularly contorted arrangement of the beds of clay, gravel, and sand, which they display in Forfarshire or in the mud cliffs of Norfolk. He told me it was so rare to see a fresh section, that he could not give a satisfactory answer to my question. Now it so happened that a railway is making from Turin to Ivrea, and although they cut through the lowest part of the terminal moraine near Mazzi, they have thought it worth while to make a tunnel through which we walked. Near the entrance I was delighted to see that curious folding of the strata which will cause the same beds to be twice pierced by a perpendicular shaft, yet with-

out the beds below having participated in the movement. I
have elsewhere speculated on the cause. What I feel to be
important is, that here as in England, when erratics and unstratified mud are in association with stratified materials,
the latter are liable to be twisted or folded in so extraordinary
a manner.

In order to appreciate the distinctive character of this
colossal moraine, you must reflect on the uniformity and
evenness of the vast plain of the Po all round it, for although really inclined from the Alps, it looks as level as
the sea; then fancy the great mounds sloping up at angles
of 20° and 30° to heights of 500, 1,000, 1,500 and 2,000 feet;
then consider that at the very extremity, as near Caluso,
there are blocks of protogine which have come 100 miles
from Mont Blanc; also that the whole assemblage of stones
is not like that which has issued from the Susa or from any
other valley, but confined to rocks such as now strictly belong
to the basin of the Dora Baltea; also that the pebbles and fragments of stone, if of serpentine or any easily striable rock, are
all striated, at least nineteen-twentieths of the whole, whereas
in a recent glacier which has only travelled ten miles, you
might only find one in twenty of the same stone striated;
and lastly, think of the narrow vomitory which has disgorged
this enormous quantity of material, the ravine above Ivrea
being as obviously the source of the whole, as is the crater
of Vesuvius the point from which its lavas have issued.
When Gastaldi read his paper to the Geological Society at
Paris, written jointly by him and Martens, Elie de Beaumont,
who had many years before visited the ground, objected entirely to their conclusion that it was a moraine, but I never
saw a stronger or more satisfactory case. But in the same
paper the authors hazarded an opinion that although the old
Alpine moraines stopped short after going a few leagues from
the Alps, yet at some former time erratics had been conveyed
to the summit of the Collina just as ' Pierre à bot ' and other
blocks had been carried by the old Rhone Glacier to the
flanks of the Jura. Now when I read this at Zürich, I
immediately recollected that in the valley of the Bormida
when I passed from Savona to Alessandria in 1828, I had
been astonished at some very huge erratics of serpentine in

the Miocene. Having never seen blocks of such enormous
dimensions in any tertiary formation, I was relieved in 1828
at finding in some spots on the Bormida projecting fragments of serpentine in place, which the erosion of the
valleys had exposed to view. I concluded that they may
not have travelled far, and when I saw some large blocks on
the Superga (in 1828) I immediately suspected that as that
hill consisted of beds of the same formation, the blocks
might have been washed out of the Miocene not far off. I
therefore now suggested this view to Gastaldi, and found
that he was by no means tenacious of his printed theory,
although he said that the blocks were many of them angular,
of very great size, and accompanied by Alpine loam. We
then examined the beds of the Superga, both those dipping
to the NW. and those to the SE., and on both sides of this
anticlinal are strata containing fragments of stone of various
kinds, some not known in the neighbouring Alps or Apen-
nines, from two to eight feet in diameter. On our ascent to
the Superga I saw a thickness of sixty feet regularly stratified of this conglomerate, in which were fragments consisting
chiefly of serpentine, but some of limestone, others of protogine granite, and one of the latter angular and eight feet in
diameter. In less than half an hour's search, I found two of
the serpentine and one of the limestone pebbles with
scratches, which would be called glacial if they were found
in a modern moraine, though not such as you would select
for examples for a museum. Still I searched this year in
some recent moraines quite as long without finding better.
As to the age of the beds, there is no doubt of their belonging to the Lower Miocene, the marine fossils of which we
collected in strata both below and above them. These
enormous blocks therefore were brought into their present
position by causes which acted in the Miocene sea. I know
of no agency but that of ice which could have quietly let
them down upon subjacent beds of undisturbed fine marl and
sand. Hence I conclude that there was floating ice in the
Lower Miocene period, and if the few scratches I saw really
imply glacial striation, the ice-rafts are probably derived
from glaciers which came down from mountains bordering
the Glacial sea; perhaps from the Alps, for that chain must

have existed before the origin of a large part of the Lower Miocene. I have kept the specimens I found of these Miocene striated stones to show Ramsay, who will be interested in hearing, that in spite of some Brazilian genera of trees and insects, and not a few palms, and some reptiles of good size, and many other fossil genera found on both sides of the Alps and supposed to imply a subtropical climate, I am not afraid to appeal to ice as the only known cause capable of stratifying these great masses in the manner in which they occur. I am sure that in the basins of the Tanaro and Bormida there are some rocks of the same age very much larger than any I have seen. Professor Angelo Sismonda was absent from Turin with the Princes of Sardinia, who I hope will become good patrons of geology. His brother the paleontologist showed me every attention.

I hope to find time before reaching Florence to tell you of some other geological matters which occupied me at Turin, and of several things seen in Switzerland not yet alluded to.

Signor Cocchi writes to say he is expecting me at La Spezzia, where I also learn that there are two active young geologists, the Marquis Doria and Signor Campanelli.

Believe me ever most affectionately yours,

CHARLES LYELL.

To LEONARD HORNER, ESQ.

Ruta, Riviera di Levante : September 25, 1857.

My dear Horner,—We are stopping to dine half way between Genoa and Sestri, having followed the beautiful road which skirts the Mediterranean among groves of olives and figs, and admiring in the gardens of the numerous villas orange trees which stand the winter, and oleanders of large size. Here and there the *Pinus maritima*, with its foliage brighter than that of most pines, and some chestnuts varied the scene. There are many hedgerows entirely made of aloes, and we saw some cactus, reminding us, as did much of the vegetation and the steep rocks and the trellised vines, of Madeira. We only saw one date palm, but that also helped the resemblance. The rocks however are not, like Madeira, volcanic, but composed of what the Swiss call ' flysch ' and

the Italians ' macigno,' a formation of vast thickness and
without fossils, except a few fucoids. I mentioned it in
my letters to you last year as being styled at Vienna the
' Wiener sandstein.'

It has had the singular fate of having been classed
during the progress of geology under the head of every
great division of strata—transition, secondary, tertiary, and
is now quite proved by Escher and others to be above the
nummulitic Eocene.

I have been seeing this flysch in my Swiss tour (in the
Vallais among other places), and always with great interest,
not only because its roofing slates look so very ancient, but
because the extreme scarcity and general dearth of fossils,
even where it is not in a metamorphic state, is so singular
in a rock of this age. It encloses near Genoa some moun-
tainous masses of pure dolomitic limestone which have also
proved as yet unfossiliferous.

There is I suppose no doubt that Hannibal with his
African elephants crossed the Alps, though so many essays
have been written to decide by which pass they got over.
Dr. Falconer has very properly started another question,
whether the Siberian elephant ever performed the same feat
long before. He found all the specimens in the museum of
Italy to which the name of *El. primigenius* was affixed, to
belong to two or three other species distinct from the
mammoth, and therefore when he at last reached Turin
and discovered that they had made the same error in regard
to several of their proboscidians found in a railway cutting,
he not only corrected their nomenclature and put the names
of *E. antiquus* and *E. meridionalis* as substitutes for their
E. primigenius, but also when he met with a real tooth of
the last-mentioned species (or mammoth) he put a query as
to whether it was really a true Piedmontese fossil.

When I had quite convinced myself that large glaciers
had once advanced many leagues from the Alps into the
plains of the Po, I felt sure that a cold climate must once
have prevailed in the highest bed of the railway cutting
which I only saw as I passed in the train, but of which
Gastaldi gave me the details. The *marmot* has been found,
the skull very entire, and not distinguishable from the living

Alpine animal. As I had seen in the present tour no less
than four individuals of this species which Dr. Debey had
procured from the loess at Aix-la-Chapelle, and another
which Mr. Morlot got from an old moraine near Lausanne,
the discovery of this creature in the plains of the Po some
miles in advance of the terminal moraines of the old southern
subalpine glaciers seemed to me to render it probable
that the mammoth might also have availed himself of the
same cold climate, and at least have extended his range as
far south as the marmot. Falconer had, I am aware, expressed himself with due caution as to the opinion that the
E. primigenius had never crossed the great chain, since he
was merely led to suspect it from negative evidence. But I
expect that the mammoth will be the rare exception in Italy,
and the same heat which melted the glaciers, vast as they
were, and caused them to terminate so abruptly in the great
plain to the north of the Po, checked also the southward
migration of the hairy northern elephant. At any rate the
publication of Falconer's paper will lead to the speedy clearing up of this question.

I had often heard that the old glacier moraines of
Switzerland had left some curious monuments of their
ancient extent in places where they have in great part
wasted away under the influence of fluvial and torrential
action, and I had planned last year a visit to the celebrated
'pyramids of Botzen' in the Tyrol, which I presume exhibit this phenomenon. Having been unable to accomplish
that point when I went through the Styrian Alps, I was
not a little pleased to meet with others which fell in my
way this year.

The first column which we saw, called the Dwarf's Tower
(or in the *patois* of Vallais, Zwerglethurm) was near Viesch,
in a small tributary valley which joins the great valley of the
Rhone. It is a single isolated pillar no less than thirty-five
feet high, of sandy mud, such as one sees in every moraine—
fine mud with sharp sand and angular pieces of various rocks
scattered irregularly through the paste, and now and then a
larger rock stuck into the same without any arrangement or
stratification. Now and then some of the small pieces of
granite, gneiss, or quartz were somewhat rounded. At the

top are two large blocks which have evidently protected the
mass below (which is about seven feet in diameter) from
pluvial action. About 100 feet higher up the same steep
slope of mica-schist covered with fir trees is a second column
about twenty-two feet high, but with a much larger block
on its top, and there are no others in the neighbourhood.

They are monuments of the complete removal by rain and
by the melting snow of what was probably an old moraine.
At any rate I saw similar columns afterwards, some of
them formed, others in the act of growth, at Stalden in the
valley of the Visp, which joins the Rhone from the south.
I measured and made drawings of them, and found them to
contain erratics of various rocks, which could not have fallen
down from the rocks above like a talus, besides pebbles of
serpentine with glacial striæ. Unfortunately the Stalden
pillars have some of them lost their cappings of large
erratics during the late earthquake of 1855, which did so
much damage to the town of Visp as well as to Brieg, and
which caused landslips and the fall of rocky masses from the
precipices bounding the Visp-thal. These same earthquakes,
by the way, may explain the origin of some of the gigantic
masses which, falling on the old glaciers of the glacial
period, were carried for more than a hundred miles into the
lower country.

James Forbes has well compared such pillars as the
Dwarf's Tower to the portions of earth or stone which the
workmen often leave to test the quantity which they have
removed. Old Time seems to have resolved to leave us now
and then a monument of the same kind, without which, in
such a case as the two pillars near Viesch, we should never
have given him credit for the extraordinary volume of matter
he has slowly and quietly carried away. At Stalden we
should have been able by observing here and there solid
fragments of the old moraine 800 feet or more above the
valley-plain to infer a once continuous mass, but I saw
nothing of the case near Viesch. There are many gorges in
the Alpine valleys from which the old glacier moraines have
been quite cleared out by the torrent, the rain, and the
melting snow, the same moraines being in great force above
and below the gorges. Just as in Auvergne, one finds a

stream of lava like that of Tartaret, above Neschers, which
has been entirely swept away in the course of ages and the
valley deepened in the granite or other rock which formed
the defile, while above and below where the valley widens,
you see a continuous flat-topped current on one or both
sides of the river, with its range of basaltic columns below
and its scoriæ above.   It is the comparatively small quantity
of old moraine adhering to the sides of the Alpine gorges
just above where the old glaciers entered the plains of the
Po which makes the extraordinary expansion of the moraines
just below the outlets or embouchures of the same old glaciers
so striking.   But it is clear that the same torrent, which has
such unlimited erosive power when rolling down a steep
descent within the Alps, would have little in comparison
when it came out into the plains of the Po.   It would make
many breaches, as it has done in the great semicircular
moraine, but it cannot destroy any of its prominent features.

Ever most affectionately,

CHARLES LYELL.

To HIS SISTER.

Florence: October 4, 1857.

Dearest Marianne,—It is wonderful to think that it
should be thirty-nine years since I was here for the first time
with you and Fanny, for that visit seems to me as recent or
more so than many things which have happened not half so
long ago.   It was the year before that only, that Mary's
uncle, Francis Horner, died, whose marble tomb, with the
medallion executed by Chantrey, we went to see at Leghorn
on Thursday, in the Protestant cemetery, surrounded by
China roses in full bloom and half hid by oleanders, which
the custode told us were full of flowers for the earlier part
of the summer.   It was evident that that monument, and
the one to Smollett, who was also buried there, are the lions
of the place, as the woman who shows it did not know who
we were, and selected them.   Mary went also at Pisa with
me to see the room where her uncle died, which happened to
be unoccupied.

It was a great treat to us at this same Pisa, to receive

your long letter about the four bairns whose different characters you have so well described. We cannot be thankful
enough that Harry[1] and his family were safe out of India
before this fearful insurrection began. When one thinks of
the warnings of Mountstuart Elphinstone, Sir Thomas
Munro, and above all Sir Charles Napier, and his words
about 'caste being another name for mutiny,' one cannot but
feel indignant at the supineness and apathy with which all
ordinary precautions were neglected at Calcutta, and at home.

I have much enjoyed the geology of the Pisan Hills, a
beautiful country. As the Professor of Geology, Savi, was
not quite well, and only able to show me the museum, he
sent an assistant with me who knew not a word of any
language but Italian. I was well pleased to find that I
could converse with him the whole day without fatigue, and
I only wish I could say the same of my German, which I
look upon as about four times as difficult to one who starts
with English, French, and Latin. I find a good deal of
scientific activity in Italy, so far as I have gone, especially
in Piedmont. We saw the Somervilles yesterday; considering that she is seventy-eight, and he a good many years older,
they are marvellously well. Both her books are requiring
new editions in England, her ' Geography,' and the ' Connection of the Sciences.'

We have taken our places in a large steamer which sails
direct from Leghorn to Messina, only touching a few hours
at Civita Vecchia (for Rome), and at Naples, and then going
on to the East.

Believe me ever your affectionate brother,

CHARLES LYELL.

To GEORGE HARTUNG, ESQ.

Naples : October 16, 1857.

My dear Hartung,—I was very glad to receive your
letter. I have already seen much in this neighbourhood;
above all, several streams of lava in motion, some going
fast, others very slow, and they have given me many new
ideas. The manner in which they cling to a slope of 27°

---

[1] His brother, Colonel Lyell.

is very striking, and they cover *continuously* an immense area, one-half of the whole cone in five and a half months, by always flowing between preceding currents. All the currents described by Scacchi of 1850–1855 are overwhelmed, the prismatic lava of 1855 concealed under new torrents which are as steep, but perhaps not so un-vesicular.

As to the walls of Somma, so far as I have yet re-examined them, they offer no difficulty whatever. They are simply thin-bedded lavas, rarely very continuous, alternating with dense beds of coarse tuff and lapilli and coarse con-glomerates, at an angle of 27°.

I begin seriously to doubt whether any of the rents, which are now dikes, imply distension as De Beaumont assumes. Scacchi's observations seem to show that they are rather the effect of a failure of support, and attended with a partial sinking. But I will explain more when I have devoted more days to the examination of this region.

I believe from what I am now seeing, and what I recollect of Etna, I shall be able a month hence to tell you that here in the Kingdom of the Two Sicilies, as in Madeira, when the beds are highly inclined they are thin bedded, one or two or five feet thick, and where the inclination is slight, as five or six degrees, they are often very thick, twenty and fifty feet thick.

I have been delayed here by the extreme irregularity of Neapolitan steamers, and shall not reach Etna for six days. There is, I hear, a small eruption going on there. I am glad of it, for I am now sure that no one can describe lava flowing so as fully to make others understand it. I assure you I scarcely expected to find so much confirmation in a few days of the upbuilding, as contrasted with the upheaval theory. Scacchi is my companion, and a scholar of his, Guiscardi. I do not believe that Scacchi would allow that one-fourth of the inclination in Madeira could with any probability be attributed to upheaval. But all that we say is, that this is the maximum of what may be. We do not say it must have been. All the movement here is bodily, as you say of the Azores.

Believe me ever most truly yours,

CHARLES LYELL.

Lady Lyell desires her kindest regards, and begs me to say she is sorry you are not with us. So am I, but every day makes me feel that you would only have been confirmed in the true doctrines, which you have gradually and not hastily embraced. I shall claim you as the first of my pupils in the volcanic line, and you will soon have seen more than I have, at the rate you proceed.

To HIS SISTER-IN-LAW, MISS SUSAN HORNER.

Avignon : December 19, 1857.

Dearest Susan,—I do not wonder that you were so charmed with Rome. It certainly combines a greater variety of attractions than any place. The sculpture struck me more than ever. As to the antiquities, much has been done since my preceding visit more than a quarter of a century before. The catacombs I never heard of when I first went there. If half a million were buried in this way for five centuries or more, the number would of course be immense, and after all, if they used the volcanic tuff for pozzolana, I can fancy its being a cheaper form of burial than ours, as the room required for the mere body in a winding-sheet was so much less than for a coffin, and the volcanic matter removed worth something.

The circus of Romulus Maxentius, entirely discovered and disclosed since I was last there, is a very grand monument of antiquity. I was also much pleased with the theatre of Tusculum, and what little I could see of the amphitheatre. The Alban Lake, also, which I went all round for geological purposes, and the sight of Old Alba. Within the walls also of Rome, much has been done to display the antiquities.

The geology of the Campagna, and especially of the Latin Hills and Hannibal's Camp, is very instructive, and there also one falls in with all kinds of Roman antiquities which Ponza explained to us. There is, moreover, a singular advantage which I did not appreciate before in the Campagna, arising from its being so unhealthy in summer and autumn, that they can only get good rents by leaving it in a pastoral condition. By this means one has a boundless space almost in a state of nature, close to a great centre of the arts and

of antiquities, and where there is much agreeable society
brought near together, and in a way to promote much social
intercourse, with good things to talk about.   All the naturalists expatiated to me on the beauty of the wild flowers in
spring.   I saw some, and several butterflies even in
November.   What a charm there is in seeing no hedges and
fences, and straight ditches and ploughed fields!   As a
woman said to us, who after settling in Nova Scotia for
seven years, returned to pay a visit to Manchester (?), 'I am
truly sorry for people who are obliged to live where they
cannot move a yard out of a straight turnpike road without
committing a trespass.'   There is some compensation, therefore, even in the malaria.   Menighini of Pisa endeavoured
to account for the increase of this pest since the Augustan
age by the gradual sinking down of the west coast, which
has submerged a considerable tract, and sent the aguish air
of the marshes up the Tiber and other rivers.   I had heard
so many theories before, that I was surprised they were not
exhausted, and this seems more likely than most of them, if
the subsidence be as certain as he affirms.

The weather during four days' excursions into the Pisan
Hills, and the hills between Volterra and Siena, was most
delightful; the air so clear, the sky so blue, and the sun,
though in December, so genial.   The vegetation in those
parts not monopolised by the olives was sufficiently unlike
our own to interest me.   Evergreen oaks, with an undergrowth of an evergreen called *Cytisus salicifolia*, and the
Cratægus pyracanthus, which we have introduced so much,
and a heath almost as high as *Erica arborea*, but what they
call by another name.   On some hills we had zones of olives,
chestnuts, and pines, one above the other.

I forgot to tell you that Mr. Gibson inquired much after
you, and paid handsome compliments to your skill in mo-
delling.   Miss Hosmer has bought a new spirited horse.
Gibson told her it would throw her, but she said she should
ride it for all that.   He says she has great perseverance, and
works very steadily.   Gibson wanted an account of his
allegorical sculptures, which accompany his statue of the
Duke of Wellington, to be printed in an Italian periodical,
but the censor actually struck out the description of the

angel carrying the Duke's soul up to heaven, as this was impossible, seeing that he was a heretic. Gibson has kept the proofs with the censor's erasures, and certainly they are a curiosity in the year 1857. He told us the story in sight of the new column raised to the Immaculate Conception, and with this in the background, no open display of superstition ought to astonish one.

Considering that no lectures on geology are authorised in the Sapienza, I was amused at the late French ambassador Count de Rayneval having made a splendid collection of tertiary fossils in the hill of the Vatican, which he and Ponza are preparing for publication. They are curious, and intermediate it seems between Miocene and Pliocene. For five years they have worked away under the Pope's window, to throw light on the earth's antiquity.

We hope to see you now in a week or ten days.

Ever yours affectionately,

CHARLES LYELL.

CHAPTER XXXII.

JANUARY 1858–SEPTEMBER 1858.

INDIAN MUTINY—DINNER AT MILMAN'S—SIR HENRY HOLLAND—RAJAH
BROOKE—BUCKLE'S 'HISTORY OF CIVILISATION'—DEATH OF DR. FLEMING
—MEETING AT THE GEOLOGICAL SOCIETY—DEATH OF ROBERT BROWN
—LAC DE BOURGET—POLITICS WITH SIGNOR BEZZI—NAPLES—EXPE-
DITION TO VESUVIUS—FESTA—MESSINA—CATANIA.

[After spending some weeks of the summer of 1858 with his
brother's family in the Berg Strasse near Darmstadt, gastric fever attacked his young nephews, and leaving his wife with the then conva-
lescent patients, Sir Charles set off at the end of August on a tour to
revisit Vesuvius and Etna, the result of which was an important paper
on the 'Origin of Mount Etna and on the Theory of Craters of Elevation,' which was read at the Royal Society. Large extracts from Sir
Charles's Journal, addressed to his wife, are given in the following
pages. The Copley Medal,[1] the highest reward of the Royal Society,
was conferred on Sir Charles Lyell this year.]

CORRESPONDENCE.

To GEORGE TICKNOR, ESQ.

53 Harley Street, London : January 23, 1858.

My dear Ticknor,—It is so long since we have had any
direct correspondence, that I shall write you a short letter,
and beg for some account of your present occupations beyond
what I gain from the interesting letters which Mrs. Ticknor
and Anna have sent, and what the newspapers told us of

---

[1] 'Nie ist wohl die Copley Medal einem Werke gewidmet worden, das
zur Verbreitung und Vervollkomnung einer Wissenschaft heilsamer gewesen
ist, als den schönen " Principles of Geology " unsers edlen Freundes Sir Charles
Lyell.'—Extract of a letter from Baron von Humboldt to Madame Pertz, early
in 1859.

the opening of the Boston Library, and your part in that affair, in which we know from other sources how much you have done.   It was fortunate for us, as we could have done no good if here, to be engaged on a tour on the Continent, when the horrors of the Indian mutiny and the anxieties of the commercial crisis were agitating people's minds all over England.

My tour was unusually profitable, first in the glaciers and then the volcanos.   I came to the conclusion that Agassiz, Guyot, and others are right in attributing a great extension to the ancient Alpine glaciers, and that floating ice which I believe has done much in Great Britain, Scandinavia, and the United States will not aid, as I formerly suspected, in explaining the transfer of Alpine erratics to the Jura.   I found Agassiz's map of the Zermatt glaciers and moraines very correct and useful.

We dined with the Milmans yesterday, both very well; their three sons also.   We met Macaulay (the new peer), looking rather older, and with a baddish cold and cough, but cheerful, and telling many a good story; the Chancellor, and Lady Cranworth, both looking in good health; the Monteagles, Sir Henry Holland, who had been far, as usual, in a short time, among other places at Jerusalem, and in sight of the Pyramids.   I think you saw when here the Herman Merivales.   We dined there last week, and met among other public characters, Rajah Brooke of Borneo. He told me that nearly all the Malays could read and write, and might be rapidly improved if they had better things to read, but the Methodist and other missionaries will only translate into the Malay language things which neither they nor he (i.e. neither the Malays nor Rajah Brooke) cared to read. What would you recommend them to give them instead? 'I would begin with the "Arabian Nights' Entertainments."'

We have just voted at the Royal Society 150*l.* to Mr. Mallet, an engineer and geologist, who has offered to go and collect physical facts relating to the great earthquake south of Naples.   I have had him with me half the morning, and expect he will do us good service.

Have you seen Mr. Buckle's book, ' History of Civilisation in England '?   I have just bought it—full of talent, and having the great merit of setting people thinking. I am prepared to find paradoxes and contradictions, as I did in the conversation of the man himself, but you will do well to get it for yourself and your Library as soon as you can. That large party here who exert a severe censorship on all who dare to think differently from the 'endowed doctrines,' as Channing called them, are up in arms, and threaten to blackball the author at the Athenæum, but I hope they will fail, or that the committee will elect him before the ballot. For so bold a book, it is having a considerable sale.

Ever most truly yours,
CHARLES LYELL.

To J. W. DAWSON, ESQ.

53 Harley Street, London : February 14, 1858.

Dear Dawson,—Dr. Fleming's loss was very sudden and unexpected by me. Less than two and a half years ago, Hugh Miller and Fleming accompanied me severally on many excursions about Edinburgh. Miller was the extinction of a great light who was making steady progress, and daring to declare his faith as he went on.

I spent more than two months in the glaciers of Switzerland (last autumn) and came to the opinion that the transportation of all the blocks in the Swiss valleys, and between the Alps and Jura, and on the Jura, and also on the plains of the Po, was effected without the agency of the sea. Agassiz, Guyot, and some who went before, were, I think, right in this. I found all the glaciers advancing from year to year.

I then went to Naples and Catania, and re-examined Vesuvius, which was in full activity, and Etna, the latter much altered, especially in the Val del Bove, since I was there in 1828.

My convictions derived from my first visits in favour of the origin of those mountains from a series of eruptions, and against the Von Buchian theory of Elevation Craters, have been fully confirmed. I am writing a paper for the Royal Society on the subject.

We are getting on well at the Geological.

It will indeed be a great point for you if Sir Edmund Head and the Government settle at Montreal.

Believe me ever most truly yours,

CHARLES LYELL.

To LEONARD HORNER, ESQ.

London: March 12, 1858.

My dear Horner,—Mary has written a journal of our proceedings to you, and I shall merely add a few words about the last meeting of the Geological Society, when Darwin read a paper on the ' Connection of Volcanic Phenomena and Elevation of Mountain Chains,' in support of my heretical doctrines; he opened upon De la Bêche, Phillips, and others (for Greenough was absent) his whole battery of the earthquakes and volcanos of the Andes, and argued that spaces at least a thousand miles long were simultaneously subject to earthquakes and volcanic eruptions, and that the elevation of the whole Pampas, Patagonia, &c., all depended on a common cause—also that the greater the contortions of strata in a mountain chain, the smaller must have been each separate and individual movement of that long series which was necessary to upheave the chain.

Phillips pronounced a panegyric upon the ' Principles of Geology,' and although he still differed, thought the actual cause doctrine had been so well put, that it had advanced the science, and formed a date or era, and that, for centuries the two opposite doctrines would divide geologists, some contending for greater pristine forces, others satisfied like Lyell and Darwin with the same intensity as Nature now employs. Fitton quizzed Phillips a little for the warmth of his eulogy, saying that he and others who had Lyell always with them, were in the habit of admiring and quarrelling with him every day, as one might do with a sister or cousin, whom one would only kiss and embrace fervently after a long absence. This seemed to be Mr. Phillips's case—coming up occasionally from the provinces. Fitton then finished this drollery by charging me with not having done justice to Hutton, who he said was for gradual elevation. I replied that most of the critics had attacked me for overrating Hutton, and that Playfair understood him as I did. Whewell concluded by considering Hopkins's mathematical calculations, to which Darwin had often referred, in reference to D.'s theory. He also said that we ought not to try and make out what Hutton *would* have taught and thought, if he had known the facts which we now know. I was much struck with the different tone in which my gradual causes were treated by all, even including De la Bêche, from that which they experienced in the same room four years ago, when Buckland, De la Bêche, Sedgwick, Whewell, and some others treated them with as much ridicule as was consistent with politeness in my presence.

Yours affectionately,

CHARLES LYELL.

To GEORGE HARTUNG, ESQ.

53 Harley Street, London: March 22, 1858.

Dear Hartung,—I have had a letter from Mr. Wollaston, who is now in Teneriffe with Mr. Lowe, formerly of Madeira, and who is now collecting plants while Wollaston collects Coleoptera in the Canaries. For six weeks they had the loan of a friend's yacht, and went to Hierro and Gomera, spending some time in exploring each entomologically and botanically; but I am sorry to say neglecting the shells except when they obtruded themselves on their attention. Wollaston says, however, they got some thirty species, mostly different specifically from those of Madeira. The weather having become more genial, they took Teneriffe in hand, and mean next to go to 'the wooded caldera of Palma,' where Wollaston expects to get plenty of beetles. He speaks with great admiration of the native and uninjured forests of Hierro, but with much disappointment of the cleared country of the valley of Tarro.

Not a word about geology; but Prof. Piazzi Smyth has just returned from an astronomical tour in Teneriffe, and published a book upon it, in which there are geological observations scattered here and there. He and his wife encamped some 9,000 feet high on the peak, or rather got the sailors of the vessel which took them out, to build a rough shelter with blocks of lava.

I am getting on with my reading up for my paper on

' Etna,' a course of reading which will be useful for Madeira.
I find that Waltershausen imagines almost all the stony beds
which you and I should consider as lavas, and which are
parallel to the tuffs and scoriæ, to be injected in the form of
dikes ! By this means he obtains proof of great upheaval.
But he is against any great catastrophe, and goes to work as
patiently as I do.   Pray let me know what you are studying
and reading and writing. As to upheaval on Etna, I saw
very few beds which required it, and what we allowed as
possible for Madeira, or one-fourth of the whole, is the utmost
which is wanted, or probable in any part of Etna.

Ever faithfully yours,

CHARLES LYELL.

To GEORGE HARTUNG, ESQ.

53 Harley Street, London : April 22, 1858.

My dear Hartung,—I was glad to hear that you had
observed in the Azores those mica-schist and other fragments,
which I believe will turn out to have been transported by
ice. But you would have followed it up with more spirit,
and traced them to their maximum height, had you thought
of the glacial theory when there. Icebergs get as far south
in mid-Atlantic in our time, and if so, thousands of them
may have floated there annually in the old glacial period.

Our present schemes for the summer are still somewhat
vague, but probably we shall leave this in July for some
place on the Rhine near Heidelberg or Darmstadt. I should
like you to join me, that I may go over the MS., especially
the illustrations of our joint paper or papers on volcanic
subjects. I also hope to get to Zürich and Berne in the
course of the autumn, and see Heer, Escher, and the Swiss
geologists, but nothing is yet fixed. As soon as I have got
off my hands what I have to say on Etna and Vesuvius, I shall
return with more knowledge to Madeira and the Canaries.
I mean to allow some upheaval in case of Etna, but exceedingly subordinate in proportion. So also in the case of
Vesuvius. Scacchi and some of the best observers in Italy
doubt whether *any* is absolutely necessary in the case of
Vesuvius. A few local and partial dislocations may of course

happen, and you may find beds vertical, but such accidents are more likely to derange the symmetry of cones than to be the cause of that symmetry. But I will not allow the extravagance of the ultra-upheavalists who call Monte Nuovo a case of *erhebung*, to provoke me into a denial that there has probably been some central distension which has modified the more dominant influence of the outpouring of lava and the outthrow cf scoriæ, &c.

Ever yours sincerely,
CHARLES LYELL.

To CHARLES BUNBURY, ESQ.

53 Harley Street, London : June 11, 1858.

My dear Bunbury,—I am sorry I have no specimen of *Equisetum Lyellii*, but I will inquire about it. I am glad you have taken up that subject, and hope you will publish on it separately—of capital mixed geologico-botanical interest.

Yesterday morning Robert Brown [2] breathed his last. They told him they might keep him alive even till Christmas possibly, Brodie, Bright, and Boott, by opium and stimulants, but he preferred not to live with a mind impaired, and so cheerfully and tranquilly, and in full possession of his intellect, gave way to the break up of nature. Every one who has been with him in his last days agrees with me in admiring the resignation with which he met his end, and the friendly way he talked and took leave of us all. Boott has been constantly with him, and looks much worn. Bennett sat up with him many nights,

I read my paper, or rather spoke it, last night at the Royal Society to a good audience.

Ever affectionately yours,
CHARLES LYELL.

---

[2] The eminent botanist.

Mary E Lyell

Engraved by G.J.Stodart from a drawing by George Richmond. R.A.

London Published by John Murray, 50, Albemarle Street 1881

JOURNAL.

To HIS WIFE.

Basle : August 24, 1858.

My dearest Mary,—Between Bensheim and Heidelberg I found myself in a second class carriage, quite full; how they talk as contrasted with first class! It is the way to absorb German with a plentiful supply of tobacco smoke in one's lungs and clothes. At Heidelberg I took first class, a handsome saloon; nine others, all English, absolutely dumb, ladies and gentlemen. All turned out at Baden-Baden, and I was then alone to Basle. What a splendid panorama of the Rheinthal! I drew back all the blinds, and could walk about the room with Bach's map spread on the carpeted floor. No map is better coloured geologically for giving a pictorial idea of a large part of Europe to the eye of one accustomed to associate scenery with the nature of the rocks. You remember, some three years ago, I was discouraged by some of our Berlin geological friends from purchasing this valuable map, because they flattered themselves that Von Decken's, and the Geological Society of Berlin, would so soon come out with a better one, on as large a scale, and this is the third journey I have profited by it, and Waltershausen told me Von Decken's had not yet appeared.

The volcanic mass of the Kaiserstuhl stands out very prominently in the valley, and is well seen from the railway. The old Rhine, long after it was extinct, covered it by its annual inundations of mud, till it was nearly potted up in loess, and the modern Rhine has never been able, when it re-excavated the great valley, to wash it clean again. So it is blindfold work, as there are no continuous sections, merely a peep at the rocks here and there.

Culoz, near Perte du Rhone: August 26, 1858.—We have had a splendid drive from Geneva along the banks of the Rhone for some sixty miles. I should like to have seen the improvements in the town of Geneva, new streets, churches, hotels built since we were last there, but I hope some day, ere very long, that we may see them together.

The arch over the river called the Perte du Rhone exists

no longer; they have broken down the limestone arch under which the river flowed in a deep narrow gorge, in order to let the wood float down freely. This scenery is in the Jura where that chain joins the Alps. What with changes of railway carriages when the travellers for Lyons and for Italy divided, and the French douane first, and then the Sardinian, and a change to a steamer on the picturesque Lac de Bourget, surrounded by magnificent precipices of limestone, something in the style of the Saltzburg Alps, there were many opportunities of losing luggage. One lady's went wrong, another left a heavy purse in our carriage, but I got it restored to her. On the lake I fell in with Signor, now Cavaliere Bezzi, returning with despatches from London in two days, as lively and wide awake as if he had slept in a bed every night, and going to cross Mount Cenis without stopping, and to see Count Cavour on his arrival. He has been active in the Sardinian Parliament, and lives on his property on the Mount Ferrat (Superga chain) near Turin, has founded several schools, some infant ones, preached against by the clergy chiefly for founding them as a heretic and Protestant, and with such effect that he lost his seat, with many others, at the last election. Count Cavour offered to bring him in for a distant county, but he said he must in that case abandon all his schools and the superintendence of his private affairs, but next spring the Government has proposed his coming in for an adjoining county. Bezzi tells me, what I am sorry to hear, that all the old nobility are alienated from the constitutional system, because in 1849 primogeniture was abrogated, 'un fait accompli' which Cavour could not alter if he would.

I remarked, 'Now you will in the next generation have only two powers, the King and People; a republic or an absolute monarchy as in France, and as your people will not yet be advanced enough for the Republic, especially if you succeed in your desire to unite all Italians, most of whom will be less advanced than yourselves in politics, you will have the " Re netto." ' He replied that ' nationality under an Italian sovereign would be worth having, even if they passed through a phase of absolute monarchy to gain it. If we could only get these foreigners out of Italy! There is only

one King towards whom there is any loyalty. All save
King Bomba and the Pope are Germans. We wish your
court had not German leanings.' I asked if he did not
suspect that the concessions now offered to the Lombardo-
Venetians by the Archduke Maximilian came from King
Leopold's and Prince Albert's advice? 'Possibly so, for they
are too enlightened to wish to see Italy governed as the
Austrians would rule her, but still it is natural that your
Prince Albert should not wish to see the Germans driven
out.' He added, 'Palmerston when hard pressed by us to
stand up for Italy on some point, coolly laid the blame on
his court (!), saying, if I can have my way on important
matters, I must yield in *small* ones, meaning by the latter
the liberal cause or some point then under discussion, of
Italy.'

I remarked, 'If an English minister who can easily control the court or resign, could be so unprincipled, why
should you believe him? I think him at heart a man who
sympathises with your enemies, whereas the Prince Consort
is by nature a liberal. Do you not think that Palmerston
and Clarendon were as free to act in the Cagliari affair, as
afterwards Lords Derby and Malmesbury? The German
leanings alleged against our court did not interfere there,
and yet you see which side they took.'

To this Bezzi replied, 'Well, it may be your aristocracy
as much or more than the court; all we Italians know is this,
that hitherto all your diplomacy has been Austrian to the
backbone in every part of Italy except Turin. Sir J. Hudson alone has been true to us, a man of enlarged views and
a good heart.' I told Bezzi how openly the French press
avowed that the union of all Italy into a compact nation
would diminish French power and influence, and although
the French and Austrians may by their mutual jealousy
preserve Sardinia independent, they would never allow a
revolution which should consolidate Italy and make her free.
So may it not be best for the Lombardo-Venetians to accept
from the Coburgs such concessions as the two rival despot-
isms would never grant? 'My dear sir, it is all too late—
the anti-German feeling has become a sentiment throughout
the best heads and hearts in Italy.' He then told me that

Savoy was a source of political weakness to Sardinia, because they did not speak Italian, and were superstitious. They refused to furnish one soldier for the Crimean war, because it was against their principles! This was an act of rebellion, and some proposed to decimate them, but La Marmora said wisely that when a whole people (who are loyal to the King) were determined, you must wink at it, however awkward as a precedent. Even with Savoy they are only five millions.

Genoa.—It is lucky I came on, as otherwise I should have had to have gone by land to Leghorn, and then by sea direct by a Tuscan boat, as I shall do now, after a delay perhaps of several days which I can well employ, not by a run to Pisa or Florence, but by regular business, having much to prepare in going over my notes made when with Waltershausen, and studying them with the two maps of Etna, and planning each day's journey. You well know that once at Naples or at Catania there will be no time for this. Instead of a night of suffering I had my memory of the Alps refreshed, and even the least good of the Alpine passes exceeds all other scenery save Madeira, from which it is so different.

Leghorn, August 29, 1858.—We got here in about fourteen hours, in a small vessel only destined in ordinary times for La Spezzia. There was not much sea, and I scarcely suffered at all and slept most of the night, but numbers were ill, the small steamer rolled so much. I asked a young man whose appearance I took a fancy to, if he were English; he said no, and that he could not talk English. We went on in French; I saw he was German by his accent. I spoke out freely about the Neapolitan panic and ignorance (for Bezzi assures me it is not a political move) which has thrown the whole commercial world, French, English, and Italian, into confusion. He by no means dissented, but I saw both he and his wife smiled now and then, and he seemed a little cautious. Next day we had some more talk, and just now at the douane, I learnt that he was the reigning Duke of Saxe-Altenberg, travelling under an assumed name with his wife and a suite, one of whom lately married, and with his young bride, a pleasing girl, spoke English and talked much with me.

On reaching this about eight o'clock this morning, and

after getting through the douane, I found a splendid English three-mast steamer going to sail *to-morrow* straight for Naples, and not touching at Civita Vecchia. But the captain is waiting in no small tribulation for a telegraphic despatch which the Neapolitan consul expects to-morrow morning, and which may derange all his and my plans and those of many others—among them a very intelligent French agent of the Rothschilds who is in this hotel with me, and who having gone to Marseilles was put into the Lazzaretto here for four days, much to his astonishment and annoyance. For the Tuscan Government was persuaded by the Neapolitan, and the Pope, to put Leghorn into quarantine. But when this same Tuscan Government had learnt that Count Cavourfus had ered positively to inflict a quarantine on Genoa, Tuscany took courage and immediately took off the nuisance here. But now comes the danger ; every one fears that in retaliation Naples may to-morrow morning put this port into the list of infected places, because she now trades freely with Genoa, &c.

Civita Vecchia : September 1, 1858.—It was rough and rolling weather when we started and many were ill, but I escaped, only went without eating till this morning when we got early to this port. I had a steady work at Leghorn, and was so rested after my march across the Alps that I can the better rough it in this steamer. The heat has not been oppressive, though of course the sun is powerful—yesterday it was cloudy. We have been five or six hours at Civita Vecchia, but I have not landed, as they talked of leaving sooner. It has enabled me to finish my notes on Mount Etna, and I shall now make a small list of essentials for Vesuvius, hoping to find Guiscardi at my service. I am told that September is thought a good month for Vesuvius, as later it is generally voted too cold.

Port of Naples : September 2.—We came here in about twenty-six hours from Civita Vecchia, the first part rough for a steamer of this size, the last half calm. So I arrive in good plight. A boat has come from Ungaro's, but it may be hours before I can profit by it, as you know how long the police are here. Vesuvius is capped with clouds, so I cannot

see how much vapour it is giving out of itself. No great
heat as yet; rather pleasant temperature in this bay.

September 2, one o'clock.—Here we are after all our precautions, and in spite of a telegraphic message from Naples to
Leghorn promising free entrance, put into quarantine for
one day, as we are told. I profit by my cabin being to myself, and have a table in it with my books, and am working
much as in Harley Street; weather cloudy and cool. Vesuvius nearly unrobed, and showing half way down between the
summit and the base, or sea-coast, the new craters of 1857-
58, smoking or fumeroling away. We are not to go to
Nisita, but stay on board.

Ungaro has come himself in a boat, and tells me my
letter arrived yesterday, and he sent it to Guiscardi. An
order came to go to Nisita, but the captain pretended he
had no coals left, so we are glad to be left opposite Vesuvius.

6 P.M.—Guiscardi has come in a boat, and had a long con-
versation, and will go with me to the Solfatara to-morrow.
Ehrenberg is at Naples, and going soon to Vesuvius. Scacchi,
Palmieri, and the others got my abstracts of Etna. All
describe the eruption of last spring as most magnificent from
Naples. I hope to get a small picture of it.

8 P.M.—We are lucky not to be in Nisita. One of the
openings of April last is emitting a glowing light, sometimes
brighter and then a dull red, rather lower than half way
from the summit of the great cone. It gives us a good idea
of the exact position of the last eruption, and of the wonderful escape the town below must have had.

As there is no moon, the lights of the houses all round
the bay look brilliant, and the lateral crater's glow is a very
distinct feature on the opposite side.

Naples: September 4, 1858.—I have paid an early visit
this morning to the Temple of Serapis, and have found the
brackish water incrustation on the walls which Babbage
speaks of, and which I missed last year, and some points
connected with it which I shall be glad to submit to
Babbage for interpretation. I got back in time to attend a
sitting of the Academy at eleven o'clock, when I saw Palmieri,
who has given me a letter which I had requested from him

to admit me and Guiscardi to sleep in the Observatory on
Monday.

Yesterday we visited (Guiscardi and I) the crater of the
Solfatara, which is full of interest. The hot vapours rush
out with a sound such as reminded me of the descriptions of
the Geysers of Iceland. The extent to which the trachyte is
turned into clay by the gases is curious, and often the felspar
crystals have disappeared and left their forms in cavities
while the felspathic base remains undecomposed.

To-morrow we go to St. Anastasia, to return at night to
visit ravines or barrancos on the outer slope of Somma.
Monday, to see the new mouth of April last, still in action,
and sleep at the Observatory. Tuesday, Atrio del Cavallo
and back. Wednesday, a great *fête*, of which more anon.
Thursday, sail by a steamer direct to Messina, only, as it is
Sicilian, will they keep their promise?

Many of the very baths which used to cure devotees in
the pagan days of the Temple of Serapis are restored by
slight repairs, and this after so long a submergence under
the sea! I saw the hot spring there for the first time to-
day.

September 7.--Just returned from a successful expedition
to Vesuvius, which in the last three days I have explored in
a variety of ways, and understand much better than before.
The first day was to visit a ravine which runs up from the
town of Somma, and which shows the structure of the
mountain to be just like that seen in the great escarpment
in the Atrio, except that there are no dikes.

Ehrenberg and Rammelsberg of Berlin were bent upon
revisiting Vesuvius the next two days with us, and we offered
them places in our carriage. We began with the Atrio.
How glad I am you were there last year! I hardly think
you could have got into it over the new lava of this year.
Horses out of the question, and the fatigue considerable; no
paths yet. A small cone stands in the Atrio just where we
came down when we descended from the summit or great
crater, and it sent a flood of lava over all the region which
we crossed on horseback going from the Observatory. It
flowed to the Fossa Vitrana.

Other small cones were formed at the base of the great

Vesuvian cone facing Naples, and though these are now dormant, and without even fumeroles, there is a constant emission of liquid matter going on from some unseen point or points below, above which are large swellings of lava, or gibbosities, as Guiscardi names them.

Palmieri believes that all the steam and gases are being let off by the great crater, which is sending out a grand body of vapour, forming in these sunny days bright and large fleecy clouds, while in every other direction there is only blue sky. He imagines that the lava is escaping from under. one or both the gibbosities, without any steam or explosive gases, tranquilly welling out, and rising up through the mass of lavas. I walked in part, and was part of the way carried in a chair by four men, to a point where the lava was issuing from a small grotto, looking as fluid as water where it first issued, and moving at a pace which you would call rapid in a river. White hot at first, in a canal four or five feet broad, then red before it had got on a yard, then in a few feet beginning to be covered by a dark scum, which thickened fast and was carried along on the surface. You then saw this superficial scoriaceous covering forced by the current into a number of crescent-shaped ropy curves, evidently caused by the greater rapidity of the motion of the central part of the stream as compared to the sides; the ropes of scoriæ farthest from the grotto were the most perfect and solid. You would at first wonder that when the outer and exposed portion of the fluid cools and congeals, it does not begin to sink, but it is a peculiarity of liquid lava, that if you throw a stone on it and press it down, it will not sink. So the plastic crust of scoriaceous matter generated a few yards from the grotto cohered a few feet farther on, and at a few yards farther was converted into those bent cables which you saw last year on the lava of 1857 in the Atrio. All this manufacture of ropy scoriæ takes place in a minute or two. I could have watched it for an hour, but the heat which it radiated was so scorching that I could only take a glance by the help of the men holding their coats before my eyes. When I took up my eye-glass it was so hot I let go, though it was really not heated enough to require that.

I was exceedingly glad I undertook this excursion to the

top of the second gibbosity, from which Guiscardi rather dis-
suaded me on account of the fatigue, but few who are not
residents at Naples ever see lava run away rapidly. We had
also a fine opportunity lower down of watching the slow
progress of lava, and saw the manner in which a small
narrow ridge is formed. Possibly those described by me in
the Val del Bove of 1852 were made in the same way on a
gigantic scale. The viscous mass moves on, swelling out
occasionally, with pieces of scoriæ rolling down in front and
from the sides, which give way in such a manner as to
undermine the top of the ridge, which continues solid, and
by the undermining and falling away of both sides, acquire
the shape of a sharp crest.

In the escarpment of the Atrio I tried to find the broad-
est of the ancient lavas, or a transverse section, showing how
wide the widest of the old streams of Somma ever were. I
could not get one above 200 feet, but this was only a rough
measurement, and I hope Guiscardi will follow it up. If
they are wider than one ever sees a stream on the flanks of
the modern cone of Vesuvius, it may be used as an argument
in favour of the upheaval theory, and therefore I was desirous
of determining the point, which has been strangely neglected.

We finished yesterday, after coming down from the Observatory, with a drive all round more than half the outer
circuit of Monte Somma, by St. Anastasia and the town of
Somma to Ottajano, where we explored a ravine in which we
were told there were dikes, but there are none ; nor have I
yet found one in the barrancos which radiate from the crest
of Somma.

September 8.—I have had the Costas, father and son,
to breakfast, *déjeuner à la fourchette* with tea and a bottle
of Capri wine. The old man, after being three months in
Calabria quite safe, was on his return knocked down by a
common robber near Naples, and because he refused to give
up his money, was stabbed in several places a few months
ago. His purse was taken and he still shows in one eye the
marks of the assassin, and is wounded in the leg, but full of
spirit and energy.

I have had three days of such exertion that, although
quite well, I am as glad of this repose to-day (though I have

to ticket innumerable specimens and pack my portmanteau),
as the Neapolitans seem to be at the *fête* of the nativity of
the Holy Virgin. The procession is to pass my window,
which is on the first floor.

Half-past three o'clock.—For an hour and a half the troops
have been passing my window; cavalry, infantry, and
artillery; Neapolitans and the Swiss guard. The latter in
red uniform like English. A fine-looking set of well-grown,
and decidedly tall men. The numbers which have passed
must be a great many thousands. Sometimes they went slow,
but sometimes 800 infantry in one minute going as fast as
they could. The cavalry managed 100 a minute trotting.
Innumerable brass bands, which with drums seem to form
nearly all the instrumental music; a great mistake, I think.
The troops, after filing one way, are now forming an unbroken
line, often four men deep, on each side of the street. Many
carriages and foot-people crowd the interval between the
ranks, and how the procession is to pass I cannot imagine.
Such a repetition of regiments, without a mixture of civilians
and priests, becomes monotonous when continued for two
hours. I have been wishing that Katharine and the two
boys were in my room now with you, dearest. At intervals
I have been ticketing my specimens, and have nearly
finished. Four men-of-war, one a steamer, and decked out
gaily with flags, twenty or thirty on each mast, opposite my
windows.

Four o'clock.—The cannon are firing to announce that
the King has left his palace. The pavement is so covered
with dust, that being never watered, it has caused some of
the fine dressed troops to be a little dusty, and I see them
wiping their coats. The cavalry raised clouds of dust when
they moved fast.

The four ships of war are firing salutes one after the other,
quite shaking the houses, though they are far enough off for
a long interval between the rolling out of the smoke and the
sound reaching us. The line of hills of Posilipo return a
splendid echo to the guns. No end to carriages and six,
and now one with eight horses, in which are the King and
Queen, and glass windows very large. The people so
excluded now, that I cannot judge of his reception, but per-

haps as it is the Virgin they are honouring, it is not a time for loyal demonstrations. The King is followed by some eight or ten six-horse carriages of state, with members of the royal family and officers of state, and several regiments of horse guards. We have now had three hours of a continuous military procession, and nothing else here at least.

The people are now closing in, and filling the middle part of the street, and seem very merry, but, for Neapolitans, not noisy.

Half-past five.—After an interlude at the *table d'hôte*, I return to my room, and find the fleet as busy as before, firing salutes. The troops still lining the streets. Some Englishmen at dinner tell me that they never heard the people make any more demonstrations of loyalty on other occasions. The priests seem to have been all in the churches, but one priest accompanied every regiment except the six Swiss ones, who, being Protestant, have none with them.

I am almost a prisoner, the streets being so lined with soldiery, and cannot help regretting that I never see Vesuvius, as every hour of the day the phenomena there are changing. They tell me the number of troops who have passed this window is just 22,000, and as they are all passing a second time, besides 100 guns, you may fancy the time they take. The infantry are actually running, and the cavalry trotting. The King and all the twenty or more carriages of state are gone by again, and the officers seem to be hurrying the men, who must be hungry enough. It is quite a curious sight to see the quick movement. All day opposite us have been thirty-two drummers and thirty trumpeters; the former were allowed to overwhelm the others by their noise. Think of the cavalry having the old flint locks still. The Swiss at least have percussion guns. They have much higher pay, but are not allowed to be buried in consecrated ground!

This rapid retreat of the soldiers twenty-four deep across the street is a far more amusing sight than the outward march. We are now beginning the sixth hour of our uninterrupted military procession. The band has moved off, but others are passing and playing. Every now and then there comes a jam of people and soldiers, and a dead halt. The

Swiss are very conspicuous with their distinct costume, more than 6,000 Ungaro says.

September 9, ten o'clock.—I must leave this soon if I learn that the ' Etna ' (steamer) is really true to her day and hour, which last is an unusual one, twelve o'clock. Guiscardi came to tell me last night that Vesuvius had given the required signs of sympathy with the two ' gibbosities ' of lava which I before described. I went out with him between eight and nine o'clock, and saw light at intervals distinctly reflected from the vapours overhanging the highest crater, showing that there is lava there also. We now agree with Palmieri that the eruption is going on in the great chimney, but the lava is escaping *without* steam from the two cones *without* craters, which are just below the windows of the Observatory.

The lava was seen by us from the Observatory in the night, radiating in narrow streams from the upper part of the two flattened domes.

Guiscardi thinks I have made great progress this time in striking out a theory to explain the connection of the pumiceous tuffs of Monte Somma and the old lavas of the same mountain. If I have done so, it is partly because I had the means of taking expensive drives to distant points and back. Guiscardi had never reached Ottajano in his life. I have set him several supplementary observations to make, to complete points required by my theory.

Messina : September 10.—When I was concluding my last letter at Naples, Signor Scacchi came in and inquired after you, and we had an animated talk over what 1 had seen of Vesuvius during this last visit. We started at half-past three o'clock instead of twelve, and in place of bringing us direct to Messina, as promised in the advertisements, the ' Etna,' an appropriate name for my steamer on this occasion, took us to Reggio, and stayed two hours discharging cargo there, and then returned here, losing thus four hours of our time. So the post for delivery of letters was closed at twelve o'clock, and I must wait to see if there is one from Dr. Carlo Gemellaro to-morrow. After all we got here in twenty-two hours. No wind, but a roll from an old storm which made three-fourths of a large number of passengers ill, but I escaped and was much amused by the conversation. All Sicilians or Neapolitans; some officers and their families returning from the procession to Our Lady of Pietra Grotta, i.e. the grotto of Posilipo, near which there is a chapel to which pilgrimages are made on September 8, and to which the royal devotee with his army go in state. There were also many priests and monks on their way back. A certain physician, young and lively, and known to many, was holding forth in our cabin, maintaining that the world of imagination and of ideas was the only thing real, true, and enduring, and all things material were fictions and a delusion, which of course he meant to provoke the others to combat. At last they came upon the ' Etna,' built last year in Glasgow, her machinery on quite a new construction, and having all her coal from Newcastle. They then appealed to Don Pietro, this learned ' medico,' to explain why there was so much coal in England, on which he went off in a more stately and exalted style than usual. ' The temperature of the English climate at a remote period in the history of our planet was ten degrees above that now prevailing in Sicily. There flourished at that time the vegetation which produced these dense strata of coal. Those plants were before the creation of animals, for it was necessary to establish plants first. Then came the herbivorous animals, afterwards the carnivorous, and lastly man was formed.' ' But, Don Pietro, you seem to require ages without end for all this.' ' Certainly, but it all admits of mathematical proof.' Some of the priests seemed intent on their breviaries, but perhaps they had one ear open to these new geological ideas, which have evidently been penetrating into those regions like Newcastle coal and Glasgow engines.

Messina, five o'clock P.M., *10th.*—I have been a two hours' excursion up the course of a fiumara which runs from the hills here, with a pharmacien, Sequenza, and found two tertiary formations, Miocene and Pliocene, between the gneiss and the sea,—a very instructive section.

Catania, September 12.—After leaving my box of fossil leaves and rocks of Vesuvius at the British consul's, I went on board the ' Diligente ' steamer at eight o'clock. The noise in the inn ' Victoria ' so great, that I mean to try a new inn, the 'Trinacria,' the rooms of which I inspected, kept by the former proprietor of the 'Victoria,' and near it. The 'Diligente' had come pretty full from Palermo, and the Catanians and Syracusans who had been at the great festa at Naples crowded into her. My commissioner had got me a good bed, but too late for a private cabin, as they were all secured at Palermo. The ladies' room crammed with women and children. The berths in the gentlemen's quite open, one over the other, not bad but public. After all were full, in came Il Barone and Baronessa, and two grown-up, lady-like daughters; no beds. One capitano after another got out and pressed the ladies to lie down on theirs. But they posted themselves in chairs with their backs to my berth, so I at least was a prisoner. Then came other ladies and gentlemen, all without beds. At last all the men gave up their berths to ladies they knew, and stood about, then got beds put on the great table and all over the floor. They contrived to give us plenty of air, and most fortunately, though the sea had been rolling for three days, the straits were as calm as a lake. In seven hours' sail we were in Catania, and not very long in getting ashore. The proprietor of this hotel came himself to help. I went to bed and learnt that no butter or milk could be had. Why not the milk of goats? 'Too late, and half-past seven o'clock; you can only buy between five and seven. It is too hot to have the goats later.'

I have been over the antiquities of Catania and examined some points in the lava of 1669 not seen last year, and had a talk with Gravina. The news came last night that Gemellaro's only daughter, aged thirty, the wife of a judge at Girgenti, was dead, leaving three children.

Calling at Tornabene's, I asked him if I should not find that Dr. Aradas could tell me all about the Gemellaros, and whether Gaetano would have to set out for Girgenti. He replied, 'I rather expect he will go, and you have no time to lose if you are determined to go to the Casa Inglese, and geologise the uppermost parts of the mountain. Go to-morrow; I will write to Dr. Giuseppe this evening.' So I am off.

Mr. Jeans has kindly lent me a beautiful hammock with which he has travelled all over Sicily, and a musquito net.

These creatures have been taking liberties with me again, after I had forgotten them since Leghorn. What a beautiful city this is, incomparably better built than Naples, something splendid in the style! One, Piazza Filippini, I had not seen,—all the columns from an old Basilica. The streets so wide, and the churches and private houses so often beautiful, and such a flood of light upon them. The elephant in the middle of the Piazza del Duomo is at this moment surrounded by a set of shrubs in full flower; a yellow one like a cytisus, a large red oleander, and others with white and blue flowers—very gay. I passed the Albergo dell' *Etna*, and that everlasting chime and bells of the church adjoining was making the usual stunning noise.

Dr. Giuseppe Gemellaro has spent 15*l*. since last year in repairing the Casa Inglese, so the 8*l*. I sent, with certain other subscriptions, will, I hope, cover this.

We have never seen the top of Etna since my arrival here, though the weather at Catania is splendid. Good night, dearest; just starting for Nicolosi.

Zafarana: September 15, 1858.—I am established here in my old quarters of nine months ago, and accompanied by Angelo, the guide of last year, who showed such resources in the Casa del Neve; which I find on Waltershausen's map is called Casa del Vescovo, and known here also by the same name. I yesterday drove to Nicolosi, seeing several things of interest on the road in the geological way. Found Dr. Giuseppe in affliction, especially as he was haunted with the idea that this blow might be the death of his brother. The necessity of making plans and preparations for my expedition with Waltershausen's map before me, and of getting provisions and guides, cheered him up considerably, and I left him less dejected and very busy preparing for my campaign in the Casa Inglese. To go there now, as I had hoped to do, is impossible, the weather is so unsettled; so I am to attack the middle heights of the Val del Bove first. The mornings are usually fine. I ascended the Monti Rossi yesterday; well worth doing, both for the crater and the scenery.

On my way here, a ride of four and a half hours, I found a new case of a stony lava on a slope of 10° and 15° and 29°, one of known date too, 1629. Also another lava with ridges

cut open, throwing some light on those of 1852. Just after we
got housed here, there came on a very heavy thunder shower
which may interfere with this evening's work and derange
plans. I find I can get on with talking with the guides
tolerably, and the map is an amazing help. I have three
mules, and a muleteer who goes on foot, making with the
guides three men. Dr. Giuseppe approved of my taking Mr.
Jeans' hammock for the Casa Inglese. He fears that the
property of the Duke of Ferdinandina in the Casa Inglese
has lapsed to those who have built and repaired it, being so
long unclaimed by the proprietor. I went to the valley of
Calanna and did some good work, but was driven back
by a soaking rain, but employed the afternoon well with the
parish priest, Sciuto, who let me copy his rough map of the
course of the lavas of 1852-3, and make extracts from the
journal he kept when the great eruption was going on for
seven or eight months.

Resolved that the weather was too unsettled for the upper
regions, and that it is expedient to descend to the sea-side
to-morrow.

Aci : Friday, 17.—Early in the morning Angelo, the
tall guide, came to tell me that before dawn they saw a
planet, quite a new one, with a long streak of light extending
from it in the east. So I suppose a fine comet has appeared.
The ride from Zafarana to Aci Reale was through endless
lavas. When I got here I re-examined the highly inclined
stony lava of the Scolazzi, all as I described it, but of
course I learnt many new facts. I then went in a boat to
see the basaltic columns of the Grotte delle Palombe. A
small Fingal's cave, in a current of lava, where it enters the
sea; a fine mass of scoriæ over the columns.

Nicolosi.—The weather has changed for the better, and
as we rode out of Aci Reale, the whole of Etna, summit and
all, appeared free from a cloud. The shape of the summit
much more divided than when you saw it. This sight made
me determine to wheel round and come here, to go up to
the Casa Inglese to-morrow. When the master of the
' Aurora' at Aci saw us return, he exclaimed, ' I thought the
gentleman would think twice before he left Aci the very day
of the Festa della Vergine ! ' In fact the band was playing

in the streets as we went out, priests were assembling, and crowds of peasants in their best attire coming in to swell a population amounting, my guides say, to 30,000. But we rode resolutely through to the woody region, not sorry to be under a milder sun, though I have really never suffered from heat. I found Dr. Giuseppe quite prepared for my change of plan. He had sent up the bed this morning.

Nicolosi : *September* 19.—The clouds again return, and I am going to Biancavilla and Bronte in place of the Casa Inglese. By choosing my time I shall succeed at last, but the first fortnight of this month was perfect, and that I should have had according to my original plan.

Ever your very affectionate husband,

CHARLES LYELL.

CHAPTER XXXIII.

JOURNAL.

To HIS WIFE.

Aderno: September 19, 1858.—My dearest Mary,—I must
have a chat with you before I go to bed after a long ride from
seven to five P.M ; only off the mule when examining the
basalts of Licodia and Biancavilla, often columnar and resting
on a sandstone the age of which I have not yet made out.
From Nicolosi to this place, through the woody region chiefly,
was a fine lesson in the aspect of lavas of every age from
one to eight centuries known, and for thousands of antecedent years of innumerable dates unsettled. I begin to have
an eye for the coming on a new lava, where the difference
would have escaped me one or two years ago. I had expected to sleep to-night in the Casa Inglese, but it has been
so enveloped in dense clouds all day that we did right in
desisting.

A procession is just passing, rather imposing ; first fifty
Franciscan monks two by two, each with a sort of Chinese
lantern lighted, and on a short pole, preceded by drums
beating in the usual Sicilian style of military devotion.
Then comes a gilt splendid carriage, in which a figure of the
Virgin, well done in the waxwork style and as large as life,
stands upright ; then a full military band playing, with
soldiers attending ; then an immense crowd of peasants.

There are also numerous banners preceding and at intervals, and fireworks and guns going off in the churches.

I think I shall soon be a good campaigner. Not liking tea and coffee without milk, and being tired of hearing of the impossibility of getting it, I ordered Giuseppe to get a clean empty bottle and a good cork. When we had been riding eight hours we came to a flock of goats, and I asked the goatherd for how much he would let me fill the bottle. He said one carline and a half—that the milk was good but not plentiful, and I should have to milk a good many of the flock. So Giuseppe set to work, and it took near half an hour (while I was geologising a cavern in the basalt) to fill the bottle with excellent milk, which I have been enjoying to-night, and shall to-morrow.

I never saw Etna look so grand as at four o'clock this afternoon. The cultivated region and the summit were in sunshine, and all the woody region concealed by a zone of fleecy clouds, above which it seemed marvellous to see a mountain. The sun was hot enough to make me hold an umbrella over my head, but this fatigue was relieved by an occasional cloud.

Nicolosi: 20.—Returned to this once more, that I may either go on with the Val del Bove or the summit, while the heat which is a little more than pleasant below, remains to cheer us above. I see clearly that after five months of very fine weather the time is come for Etna to be rarely fine for two whole days. But I have the great advantage of being here one month earlier than when I encountered the snow-storm last year. Dr. Giuseppe was glad to see me, and I hope my next date will be the Casa Inglese. The provisions are all prepared so far as I can see to them, but as I am to have daily communication with Nicolosi, I hope to repair omissions. As I have not of course tasted butter since I left Messina, my great point is to see to the milk. They take great care about wine, but cannot be brought to believe that coffee and tea require milk. Yet most of them think goats' milk better than that of cows.

Casa Inglese: September 21.—Got off with two guides and two muleteers and four mules at half-past seven, in bright sunshine, from Nicolosi, and after a beautiful sunny

ride of three hours through wooded craters, protected from
the heat by my umbrella, was gradually enveloped in clouds.
I saw a lava stream, where the oaks had been surrounded
by lava which had taken the form both of upright and prostrate trunks, surrounding them with tuff, and the wood
being burnt up they are now cylinders of scoriaceous lava.
After a couple of hours we got above the clouds, when about
8,000 feet high, but not till my hands were numbed, for I
could not believe for a long time in the necessity of my putting on a cloak. After reaching this place I set out with
Angelo for the top of Etna, leaving Giuseppe to cook. We
had now and then a small drifting cloud, but on the whole
splendid sunshine. I saw the spot at the foot of the great
cone where the Catanians, as I mentioned in 1830, quarried
ice from under a current of lava. My guide saw the same
done six years ago, while the eruption of 1852 was going
on in August and September ; the sand and lava ten feet
thick, and four feet of ice below (once snow), and bottom not
seen. When there is a scarcity of snow they send to this
reservoir. Not far above the ice I warmed my hands at a
fumerole, where the steam and some sulphuretted hydrogen
were given off at such a heat that I was obliged to be careful how far in I put my fingers. This welcome heat enabled
me to write. The bank of clouds below us was such as
aeronauts describe, finely lighted, some parts representing
even plains, others alpine heights one beyond another, some
nearer masses moving apparently much more rapidly than
others. When we reached the edge of the crater, the whole
of Sicily was hidden except the higher part of Etna between
us and the Montagnuoli. But Lipari and Stromboli stood
out in the sea very conspicuously.

I made a rough sketch of the two craters ; the smaller
one has lately, I believe, fallen, and shows a section of some
of the horizontal beds of lava with which it has been filled
up nearly to the top. It seemed to me about 200 yards
diameter and perhaps 150 yards deep, and the great crater,
separated by a very narrow partition, vertical on one side
and steeply sloping on the other, seemed three or four times
as wide and deep. It was a considerable exertion climbing
and going half round it, after a seven hours' ride, and this

makes the Casa Inglese, which is the roughest place I was ever in, seem a hospitable mansion, as it saves our returning. The wind is whistling round, and somewhat through it, but Dr. Giuseppe, I hear, has made it weather-tight. There is no chimney, and we have charcoal burners, but if the wind always blows like this, I am at any rate guaranteed from asphyxia. I was determined to write to you before trying Mr. Jeans' hammock. The two guides have an adjoining room. Our muleteer has taken back the mules, and I have offered him a reward if he brings me up milk to-morrow, which the others promised to obtain on the way, and failed.

September 22.—Two experiments at once—this house, and a hammock to sleep in for the first time—was too much, for this house is the roughest I was ever in, since we cannot let in the light by the one window, which is without glass, without admitting a blast which in a minute annihilates an hour's warming of the charcoal apparatus.

The well-swung hammock pitched like a cutter in a short sea when I got in and out, and rolled like our old friend the ' Britannia ' steamer when I was restless, so that 1 lay awake wondering whether I should be sea-sick, withal a smell of an occasional whiff of sulphur from the highest cone. Having come down from that cone hungry, I had eaten too heartily just before going to hammock, and this was perhaps the chief reason of a sleepless night, just as I have got to where the cold protects us from musquitos. I was glad when daybreak was announced, although accompanied by Giuseppe's anticipation of snow. The sun, however, conquered, and I got a fair day's work, breaking ground on those enormous precipices below the flattened dome-shaped area, on which the principal cone with its two craters rises, and on which also this wretched remnant of the earthquake-ruined edifice which the English army built, stands.

There were clouds below all day, but we were generally in sunshine.

To-night I have been taking various measures against cold, for having tried with the guides to plan all manner of ways of exploring certain parts of Etna, without coming and staying here two or three days, we find it an unavoidable evil.

Everything one touches is like ice just now. The floor so uneven that a touch causes a chair to fall, and it is strewed with sand like the lavas which surround the volcano.

September 23.—Before I was up, about dawn, the arrival of five English officers of Captain Clifford's ship the 'Centaur' was announced. They had encountered rain in the night, and though they went to the summit, saw nothing but the rim of the crater. I was much pleased with their chief, the second Lieutenant de Kantzow, of Swedish origin he told me, at least the name,—well read in my works, and had tried hard, but in vain, on several occasions, to get near the lava of Vesuvius during the late eruption. The heat was too great. He thought me singularly lucky in what I saw, and I must remember this good fortune when I am disposed to abuse the quarantine regulations for causing me to have this struggle with the elements up here. However, I had a much better night, managing by dint of clothing to warn off the great enemy cold.

The compact lava I found yesterday is one great reward for coming up here, as I can so easily refer to the place. It is perhaps my best case, and the Royal Society will let me add it with the new date.

To-day I have explored the Montagnuoli, and feel sure it is a simple case of a double cone, the newer one thrown up in the common way. How Abich could think otherwise I wonder much. To-morrow I hope to descend into the Val del Bove. The mists to-day have interfered more than any previous one with my geologising.

Casa Inglese : *September* 24.—Giuseppe, the son of the inn-keeper of Nicolosi, was amusingly put out by the roughness of the quarters here, where he had never slept before, and formed all sorts of schemes for enabling me to explore the precipices of the upper region of the Val del Bove without braving the rudeness and coldness of things up here. I think the guides' room the warmest and lightest of the two, and have therefore been sitting there nearly half the day with Angelo, who is alone here. For I sent Giuseppe round by Nicolosi to Zafarana, meaning him to bring the mules to the foot of the great precipice, weather permitting, and so to go on to my old quarters at Zafarana.

But we had rain and snow whitening the ground, and ever since a regular Scotch mist, and instead of the grand scenery of this spot, you would fancy yourself in shooting quarters on some Highland moor. I have at any rate had a day of rest from fatigue, whether of body or mind. Got a good read of Hoffmann, and however dry he and his editor Von Decken may be, the book interests one on the spot, as the rocks he describes in detail are what I have been collecting, and the places so many of them what I have seen or have to see. But the most useful part of the day's business has been the planning each day's excursion with mules till the day I must leave for Messina, availing myself of Angelo's local knowledge of distances, inns, and places and sights, and the power of mules. It is a most difficult country, Etna,— especially the lavas so often interposing a barrier to mules, and an immense delay to men on foot. I have written all out, and allowed for some rainy days, and three or four in Catania. Angelo, a hardy peasant, is only diverted with Giuseppe's dismay, and it is amusing how little the education of the one makes up for the superior mother-wit of the other. In most of my future excursions I mean to dispense with Giuseppe and have a muleteer under Angelo, who gives me most information when alone with me. Lieutenant de Kantzow, to whom I had presented some milk for his tea, presented me with half a bottle of good brandy. It has done me good by spoonfuls, as the wine here is too acid for me. By putting on all my clothes double I have kept warm, and have often thought of Samuel Rogers' maxim, that a man who suffers from cold must be a fool or a beggar. If it were some tropical volcano, how much more difficult to contend against the enemy!

26.—Yesterday I made the grand descent of the Balzo di Trifoglietto, 4,000 feet, and really without much difficulty, and in about three hours, spending a great deal of that time in observing the interesting fact, that in the middle of this cliff, the head of the Val del Bove, and just below the great cone, the strata are perfectly horizontal—a fact first made out by Waltershausen and I think first explained by me. I had made up my mind that possibly Giuseppe, from want of courage or from taking a different view of the state of the

weather, concluding I should not descend, might not bring
the mules to the foot of the precipice. It turned out that
he had come for me the day before, in spite of the bad
weather, and had returned yesterday, but lost his way in the
mist. So we found no one; but after walking an hour and
a half, I told Angelo to go to Zafarana, get the mules, and
return to my rescue. There was time for all this before daylight was over. Angelo proposed to give one more loud
halloo, which we gave, and were answered by splendid echoes
from the surrounding vertical walls of the grand amphitheatre. But Angelo was positive that there was more than
an echo. I could not believe it, though he turned out to be
in the right, and after nearly an hour the lost Giuseppe
found us, and I got to Zafarana between three and four,
just 7,500 feet below my station of the morning—a most
agreeable change of climate, bed, and all things. To-day I
have come to Giarre, having re-examined on my way the
Cava Grande case, of steeply inclined lava.

I have had a long talk with Dr. Mercurio. He tells me
there are *two* comets, but it is too cloudy to see them to-
night. I was glad to be free to-day from that chattering
Giuseppe. Angelo and Delfo the muleteer are now my
companions.

Aci Reale: September 27.—I have satisfied myself that
Don Carlo Gemellaro is right, and that the basaltic columns
of the Grotte delle Palombe belong to a lava, which instead of
being the oldest, is newer than all the currents of Aci Reale.
I found a cliff 600 feet high or more, inclined at an angle of
about 50°, and cascading over it reduced the slope to 23°.
So that Waltershausen is wrong in his view, suggested, I
suspect, by the old superstition for which Von Buch has in
part to answer, that columnar basalt is an *older* production
than ordinary lava.

Before leaving Giarre, I revisited the bed of the torrent
where I found last year a current of columnar basalt
reposing on the old alluvium of Giarre, and shall be able
to make the diagram given in my paper more exact and
instructive.

Catania.—Cal'ed on the British consul, and had the
happiness of receiving two of your letters. . . .

Casa Inglese, Etna: October 1, 1858.—Here I am once more in these elevated quarters, about 10,000 feet above the sea, having fine weather, and having flattered myself that I shall get off with one night, making to-morrow the descent of the Montagnuoli, as I have already done the loftier but not equally precipitous height of the Giannicola. Yesterday morning I had appointed Gravina between seven and eight o'clock to start with me. The news came that Angelo, having out of his punctilious honesty preserved the fag end of a ham used in the last excursion, had been stopped by the police at the entrance of Catania (coming from Nicolosi), with the three mules. So I had to get Gravina to apply to the chief of the police, and while he was so employed got my breakfast and saved my character for punctuality.

I had a good day's geology at Fasano, which you may remember on the road to Nicolosi, St. Paul, and Catera in that neighbourhood, Gaetano Gemellaro meeting us, and he and Gravina wanting my opinion on a geological point on which they had been contending. I forgot to mention that the day before Aradas had given me a list of 150 shells of that Newer Pliocene clay on which Etna rests. Nine-tenths of them agree with Mediterranean species; but as he means to lend me the whole collection, which I hope to get Deshayes to overhaul, it will be a valuable addition to my paper, as giving a precise paleontological date to the volcano, the same in truth as I gave in 1828, which is satisfactory. In truth, the first twenty-five species of shells would generally give the result which 150 afterwards confirm.

Zafarana.—I came down the heights under the Montagnuoli, about 3,000 feet, without fatigue seeing the rocks on the way. It would take weeks to study such countless dikes and lavas, but I read off not a few of the principal features.

We were met at the foot of the hill as near the appointed time as could be expected, and I now feel that the most difficult part of the task is accomplished, as I have verified the fact of the inclination of the lower, the horizontality of the middle, and the unconformable southern dip of the uppermost part of the great 3,000 and 4,000 feet high precipice.

October 3.—I have accomplished a ride I always wished to make, and which Waltershausen, not having knowledge or faith enough in mules, believed I should never make out. I got from here almost without stirring from my saddle, up the external valley, one outside of the Val del Bove, called Cava Secen, and so ascended gradually till 1 reached the narrow rim or knife-edge which divides the Val del Bove from the outer slope. I wanted to see among other points whether the junction of this deep valley (of Tripodo) with the rim of the Caldera or Val del Bove was marked by a depression or notch, or what the Americans would call in reference to the Alleghanies ' a wind-gap,' for you probably remember there is not a single ' water-gap ' like the Barranco de las Angustias in Palma, allowing a river to pass out of the Val del Bove. It is strange that no one in Catania could help me as to this point, and yet there is not a more marked feature in the geography of Etna than the deep indentation, narrow, and I should guess 800 feet deep, if not much more, which marks the situation and the unbroken rim of the water-shed between this Val di Tripodo and the Val del Bove, or a narrow ravine which descends inside into the great Caldera. My guide, perhaps the most experienced on Etna, never was there before. It is a most picturesque spot. On the one side is a deep Alpine style of valley, the rocks covered with old beech trees, on the other a look into the great gulf or cauldron-shaped amphitheatre, with Etna steaming away and the great range of precipices below it, and on the left, between the point of view and the Montagnuole, a series of five or six vast promontories one behind the other, about 3,000 feet high each, without a shrub, but pinnacled with dikes, some standing up fifty feet or more. The whole scene of the valley was shown in sunshine for half an hour, and the crater of Etna; then the clouds gathered rapidly, and thunder soon began, but we rode back in sunshine, soon escaping from the mists.

Nothing amused me more than going to see what such a torrent as that of Tripodo did, when a very powerful lava current, such as that of 1792, has blocked up its mouth and drunk up all its waters. The lava cannot absorb its sand boulders and mud. So the torrent has been working away for eighty years, and has already filled up a lake-like cavity, and redeemed the space of a large field from the sterile domain of the lava, and when it once gets its alluvium high enough to overtop the lava, the side of which has been probably thirty feet high, it will get on fast with its fertilising soil, leaving occasional black ridges projecting above.

Zafarana: October 4.—I got off by six o'clock this morning, Etna and the whole region without a cloud, and secured six hours' good survey of the centre of the valley, and the craters and lava of 1852, having now satisfied myself that Gaetano Gemellaro was quite out about the rock he still will call Musarna, and his uncle Giuseppe very wide indeed of the mark, in laying down his three points of eruption, and in some respects the direction of the lava.

The views this morning were very fine. Just at twelve o'clock the first white speck of cloud appeared on Mont Zoccolaro, and it soon began to roll into the great valley, and filled it with mist in an hour. We rode down out of it and reached this soon after two o'clock in sunshine. I have now done what I intended of the higher and difficult region, and am now going round the base to see all the tertiary and secondary strata which peep out from below the ever encroaching envelope of lava. I hope by this means to be able to speculate on the nature of the pedestal of Pindar's 'pillar of heaven.' I must take care of the mules now they are in the plains. I feel here, as in Madeira, that a good mule is like presenting an old geologist with a young pair of legs. They would mount up the side of any English stone quarry where a man can get, and you need not be merely taken to the quarry's mouth.

The comet is most splendid on a moonless clear night in the north-west. I hardly think that that of 1811 was brighter, but the tail was much longer.

Randazzo: October 6, 1858.—After leaving Giarre, I wanted to see a small outcrop of tertiary between Piedemonte and the sea, and not having found Dr. Mercurio was looking out for an informant, when I saw by the road-side an architect whom I employed last year to copy a map of the rivers coming from the Val del Bove which he had been employed to make by Waltershausen. He was one of the few men in Sicily who could exactly tell me where and how to go

to the point with my mules. He was amused to see Waltershausen's map. His name is Vincenzo. I spent a pleasant
hour on a hill-side collecting old friends, *Dentalium elephan-
tinum, Venus dysera, Pecten jacobœus, Limopsis, Pecten varius,*
and many others, the guides and muleteer also helping very
well. We then went to Linguagrossa, bad quarters, but
luckily Don Gallo of Giarre had set me up with provisions.
At such places (and it is the same here) they are thankful if
you want nothing but a bed. If you insist on the whole room
you pay for *two* beds, only two carlines each ! The cone of
Mojo which I passed to-day, the most northern dependency of
Etna, reminded me of Tartaret by its situation in the middle
of the river's ' thalweg,' and I find many remembrances of
Auvergne now that I am on the borders of the volcanic and
Eocene formations.

I passed a good deal of heath (*Calluna vulgaris*) to-day
on the lavas, and abundance of *Cyclamen Europœus*; also a
short-stalked purple *mallow* (I think), very handsome ; the
yellow crocus large ; the *Colchicum autumnale.* A good deal
of *Euphorbia,* like English, I think, but larger, and a second
species ; and the ' weebo ' of Forfarshire, and ' traveller's joy,'
and hawkweeds and others, all almost entirely common
British weeds, such as *Euphrasia,* &c.

Bronte : October 7.— Have had another fine day for sur-
veying the northern and north-western side of the volcano.
Saw the place where a crowd assembled in 1842 to see the
lava flow into a great artificial reservoir of water. The
melted current of stone came forward with a front thirty or
more feet high, and falling suddenly into the water produced for awhile no effect whatever, as if, like white, hot
metal in Butigny's experiment, it required to cool down
before it could cause explosion. At length it went off
suddenly, and everybody but one or two, fifty or more in
number, were killed, asphyxia it is said, for no wound was
found on anybody. Had it not been for some heavy rain
many hundreds would have been there.

Catania.—I made out my journey well yesterday evening, and cannot sufficiently congratulate myself that I
should have been so many days for eight hours a day on
mule-back on such roads and crags, and without the smallest

accident. The chief difficulty is the food, but I have learnt by degrees how to manage.

I feel most thankful for the privilege I have had in going over and checking my own and Waltershausen's observations previously to printing my paper, which will gain greatly in that kind of minute accuracy which is desirable when one's facts bear on controverted questions, with the disputing of which the *amour propre* and reputation of living authorities is most sensitively concerned, and I have had the pleasure of being greatly strengthened in my convictions.   Your very affectionate husband,

CHARLES LYELL.

CORRESPONDENCE.

To MRS. BUNBURY.

Casa Inglese, Etna : October 1, 1858.

My dearest Frances,—The cold in this habitation, about 10,000 feet above the sea-level, is uncomfortable, but sleeping here, which I am doing for the fourth time, having returned to it a second time, is the only way of examining great precipices which are a key to the structure of Etna.   I have found striking confirmatory examples of the fact that lavas form sheets of excessively compact rock on slopes of 35°, and they can form continuous stony masses at an angle of 40° without vesicles ! It is strange that this should have been denied on such scanty grounds.

I should not think that Etna would afford a rich botanising ground in comparison with its size and height.   The eruptions destroy plants, and the single ones which can live on loose sand, or can stand the barren lavas, such as broom and *Hypericum*, and some others, monopolise large tracts. The *Cyclamen Europæus* is abundant, and like the *Crocus* (or *Colchicum*) *autumnalis*, very ornamental, and the yellow crocus, but that is rare.   Also another purple flower which I must learn the name of—not English, I feel sure.   Heer, when he decided that one of my fossil plants from the tuff was *Pistaccia lentiscus*, said it was not in the floras of Etna which he had seen in print, but I saw it yesterday so abundant that I cannot suppose it had all strayed from gardens.

There are but few travellers here this year, so I am sorry two are coming up to sleep here to-night; but they can have a separate room. I have not yet seen the comet, which you seem to have a good view of, but my guide announced it ten days ago 'as a new planet, one he had never observed before, with a streak of light proceeding from it,'—not a bad description for a most illiterate peasant.

There is a great pleasure in reiterating one's observations on one limited area like Etna, and considering the volcano in all its bearings and its age as proved paleontologically, its structure, the relative date of its valleys, and so on.

In consequence of the death of Carlo Gemellaro's only daughter, I have got little help from him and his son this time, but others have aided, and I feel very independent with my guide of last year, who knows enough Italian to get on pretty well with me. My correspondence and talk has been so much in Italian, that had the Sicilian spoken the real language I should have made great progress. To hear my guides talk, you would suppose that the staple of Sicilian was 'da bano,' which I am told means 'dall' altra parte,' but if it meant nothing more, it could not by any ingenuity be introduced so often. When they talk together I hardly catch a word, yet their newspapers are like those of Tuscany or Rome as to diction.

The clouds have rather interfered with some of my work in the higher region, but all Etna has been clear to-day, and since I began this campaign I have only lost one day by rain, and that was when last up here on the summit. And yet when I came they had had five months of fine weather.

Zafarana.—I have descended safely into the region of vines from 10,000 to 1,800 feet above the sea, and have studied another part of the great precipice on a fine day, and found *all right*; but I must confess that the lower third out of 3,000 feet is such a wonderful instance that dikes have intruded in such numbers and of such thickness as to make the beds they invade seem to play a subordinate part, that other theorists may be expected to find facts to suit their doctrines.

Zafarana: October 3.—1 have told Mary of a ride which
perhaps delighted me the more as it was one never recommended to me, and which I cannot find out from the guides
anyone has been at, though of course Waltershausen got
to the point when he made his map. It seemed to me the
most beautiful scenery I had yet seen, the dividing ridge as
sharp as the crater of Etna itself, between the great Val del
Bove, which you look down into in its grandest part, and the
Val di Tripodo, the only considerable valley which furrows
the side of Etna, and which is well wooded, and contrasts
finely with the savage scenery of the Val del Bove, now
almost without a spot of green. I won't say it beats the
north of Madeira, but no one has seen Etna who has not
been there.

October 4.—I have had another fine ride into the centre of
the Val del Bove, to study the effects of the grand eruption
of 1852. The two craters after rain still send out such
volumes of steam that they look as if they were beginning
a new eruption, but without such fuel they scarce give any
signs of life.

I must conclude this with my love to your husband, who
will, I hope, tell me by the end of the month what he has
been doing in his study in the way of fossil botany since I
saw him. I little thought when I wrote to Sir Henry Bunbury about the Casa Inglese, that I should ever see it again
or sleep in it. The lawyers here say that they who built it,
and have repaired it, have the property as squatters, and
neither the Prince of Palermo nor his assignees have any
objection. But the singularity of the case is that no one
can say who it is that has squatted there.

The Gemellaros say the English army, others say the
Gemellaro family, and so stands the case. Dr. Giuseppe, who
has an income so small it would astonish an English apothe-
cary, has spent 15*l.* this year in repairs, and more than half
out of the subscription I sent him.

Believe me ever yours affectionately,

CHARLES LYELL.

To His Sister.

Messina, Sicily: October 16, 1858.

My dearest Sophy,—No excursion could have been more
successful in a geological point of view than my re-examina-
tion of Mount Etna, and the strong and sure-footed mules of
that country gave me the power of scaling such heights as
you climbed so successfully in Switzerland by means of the
same power. The beautiful climate of autumn in this island
makes up for much of the drawbacks in the way of bad and
dirty inns, or rooms with beds where they do not profess to
have inns. But nothing but a considerable scientific excitement could carry one through. I was often afraid I should
not keep well where the diet was so strange. They cannot
understand a man caring for milk and butter, but here at
Messina I see these rarities again. Night after night I
stopped with my guides and muleteer at large towns, with
grand churches and priests and friars without end, and processions and tolling of church bells, and monasteries, as at
Linguagrossa, Randazzo, Piedemonte, Bronte, where there
was not a single inn. You find at last a dirty hole where
they show you one or two rooms, in which there are two or
more beds, but nothing else, and before the white sheets are
brought, they look like pieces of furniture less inviting to
lie down upon than those lavas of Etna which are liberally
sprinkled with volcanic sand. You then bargain that no one is
to pass through your room and that no one is to sleep in any
other bed in it. You accordingly pay for the other beds;
a mere trifle. You promise something extra for water and
a towel, and send out your guide to try and buy something
for food—bread and eggs at any rate, and you may have
brought with you some tea, a teapot, salt, sugar, and
Bologna sausage. Sometimes you have a landlady who,
having never seen an Englishman before, can scarcely be
kept by lock or bolt from prying into the room and watching all your proceedings, to learn what you can want with
basin, towels, and other articles, never I suppose asked for
before. But the worst trial is the dirt of your own servants,
guides, and muleteers, who during the ten days they are with
you never wash or change raiment, if indeed they ever did

before or after, but if you have fine weather and scenery in
the daytime, and are laying in a store of scientific facts, and
tired enough at night to be sure to sleep, and hungry enough
to be glad of bread and eggs, and above all, if you have no
one to blame but yourself when you feel you are in a scrape
so far as fare and comfort are concerned, it is extraordinary how one comes through it all with health and
spirits. I must say it often did me good to think how much
worse our officers fared in the Crimea, and under an Indian
sun, to say nothing of being shot at to boot. I usually
held my umbrella over my head five or six hours of the day,
except when in the Casa Inglese, where we had snow and a
difficult fight with the cold. I slept there four nights, and
was glad when they were over, but they repaid me amply.

I made some pleasant excursions from Catania with the
geologists of that city, and am going to-day with a M.
Sequanza of this place to explore a valley near the town,
where we shall find a great many fossils, and have splendid
views of Messina, the Straits, and deep blue sea, and opposite coast of Calabria.

Believe me, my dearest Sophy, your affectionate brother,
CHARLES LYELL.

CHAPTER XXXIV.

FEBRUARY 1859–NOVEMBER 1860.

AGASSIZ MUSEUM—OWEN ON THE 'GORILLA'—MR. PRESCOTT'S DEATH—LORD
LYNDHURST—MANSEL'S 'BAMPTON LECTURES'—BRITISH ASSOCIATION,
ABERDEEN—DARWIN'S 'ORIGIN OF SPECIES'—BIRTHDAY PARTY—DEATH
OF MACAULAY—DAWSON'S 'ARCHAIA' ON EVOLUTION—FLINT IMPLE-
MENTS IN THE VALLEY OF THE SOMME — DEATH OF CHEVALIER
BUNSEN.

[He visited Holland, Paris, and Le Puy in the course of 1859, and
in September presided over the Geological Section at the British
Association, Aberdeen.

In 1860 he was in Germany, visiting Rudolstadt and Coburg, and
in the autumn went to the Oxford British Association. In these
years he turned his attention especially to the antiquity of man.]

CORRESPONDENCE.

To GEORGE TICKNOR, ESQ.

53 Harley Street, London: February 16, 1859.

My dear Ticknor,—I was glad to learn from Prescott's
last letter, that Agassiz was stirring about a museum, and
that at a meeting at James Lawrence's, the Governor had
promised aid from the State. With the splendid bequest
left by Mr. Grey, I hope something grand will come of this.
There is no getting on without such collections. Sir Philip
Egerton told me the other day, that he must go to Vienna
to compare some fossil fish with the recent collection, so
much more complete there than in London or Paris. I
expect we shall have to go to Boston instead, if once Agassiz
has free scope in that particular branch.

I saw Mr. Harness (the Rev.) just now, who was remarking to me that although Milman is very well, he is

much more bent, and therefore older looking than ever, but he is in good spirits. These evening services in St. Paul's have given him much to arrange; rather fatiguing, and some preachers inflict more than an hour.

The Italians here have been annoyed at Mazzini for having expressed no sympathy at the alliance of Sardinia and France, and the promised aid, and they said to him that if they could get help from the devil himself, they ought to jump at it. His answer was, ' You will have the devil and the Austrians too.' My friend Dr. Falconer obtained on his way through Lyons a fine cast of a species of fossil Rhinoceros, which we have in the valley of the Thames also. At Naples the police voted it to be an infernal machine, and took it away. It was five weeks before (after much trouble) it was restored, but some MS. on osteological subjects in Falconer's not very legible handwriting, was submitted to certain priests, who still think it treason. I fear that when he gets to Palermo he will find the prisons full of newly arrested men of family. Affairs there are much as in 1847–48.

Charles Murray is returned from Persia, and we are to meet him at dinner at Horner's on Monday.

My three nephews and niece are quite well and in spirits, after their summer's illness. Susan Horner's translation of Colletta[1] reads well, and has come out at an opportune moment, when so many are thinking of Italy.

Owen has been reading a paper to prove how much nearer the Gorilla comes to man, osteologically, than does the Chimpanzee, but he seems not to have more, if so much intelligence as the elephant; so whether the Lamarckians will profit much by the fact I cannot foresee, though they may do so if the Negro is more like the Gorilla than the white man. I am rather suprised at the popularity of the doctrine of a chain of beings leading up to man. Agassiz is very fond of it, and strains a point for it, as do some others, while they protest against Lamarck's transmutation, but they are I suspect drifting towards the same goal without knowing it.

Ever affectionately yours,

CHARLES LYELL.

[1] *History of the Kingdom of Naples,* by General Colletta. Translated by S. Horner.

To GEORGE TICKNOR, ESQ.

53 Harley Street, London : March 11, 1859.

My dear Ticknor,—Mary is writing I find by to-day's post on a subject [2] which has been so much occupying our thoughts, as it has yours, and on which so much has already been well said and felt, on both sides of the Atlantic. Mr. Prescott's correspondence with Mary was so frequent, that we cannot see a mail come in from America without sorrow, to think that there can be no letter from him. I am sure there is no one of his friends to whom his death will be so great and permanent a loss as to you. The universality of the expression of grief and regret has not surprised me, but I have felt some consolation at the unusual amount of ability and eloquence, and upon the whole good taste, of the various speeches and biographical sketches extemporised on this occasion, especially in the United States.

No impartial reader can peruse them without feeling that from such a soil and in such an atmosphere great literary men must continue to spring up. In spite of the warning given by Prescott's attack a year ago, I was quite unprepared for his being taken away from us so soon. A letter which he wrote to me five days before his death was in his usual cheerful and sprightly tone.

I have been amused at meeting and conversing with two notabilities since I last wrote to you, whose minds preserve remarkable vigour and liveliness at a great age, Lord Lyndhurst eighty-seven and Brougham eighty. The first I met at a dinner at the Palace. I had not seen him to talk with him for thirty years, when I passed several days in Trinity Hall, Cambridge, where Copley, the Attorney-General, was staying. He alluded to those days as 'jolly times,' and then questioned me as to the latest geological discoveries respecting the antiquity of the human race. I told him among other things, of Horner's paper on ' Egypt,' proving that pottery was made there 13,000 years ago, and told him I would lend him the paper. On receiving it a few days later, he wrote to me to say that he had found out that Bunsen in some

---

[2] The death of W. H. Prescott, the historian.

paper just published had arrived at analogous results from different data and by another process; a discovery which amused me, for in that short interval he (Lord Lyndhurst) had been occupied in addressing the House of Lords on the removal of the Royal Academy to Burlington House, giving a sketch of their history, legal rights, &c. But though his mind is clear and lively, he can only get upstairs, or even go from the dining to the drawing-room, with the help of an attendant. It was pleasing to see the Queen paying the attention due to his years, and introducing the Princess Alice to him.

Brougham is upright and strong on his legs, and amused me by almost confessing his sympathy with a new pamphlet on the ' Reform Bill,' by John Austin, once as great a Radical as ever Harry Brougham was, and now as much of an alarmist as any Tory can be. We met him at Milman's. Lord Lansdowne was to have been there, but had to go to the Duchess of Cambridge instead. He is getting a little deafer, otherwise very well on the whole. We dine with him to-day. Lord Stanhope and the Bishop of London were at Milman's, and Lord Shelburne, the day we met Brougham, and they discussed the new Reform Bill very freely. The Bishop says that the artisans are so much more intelligent than the small tradesmen, that those of the former class who have any property would make a far better constituency than the tradesmen, to whom we are in danger of transferring so much power. I fully believe this, and think the scheme of votes for those who have 60*l.* in saving banks a good idea.

Have you looked at Mansel's ' Bampton Lectures ' on the ' Limits of Religious Thought '? There were many fruitless discussions among the dons of Oxford, how to force the young men by various pains and penalties to attend the University Church, which was nearly empty, but there were no precedents for such proceedings. At last some original thinker suggested, that possibly if they named some good preacher it might remedy the evil. So they made inquiries for some young men of ability, and found this Mansel, who forthwith filled St. Mary's to overflowing, and when the lectures were printed they soon reached a second edition. A friend of mine, Huxley, who will soon take rank as one of the first

naturalists we have ever produced, begged me to read these
sermons as first rate, ' although, regarding the author as a
Churchman, you will probably compare him, as I did, to the
drunken fellow in Hogarth's Contested Election, who is sawing
through the signpost of the other party's public-house, forgetting that he is sitting at the outer end of it. But read
them as a piece of clear and unanswerable reasoning.' Soon
after I had seen them, I was recommended by Sir Edward
Ryan to read a powerful article in the last 'National,' in
answer to Mansel, by Martineau ; and certainly it is worth
reading, and shows among other things, in an episode devoted
to Butler's ' Analogy,' how much more comfortable and con-
solatory is the system of creation, or the divine dispensation,
when viewed from a Unitarian than from an orthodox point
of view. At length, after expending much admiration and
adulation on their new defender of the faith, the Oxonians
have become alarmed, and Milman told me that one of them
had written to Hampden ' You are avenged ; ' while Dr. Jeune
had exclaimed, ' To think that I should have lived to hear
Atheism preached from the University pulpit, and the
member for Oxford recommend the worship of Jupiter ! '
    You will understand, I daresay, the last hit better than me,
for I have not read Gladstone's Homeric lucubrations.
    I must conclude, as I have to go to Owen's lecture on
' Fish.' He assured us yesterday that every act of thought,
even in a reverie, is so far material that we are the worse for
it, and require reparation through the digestive organs.
There is something *burnt*, or a loss of oxygen, by every, even
the slightest effort, of the brain in thinking.
                    Ever affectionately yours,
                            CHARLES LYELL.

                To MRS. BUNBURY.

                            Leyden : April 22, 1859.
    My dearest Frances,—We have just returned from an
excursion to the sand dunes of Catwyk, at the mouth of
the Rhine, collecting shells, star-fishes, spatangi, seaweeds,
and other treasures on the sands over which the waves were
rolling on a very bright day. We then visited the sluices by
which the water of the Rhine is let off at low tide, so as to

drain the inland country, and sever out a passage through the dunes which used to intercept the water of the river before Napoleon I. made these sluices; having cut a canal through the dunes. On the sand hills we gathered abundance of wild pansy, a small *Myosotis, Stellaria,* groundsel, and a few other plants in flower. We saw in the gardens and suburbs of several villages on our way large patches of brilliant tulips, of all colours, arranged in groups.

In the Museum here we saw parts of a crab called *Inachus macrocheirus,* from Japan, which they calculate is eight feet long when entire. I had thought that the Devonian crustacea, *Pterygotus,* &c. which they say were seven feet long, exceeded any of the creatures of these degenerate days, but one never sees a rich collection like this without being convinced how little one knows of the present inhabitants of the globe.

I rather think it is thirty-six years since I was last in Holland. I am sure the Hague must have doubled at least in size and population. It is a very handsome city now.

Ever affectionately yours,

CHARLES LYELL.

To MRS. HORNER.

Aberdeen :[3] September 15, 1859.

Dearest Mamma,—As you asked me to write I will send a short letter, and send it to-day unless interrupted in this busy time.

Dr. Fowler, at ninety-four, looks well enough, but having eaten turtle soup and melon too close to the rind, and other imprudences, is not quite well to-day. I went in the Provost's carriage to dine at Banchory, to meet the Prince, Sir Roderick Murchison, General and Mrs. Sabine, Sir David and Lady Brewster, Lord Rosse, Duke of Richmond, and several local managers of the Aberdeen meeting. I sat next General Sir Charles Grey, whom I always find agreeable.

At the evening meeting I was on the platform next Lord Ashburton, who introduced himself, as he says he does to all his friends (since his return from the East with a beard), as not a soul knows him. He is quite changed by it. The Prince's speech was well delivered. We shall send it

[3] At the British Association.

to you when we get it, perhaps to-day. In consequence of
General Grey talking with me about the flint hatchets of
Amiens, he for one determined to hear my opening address.
I had thought it best not to ask the Prince to be there, lest it
should interfere with other arrangements, though I had
two talks with him at Banchory. But next morning
Sabine came with a message that H.R.H. the President
requested me to defer my opening speech till he could
attend at twelve o'clock, so I got on Prof. Nichol first, on
the ' Geology of Aberdeenshire.' When the Prince came,
the room, which had been gradually filling, must have
contained about 800. The Prince, the evening before, had
2,200, and many ladies had been refused tickets.

Young Geikie[4] has read the best paper to my mind yet
presented to our section, on the 'Age of the various Trap Rocks
of Scotland.' He finished by endeavouring to prove the
top of Arthur's Seat to be tertiary ! Of the young men he is
certainly the coming geologist and writer. I am glad Horner
likes his book. I expect he will one day be a leader in the
Ordnance Survey.

My address, which I hope to send you to-morrow in the
newspaper, will be a full answer to Horner's very natural
suggestion about the Denise fossil man. We have talked
with Symonds of Pendock (who was escorting a daughter
of the late Hugh Miller,) with Sir Richard Griffith, Profs.
Ramsay, Huxley, Harkness, Lord Monteagle, Mr. Vernon
Harcourt, Prof. Phillips, Sir Philip Egerton, Sir James
Clark, Dr. Allen Thomson, Dr. Gould, Sir William Jardine,
Prof. Owen, Mr. Hopkins, &c.

Sedgwick was to have dined at Banchory, and slept there,
but wrote on the day to say he could not come, having a cold,
but hoped to be here on Saturday.

Sir James Clark wanted Mary to go to a ball at Balmoral
on Monday, but she will stay with me, and perhaps go with
the Clarks to the luncheon, to which I shall probably go at
Balmoral on Thursday the 22nd.

With love to all, ever affectionately yours,

CHARLES LYELL.

[4] Professor Archibald Geikie, Director of the Geological Survey of Scotland.

[After the meeting of the British Association at Aberdeen, Sir Charles Lyell asked Mr. Symonds to accompany him to Elgin to investigate the history of the *Elgin Sandstones*, which contained the fossil remains of the supposed Old Red reptile, Telerpeton. The late Sir William Jardine, Professor Harkness, and Lord Enniskillen also went with them. After many days' examination, Mr. Symonds came to the determination that the Telerpeton beds were New Red—not Old Red—and it turned out that he was right eventually.]

To Charles Darwin, Esq.

Drumkilbo, Meigle, Perthshire : October 3, 1859.

My dear Darwin,—I have just finished your volume,[5] and right glad I am that I did my best with Hooker to persuade you to publish it without waiting for a time which probably could never have arrived, though you lived to the age of a hundred, when you had prepared all your facts on which you ground so many grand generalisations.

It is a splendid case of close reasoning and long sustained argument throughout so many pages, the condensation immense, too great perhaps for the uninitiated, but an effective and important preliminary statement, which will admit, even before your detailed proofs appear, of some occasional useful exemplifications, such as your pigeons and cirripedes, of which you make such excellent use.

I mean that when, as I fully expect, a new edition is soon called for, you may here and there insert an actual case, to relieve the vast number of abstract propositions. So far as I am concerned, I am so well prepared to take your statements of facts for granted, that I do not think the *pièces justificatives* when published will make much difference, and I have long seen most clearly that if any concession is made, all that you claim in your concluding pages will follow.

It is this which has made me so long hesitate, always feeling that the case of Man and his Races and of other animals, and that of plants, is one and the same, and that if a *vera causa* be admitted for one instant, of a purely unknown and imaginary one, such as the word 'creation,' all the consequences must follow.

[5] On the *Origin of Species*.

I fear I have not time to-day, as I am just leaving this place, to indulge in a variety of comments, and to say how much I was delighted with Oceanic Islands—Rudimentary Organs—Embryology—the Genealogical Key to the Natural System—Geographical Distribution; and if I went on I should be copying the heads of all your chapters.

With my hearty congratulations to you on your great work,
Believe me ever very affectionately yours,
CHARLES LYELL.

To DR. JOSEPH HOOKER.

London : November 13, 1859.

My dear Hooker,—Your generalisations on species are worth more to me in proportion to the vast numbers of individual forms and varieties with which you deal, and to the more extensive grasp which the botanist has already got of the earth's flora, than the zoologist has of the fauna. I therefore wish all the reasoning in the old New Zealand Essay, in the review of De Candolle, Introduction to 'Flora Indica,' and in your new book, could be given in one 8vo. volume, referring for details to the Essays.

Reviews, even when not anonymous, are so apt to be classed with ephemeral criticisms, in which the usually unknown and half responsible critic writes off-hand on the book he is treating of, that it is not the best channel for recording with a date the results you had then come to, after collecting personally a greater body of evidence than any other naturalist has brought to bear on a momentous question.

Ever affectionately yours,
CHARLES LYELL.

To HIS SISTER.

53 Harley Street, London: November 15, 1859.

My dearest Caroline,—I have to thank you for your kind present of a purse, which will be soon useful, as Mary says I wear mine out very fast. I wish you could have seen the boys romping for nearly five hours, and very well behaved, drinking my health at dinner without any *mauvaise honte*, and keeping every one alive. Lady Bell is wonderfully well, and danced in the Haymakers. From her to Rosamond [6] there was a capital gradation of ages. Next week we are to have the Hookers staying with us. He is finishing the printing of an Essay on the ' Flora of Australia,' in which the great question of the mutability of species is treated of, and as he has for years been discussing this great problem with Charles Darwin, and goes nearly as far as he does, I long to read it before I have my say in the new edition of my Manual. Sir James Clark is rather better, but has suffered much. He never missed a birthday dinner of the Prince of Wales before, and the Prince wrote him a very good letter on the occasion, very natural and kind. The Queen has also been very attentive and concerned about him. You should have seen Leonard introduce me to my table of presents. He was as good as an auctioneer. He pronounced a cup and saucer of Sèvres manufacture, much to Lady Bell's amusement, who was the donor, as decidedly the handsomest thing on the table ; praised Susan's travelling letter-case as most useful, and so of the rest, informing me who had sent each, which he got up for his own curiosity and my edification, and by no means to show off, for he was unconscious of the amusement he afforded the company.

Believe me your affectionate brother,

CHARLES LYELL.

To DR. JOSEPH HOOKER.

53 Harley Street, London : December 19, 1859.

My dear Hooker,—I have just finished the reading of your splendid Essay [7] on the ' Origin of Species ' as illustrated by your wide botanical experience, and think it goes very far to raise the variety-making hypothesis to the rank of a theory, as accounting for the manner in which new species enter the world. Certainly De Candolle's book was like the old doctrine of those who only called in spontaneous generation for explaining those cases where they were unable to trace the origin to an egg or seed. Nevertheless the extent to which he granted nearly half of what he really believed

---

[6] His young niece of three years old.

[7] Introductory Essay to the Tasmanian Flora.

to be true species to have been derivations in the way of varieties, was calculated to lead philosophical and logical minds to go the whole length of transmutation.

I thought your way of putting it very clear, and the style luminous; the acknowledgment of Robert Brown handsomely done.

The number of grand generalisations is really stupendous when one considers the vast number of species to which they relate. Such as the excess of unstable over stable forms; the limitation of genera and orders owing to extinction, and of species by destruction of varieties; the non-reversion to wild stocks, which struck me as very new and important; the centripetal tendency of hybridisation, species being realities even under the new view. The equality of distribution of the Acot-, Mono-, and Dicotyledons is very wonderful if the dicotyledonous Angiosperms are geologically so modern.

The gymnogens ought, like the marsupial quadrupeds, to have kept some one country to themselves from the oolitic period in order to make the case of plants parallel to that of animals. Though it must be owned that we do not yet know whether in tertiary times there may not have been a rich placental fauna in the Australian area.

The first two notes of page vii. are very interesting, and show what grand speculations and results ' the creation by variation ' is capable of suggesting, and one day of establishing.

The facts and views in page xvii. are wonderfully suggestive and grand, as in all about the glacial migrations. The geological chapter is only too short.

I read your Essay when staying with the Van de Weyers, and he ordered it, and will I am sure appreciate it, in a way that few of *our* literary men can.

Ever affectionately yours,
CHARLES LYELL.

To CHARLES BUNBURY, ESQ.

53 Harley Street, London: January 3, 1860.

My dear Bunbury,—I am sorry to hear of your cold, and that you have not been able to get out these fine frosty days.

You asked me whether anything new had turned up about the Bovey Tracey beds. The very day your letter reached me with this query, Pengelly came to town with a fresh store of specimens. Among these the *Glyptostrobus Europœus,* with fruit now, as well as innumerable leaves, was conspicuous. They have come upon another bed in which a large palm-like looking plant, sometimes two or three feet long, and with a somewhat fan-shaped arrangement of the flabellaria-like leaves, abound, but I could find no point from which the leaves radiated, and we had no botanist to help us. We thought it most like one of Heer's plates of *Manicaria.* It is at any rate an outlandish form for Britain.

Did I tell you of Hooker's ascent of Lebanon, 11,000 feet high, and his finding, though now there is no perpetual snow there, old glacier moraines descending 4,000 feet down from the summit, of which he has made some very graphic sketches? All the cedars of Lebanon grow exclusively on these moraines.

Ever affectionately yours,
Charles Lyell.

To George Ticknor, Esq.

53 Harley Street, London : January 9, 1860.

My dear Ticknor,—I have been so absorbed in preparation for a new edition of my ' Geology ' that I have really had no ideas to exchange, except on those matters which the initiated are discussing, or that question which my friend Charles Darwin's book has brought before the British reading public, both scientific, literary, and theological ; whether, as Dean Milman expresses it, Lyell and his friend have come from tadpoles, against which the Dean, after reading the book on the ' Origin of Species,' vehemently protests, saying that the production of such a book is in itself enough to refute the possibility of such an origin.

Nevertheless it is easier to say and feel this, than to gainsay the continually increasing body of evidence to which I shall try to add some arguments not yet advanced.

Of all the small books of a readable kind which have

come out in our time, you will find it, I expect, the one which takes the longest to read and digest.

Milman has been much overcome by the sudden death of Macaulay—ten years younger than himself. I wish he were not pall-bearer, as I think it is too much for him. The last time we met him, in July I think, Macaulay was in excellent spirits. It was at Lord Stanhope's. The Duke d'Aumale and Motley were there among others, and there was much talk about the true English origin of most Americanisms. Lord Stanhope is also, I hear, to be a pall-bearer. I have heard nothing positive about the state of Macaulay's historical MS., but the general story runs that one volume at least is finished.

I have been much occupied with another geological subject, besides that which your niece, Ellen Twisleton, irreverently calls, the proving her to be first cousin to a turnip (a violet she should have said). I mean the antiquity of man as implied by the flint hatchets of Amiens, undoubtedly contemporaneous with the mammoth, and also the human skeletons of certain caves near Liége which I believe to be of corresponding age. I regard the Pyramids as things of yesterday in comparison of these relics. I obtained sixty-five recently dug up, and Sir George Grey, of the Cape, and formerly Governor of New Zealand, recognises among them spear-heads like those of Australia, and hatchets and instruments such as the Papuans use for digging up roots, all so like as to confirm the saying you used to quote, of ' Man being a creature of few tricks.'

We hear that the open speaking out against the Government at Vienna forms a remarkable feature to the state of things before the war. The Concordat irritates the low-church party, the army complains that they were not well led in Italy, and the Hungarians are of course more discontented than the German population.

Among the stories I hear is that when the two Emperors took a ride alone after the peace of Villafranca, they went to a rock where there is a famous echo, and Louis Napoleon called out Eugénie, to which the echo replied, ' Génie,' and he proposed to the Austrian to call out the name of his wife Elizabeth, to which the echo answered ' Bête.'

To return to Darwin's book, Twisleton, who has called
since I wrote the above, being up here for Lord Macaulay's
funeral, told me he had been much taken with the new theory,
and stated some objections, and ended with asking me what
I thought Agassiz would say to it, after he had nailed his
colours to the mast in his recent work on 'Classification.'
Now I should like much if you will learn what Agassiz does
think and say, and if he has already written anything, please
send it to me.   Asa Gray, among your scientific men of
note, is, I think, the one who comes nearest in his opinions
to Darwin.   I confess that Agassiz's last work drove me far
over into Darwin's camp, or the Lamarckian view, for when
he attributed the original of every race of man to an independent starting point, or act of creation, and not satisfied
with that, created whole 'nations' at a time, every individual out of 'earth, air, and water' as Hooker styles it,
the miracles really became to me so much in the way of
S. Antonio of Padua, or that Spanish saint whose name I
forget, that I could not help thinking Lamarck must be
right, for the rejection of his system led to such license in
the cutting of knots. . .

January 10.—Called this morning at Edward Romilly's ;
all talking of Macaulay's funeral.   Edward Romilly said he
never heard Macaulay say anything with humour in it.   I
told him he once, when we were talking of novels, said to
me, 'I suppose if you were ever to write one, you would
make the lovers meet on the "millstone grit." '   E. Romilly
declared there was more humour in that than in all he had
ever heard him say.   As to my flint implements of the age
of the mammoth, Macaulay was very clear I must be mistaken.   He said they must have got into the drift by some
accident.

With love to Mrs. Ticknor and Anna, believe me affectionately yours,

CHARLES LYELL.

To Principal Dawson.

53 Harley Street, London : May 15, 1860.

My dear Dawson,—I ought to have thanked you sooner for your handsome present of 'Archaia,' which I read through with great interest.

I thought some parts very eloquent, but you well know I am one of those who despair of anyone being able to reconcile the modern facts of geology and of many other sciences with the old cosmogonies handed down to us by the unknown authors of the early chapters of Genesis. A great part of your book, however, may be read with no small profit and pleasure, without reference to such matters, and I was glad it was written before Darwin's book came out, and after Agassiz.

You have truly remarked that the latter, by referring varieties and races to separate creations, leads practically to Lamarck's transmutations. Indeed the license in which, in his work on 'Classification,' he indulges in multiplying the miracle of creation whenever he has the slightest difficulty of making out how a bird or a fish could have migrated to some distant point from its first or other habitat, prepared many to embrace Darwin's and Lamarck's hypothesis.

The argument that a plurality of original stocks, demanded by Agassiz, requires more than a sufficient cause, is unanswerable. But Agassiz honestly felt that if he had to allow that the Negro and the European came from one stock, he should go more than half over to the transmutationists. This he candidly confessed in one of his reviews.

Kenrick was the first I remember to argue that the antediluvians who lived five hundred to one thousand years could not have been of one species. Such a difference in regard to longevity would far exceed the other peculiarities of the Negro. If I believed (as I certainly do not) that any race of the genus Homo ever differed from us as much, I should think that a good specific divergence could take place in a much shorter time than Darwin would deem possible.

I by no means deny that progressive development may be inferred from the manner in which the mammalia increase and rise higher in grade as we trace them down to times nearer our own. Hugh Miller may have pushed it too far, and few allow enough for our ignorance of the ancient inhabitants of the land, yet there is no doubt much truth in the theory of advance, and you ought to have given Darwin credit for not having insisted more upon this class of facts. The truth is that what with occasional proofs of degradation of some types and of persistency without improvement in others, he did not feel justified in going as far as his orthodox opponents. He would otherwise have rejoiced in believing that the rise from the sponge to the cuttlefish, and thence through fish, reptile, and bird to marsupial had occurred, and from that to the intelligence of the Gyrencephala, and from the Chimpanzee to the Bushman, and at length to naked Britons, all by a law of creation ending with the development into an Anglo-Saxon. This successive evolution of sensation, instinct, intelligence, reason, which is such a popular creed with those who shrink from transmutation, is the direct way which leads to Lamarckianism—possibly the road of truth, but they who travel by it hardly I think, see the natural consequences or the goal to which they are approximating. I wish you had shown more appreciation in your review of the number of very distinct sets of phenomena of which Darwin's hypothesis (I claim no higher name for it) offers a solution, and for which no other scientific hypothesis hitherto advanced affords any. With *limited* variability, which is an arbitrary assumption after all, we can explain nothing.

His tables of large and small genera will show you that it is only when you have thousands of species to deal with that you can strike an average, and what you say of Asters will, I think, be denied by botanists.

Hooker's 'Introduction to the Flora of Australia' will interest you much.

Ever most truly yours,
CHARLES LYELL.

To MISS HORNER.

London : June 22, 1860.

Dearest Susan,—I have a promise that ' the bust of the late Mrs. Jameson, by Gibson, R.A., shall be placed in the corridors to be appropriated to memorial sculpture on the grounds of the Royal Commission of 1851 at South Kensington.'

Also, that if the place for it is not finished or ready, it shall provisionally be put into the room of sculpture.

The Prince wishes, in this and other cases, that there should be inscribed on the pedestal, not only the name, but some statement of the merits of the individual commemorated, which he thinks too much neglected in our public statues, the people requiring instruction.[8]

At first they will put all such statues and busts together as they accept them ; but eventually they will be classified, not according to *sex*, but their 'specialities,' as the French would say—historians, poets, artists, &c.

He asked what division Mrs. Jameson would best come into. I said ' *belles lettres*, relating to the fine arts ; ' but that I would consult her friends. At any rate that will be for future arrangement.

You may write to Gibson now.

Yours affectionately,

CHARLES LYELL.

To Sir Charles Bunbury.

53 Harley Street, London, W. : July 4, 1860.

My dear Bunbury,—Baron Anca, a Sicilian, has just found in a cave near Palermo, or rather has just brought to Falconer, remains of the living *Elephas Africanus* in a fossil state, associated with the living African spotted hyæna, and extinct *El. antiquus* and *Hippopotamus Siculus*, all from a cave near Palermo. He brought them to me in this room. They show that Sicily must have been united with Africa, and you perhaps remember that Admiral Smyth discovered that a shallow hundred fathom (?) bottom ran from Tunis to Sicily, while on each side of this submarine ridge there was

---

[8] Miss Horner modelled a medallion of her friend Mrs. Jameson's head from a cast taken after death, and this aided Mr. Gibson in making the marble bust at Rome which stands in Kensington Museum. The inscription on the pedestal was written by Miss Horner.

very deep sea, a thousand fathom deep on the west side, and never yet fathomed on the east.

Lord Ducie told me a year ago of a hippopotamus having been found in Malta, and the other day at Oxford he showed me a letter from Captain Spratt, giving an account of a cave in Malta in which he had found an elephant (fossil) and a gigantic mole.

Both Anca and Falconer seem to believe that man was contemporaneous with this land connection of Africa and Sicily.

I was not able to attend the section of Zoology and Botany [9] (Henslow in the chair), when first Owen and Huxley, and on a later day the Bishop of Oxford and Huxley, had a spar, and on the latter occasion young Lubbock and Joseph Hooker declared their adhesion to Darwin's theory.

Owen and Huxley discussed the osteological and cerebral distinction of Man and the higher Apes, Huxley contesting seven of Owen's propositions laid down in his lecture at Cambridge as untrue and unsound in fact.

The Bishop of Oxford asked whether Huxley was related by his grandfather's or grandmother's side to an Ape. Huxley replied (I heard several varying versions of this shindy), 'that if he had his choice of an ancestor, whether it should be an ape, or one who having received a scholastic education, should use his logic to mislead an untutored public, and should treat not with argument but with ridicule the facts and reasoning adduced in support of a grave and serious philosophical question, he would not hesitate for a moment to prefer the ape.' Many blamed Huxley for his irreverent freedom; but still more of those I heard talk of it, and among them Falconer, assures me the Vice-Chancellor Jeune (a liberal) declared that the Bishop got no more than he deserved. The Bishop had been much applauded in the section, but before it was over the crowded section (numbers could not get in) were quite turned the other way, especially by Hooker.

Mr. C. Moore interested me much in our section, by the result of a collector's feat which I never heard equalled. He carted away from a fissure near Bristol two tons of the

[9] At the British Association at Oxford.

detritus of the triassic bone-bed which had been accumulated
in the said upfilled rent. He had it conveyed to his house
twenty miles distant, took two years to examine it, and
found in it forty-five thousand teeth of the genus *Acrodus*,
counted by the pint measure innumerable other fish and
reptiles' teeth and bones, many shells, and nineteen teeth,
besides a few vertebræ, of *Microlestes* and two other mammalia
genera of triassic age!

I hope to hear soon that you are returning to your scientific reading and investigations, and that you will pay well
for what we call 'factorship' in Scotland, rather than let
your valuable time be seriously absorbed by such superintendence and agency as can be purchased.

My love to Frances, and believe me affectionately yours,

CHARLES LYELL.

To the REV. CHARLES KINGSLEY.

53 Harley Street, London : September 23, 1860.

My dear Sir,—On my return from the Continent, I find
here your excellent sermon on the Prayer for rain, sent to me
I presume by your direction, and for which I return you many
thanks. Two weeks ago, I happened to remark to a stranger
who was sitting next me at a *table d'hôte*, at Rudolstadt in
Thuringia, that I feared the rains must have been doing a
great deal of mischief. He turned out to be a scientific
man from Berlin, and replied, 'I should think they were
much needed to replenish the springs after three years of
drought.'

I immediately felt that I had made an idle and thought-
less speech. Some thirty years ago I was told at Bonn of
two processions of peasants who had climbed to the top of
the Petersberg, one composed of vine dressers, who were
intending to return thanks for sunshine and pray for its
continuance, the others from a corn district, wanting the
drought to cease and rain to fall. Each were eager to get
possession of the shrine of St. Peter's chapel before the other
to secure the saint's good offices, so they came to blows with
fists and sticks, much to the amusement of the Protestant
heretics at Bonn, who I hope did not by such prayers as you

allude to commit the same solecism occasionally, only less coarsely carried out into action.

Have you read Freeman Clarke on 'Prayer' (Boston, United States), who states more fairly than any author I have read the philosophical difficulty, and, though he cannot clear it away, treats the subject more ably than perhaps anyone has done yet? Horace Bushnell, on 'Nature and the Supernatural' (1859, New York), has also some splendid chapters on the relation of God's free will to the immutable laws of Nature, but the book is very unequal. I hope we shall have the pleasure of seeing you when you are next in town.

Believe me ever truly yours,

CHARLES LYELL.

To the REV. W. S. SYMONDS.

London : October 1, 1860.

My dear Symonds,—I should have written to you a day or two ago had I not been somewhat overwhelmed on my return with arrears of correspondence, and until I had seen Dr. Falconer and learnt whether he and Prestwich had made as much progress as I had hoped on the great question at issue about the relations of certain elephant beds and the glacial epoch. I am disappointed at finding matters so much where they were, here at least, for I hope I see my way rather more clearly in reference to Picardy, Belgium, and the Rhine, where I have been examining both into the question of the antiquity of man, and that of the supposed return of a warmer climate than we have now, after the era of glacial cold.

Some of the evidence relied on for the latter opinion, such as the *Elephas Africanus* in modern drift, has decidedly given way.

My idea of going to South Wales, and taking your district on my way, and getting the benefit of your co-operation, was dependent on some progress having first been made by Prestwich, Falconer, and Colonel Wood in regard to the age of the South Wales caves, with not only *Elephas primigenius*, *Rhinoceros tichorhinus*, but also some of them with the other elephants and rhinoceroses (*E. antiquus*, and *R. leptorhinus*,

now called by Falconer *R. hemitœchus*), the age of these
relatively to the glaciers, glaciation, and submergence of North
Wales, and the deposition of the northern drift. Interesting
as this question is, it is only one of many which I have to go
into, such as the new discoveries of human remains coeval
with mammoth, Darwin's theory of species, which affects
certain passages even of the 'Manual,' and I know not how
many other matters which in my travels of the last five years,
since the last edition was published, have accumulated upon
me, and on which I have to pronounce some opinion. I must
therefore stay in town and unpack my boxes, and only thank
you heartily for your kind invitation, for which my wife also
desires me to send her thanks to you and Mrs. Symonds.

Ever sincerely yours,
CHARLES LYELL.

To PRINCIPAL DAWSON, *Montreal.*

London : October 27, 1860.

My dear Dawson,—I received a letter a few days ago
from Sir Charles Bunbury, in which he called my attention
to your paper in the February 1859 No. of our Journal,
as 'admirable and of great importance, one of the most
material additions to our knowledge of vegetable structures
in coal that we have had for a long time.' Although I
had been much struck with it when it was first read, I
had not fully appreciated it till I reperused it in print. The
doctrine of the mineral charcoal having been formed by
plants decaying in the air is a grand step, and seems to me
very unanswerable. What you say also of the Sigillariæ
and Calamites, and their not having been of lax and soft tissues
but of slow growth, is most interesting ; also the number
of generations of Sigillariæ in one foot of coal. In short the
whole paper teems with grand results, and makes us wish
for your continuation in giving us the history of the other
seventy-six successive coal-beds of the South Joggins.

Another *Stereognathus ooliticus* jaw has turned up in the
Stonesfield slate. An animal about the size of a hedgehog,
but I fear we shall learn nothing new from it.

Falconer has made out clearly the former existence of

a small elephant, the size of a Shetland pony, in the small island of Malta. They have also found a hippopotamus there, and a gigantic dormouse the size of a rat, but I fear Falconer's having to go to Sicily for his health will prevent our having a printed announcement of these discoveries.

The African elephant, living species, has turned up in the Sicilian caves, together with the existing Cape hyæna (*H. crocuta* I think it is called), showing the comparatively recent land-connection of Sicily and the African continent, between which there is a very shallow sea 200 or 300 feet deep.

I abandon the Old Red reptile, which will gratify the progressionists, some of whom still feel inclined to adhere to it. The Telerpeton I mean. If Darwin's theory is ever established, it will be by the facts and arguments of the progressionists such as Agassiz, whose development doctrines go three parts of the way, though they don't seem to see it.

Ever most truly yours,
CHARLES LYELL.

To HIS SISTER.

53 Harley Street, London: November 15, 1860.

My dearest Marianne,—I have to thank you for a letter and your good wishes on my birthday, and those of the rest of the family north of the Tweed. Our party [1] went off very pleasantly, as it was sure to do, for we had no one here that we did not like, and scarcely any one of whom I may not say that we were fond, from Rosamond upwards to grandpapa, and Sir Edward Ryan. The impromptu charade of the Catacombs was most entertaining, and the final scene especially. The children, who all acted, were wide awake to the last. I wish you and all of you could have seen them, and the party.

Our new picture, which Susan got for us at Dresden,[2] and of which I presume you heard, was much admired.

[1] On his birthday.
[2] A copy of the Madonna di San Sisto.

z 2

Both Gibson the sculptor and Boxall the painter admire it as a copy exceedingly.

The other day Mr. Wollaston, the entomologist, told me that he had just received some insects, Coleoptera, from St. Helena. Not only the species but the genera so entirely different from all the rest of the world. He said, when we were talking of the extinctions fast going on, that the large Copper, *Lycæna dispar*, is believed to be fairly blotted out of creation. It was only British, and you saw what they had done to Whittlesea Mere. I hope our specimens are in good preservation, for what with their beauty and rarity, they are getting up to a fabulous price, which shows how they are valued. I am sorry I did not find time when last at Drumkilbo to go over the Coleoptera and Lepidoptera at least. Some of the St. Helena Curculionidæ seem to be the most abnormal types.

I have just got Bunsen's ' Egypt's Place in Universal History ' from Mudie's, and am much interested in it. Max Müller's Essay on ' Comparative Mythology ' in the Oxford Essays for 1856 is a splendid article.

It was like old times again to meet the Heads [3] at the Milmans', but the loss of their son makes a great blank. There was no young man of all my acquaintance as able and enthusiastic as a geologist. The Heads will return in February to Canada, but I hope for no long time. Joseph Hooker is expected this week to return from Syria. Faraday was calling here to-day, to explain why he could not romp with the boys as he did last year. He is very well, and going to lecture at Christmas, and Leonard is to hear him. With love to all,

Believe me affectionately yours,
CHARLES LYELL.

To GEORGE TICKNOR, ESQ.

53 Harley Street, London   November 29, 1860.

My dear Ticknor,—I was glad to get the news in Mrs. Ticknor's last letter of the opening of the Museum and of Agassiz's doings. I fully expect it will soon be a model

[3] Sir Edmund Head, Governor-General of Canada, and Lady Head.

collection. I shall be curious to see a second reply of
Agassiz to Darwin, which I understand is coming. Murray
has sold all that remained, and more, of my friend's ' Origin
of Species;' 4,250 copies printed, and only out about a year,
and he must now prepare a new edition.

Whatever faith we may settle down into, opinions can
never go back exactly to what they were before Darwin came
out. The Oxford Professor of Geology, J. Phillips, has
fought Darwin by citing me in pages out of my ' Principles,'
but I must modify what I said in a new edition. Agassiz
helped Darwin and the Lamarckians by going so far in his
' Classification,' not hesitating to call in the creative power to
make new species out of nothing whenever the slightest
difficulty occurs of making out how a variety got to some
distant part of the globe. Asa Gray's articles, all of which
I have procured, appear to me the ablest, and on the whole
grappling with the subject, both as a naturalist and meta-
physician, better than anyone else on either side of the
Atlantic.

I have been very busy with the proofs afforded by the
flint implements found in the drift of the valley of the
Somme at Amiens and Abbeville, and more recently in the
valley of the Seine at Paris, of the high antiquity of man.
That the human race goes back to the time of the mammoth
and rhinoceros (Siberian) and not a few other extinct mammalia is perfectly clear, and when the physical geography
was different—I presume when England was joined to
France.

This will give time for the formation of many races from
one, and enable us to dispense with the separate creation of
several distinct starting-points, to make up for unorthodox
conclusions about ' preadamite man,' of which I see some
writers are freely talking. How are you getting on with
your ' Life of Prescott '? faster I hope than I am with my new
edition of my ' Geology.' I am afraid there is no chance of
Baron Bunsen's recovery; but when we saw him two months
ago he was full of vigour and animation. His date of 10,000
years B.C. for Noah's flood must astonish some of the
orthodox in Boston. This reminds me of Max Müller's
Essay on ' Comparative Mythology ' in the Oxford Essays for

1856, which appears to me in the philological part very excellent. The argument for the existence of some aborigine language, whether it be called Arian or by any other name, seems conclusive, and it must go a far way back, as they branched off into such distant and ancient nations. Bunsen's testimony that there is no tradition of the Arian deluge in Egyptian history and mythology is striking.

Ever affectionately yours,

CHARLES LYELL.

November 30.—The newspaper to-day brings the news of Bunsen's death. I was not prepared for it so soon. He was so beloved by his family, that it will be a great blow.

CHAPTER XXXV.

APRIL 1861—DECEMBER 1862.

EXCURSION TO BEDFORD, WHERE FLINT IMPLEMENTS HAD BEEN FOUND—
OFFER OF CANDIDATURE OF M.P. FOR UNIVERSITY OF LONDON—KREUTZ-
NACH—FOSSIL BOTANY—MEETING AT PHILOSOPHICAL CLUB—DEATH OF
PRINCE ALBERT — LETTER ON THE PRINCE CONSORT — ANNIVERSARY
DINNER AT GEOLOGICAL SOCIETY—CURTIS, THE ENTOMOLOGIST.

[Various offers of a flattering kind were made to him at different times to accept offices; a Trusteeship of the British Museum, President of the Royal Society, &c. : and during the year 1861, under the proposed Reform Bill for a representative in Parliament for the University of London, the candidature was offered to Sir Charles Lyell. But he declined all, and resolved that he would devote himself to the end of his life to his favourite science, which was daily opening up more interesting matter for study and research. The first symptoms of ill-health declared themselves in this year, and he was advised to spend some weeks in Kissingen in Bavaria, whither he went in June, accompanied by his wife and nephew.

In January 1862 he was elected Corresponding Member of the Institute of France. This year was saddened by the death of his wife's mother at Florence, whither he and Lady Lyell went in May, and later in the year they visited various places on the south coast of England.

In 1863 he visited the Caves of Liége and Maestricht, and in the autumn made a tour in Wales, accompanied by the Rev. W. S. Symonds.

The 'Antiquity of Man' was published in February of this year. An order of scientific merit was conferred on him by the King of Prussia, the present Emperor. The order was instituted by the former King (Frederick William IV.) and Alexander von Humboldt was its first chancellor.]

CORRESPONDENCE.

To SIR CHARLES BUNBURY.

London : April 26, 1861.

My dear Bunbury,—I am laid up for a day or two after an excursion to Bedford with Prestwich and Evans, to see a section where a Mr. Wyatt, editor of the Bedford provincial newspaper, has just found two fine hatchets of the true Amiens and Hoxne type. They occurred in working a gravel pit at Beddingham, which I visited more than thirty years ago, when I stayed with Admiral Smyth, then residing at Bedford. That part of the pit, twenty-five feet deep and of large extent, which I then explored, is a wood of tall larches which have grown fast and do credit to the interval.

As the hatchets occurred at the bottom of the old fluviatile gravel which contains Cyclas, Lymnæa, Helix, &c., of recent species, they prove that man was in the country before any mammalia of which the bones are buried in said gravel. I saw a specimen of *Elephas antiquus* found in the town of Bedford at the same level as the *E. primigenius* of Beddingham. At the latter place is also the tichorhine rhinoceros. I was assured that the hippopotamus major was also found in another locality ; but I will not answer for that, but am prepared for anything after my last visit to Paris and revisit to Abbeville. For I am now prepared to believe that all the animals above mentioned were in the country after man, and by digging a pit myself at Abbeville, I obtained five implements below a fluvio-marine bed containing the *Cyrena fluminalis,* alias *C. consobrina* of the Nile. The relation of man to the close of the glacial period is a point on which I have not yet made up my mind, but I suspect those beds in France may have begun before the northern drift of England was finished. The late discoveries at Herne Bay and Reculvers convince me that man inhabited England when the Thames was a tributary of the Rhine.

But enough of these very modern affairs. What you tell me of those plants of more respectable antiquity, and of Lesquereux's observations,[1] interest me much. I am surprised

[1] On the Coal-plants of Pennsylvania.

that he can make out, even by the omission of some conspicuous plants, any kind of resemblance between the
American swamps and European peat mosses, except by
going very far north in the Transatlantic Continent. No
doubt all the coal-plants of Europe and America have been
formed in very analogous habitations. Dawson tells me that
he has found such a number of different species of Trigono-
carpum in that part of the Nova Scotia coal, where there is
a similar abundance and variety of Sigillaria, and where
Coniferæ are scarce, and has found so many Trigonocarpa in
hollow Sigillariæ, that he is half inclined to guess that said
fruits belong to said genus.

Ever affectionately yours,
CHARLES LYELL.

To HIS BROTHER.

53 Harley Street, London : April 23, 1861.

My dear Tom,—Under the new or proposed Reform Bill,
the University of London, with a constituency of about 700,
is to have a member, and a strong party, wishing to
have a new literary or scientific element introduced into
the House, as representing academical interests, offered to
bring forward Mr. Grote, formerly M.P., but being now im-
mersed in finishing his ' History of Greece,' he declined
about a fortnight ago, and they then offered it to me.

I was obliged to consider the affair just as if the Bill
was carried, and as if (which I really believe) the committee
could carry me. I declined, though there is no place in
the House I would so soon hold ; and as the party who made
the offer are by no means the extreme Liberals, I should
have fully sympathised in their political as well as educational
views. But my happiness is in carrying on my Geology, and
I believe I shall do more good in that line, as I have on
hand years of unpublished travels, and reading, and thinking.
There is a great demand for a new edition of the ' Manual,'
now out of print, and the ninth edition of the ' Principles '
(5,000 copies) nearly exhausted. I have also a new book
on the stocks.

We don't talk of the overture, as I would not cheapen

the honour.   After all, will Lord John carry his Bill?   They (the University of Edinburgh) offered me a degree at Lord Brougham's installation, but I grudged being absent five days from town, but was half sorry to refuse to go.   Here is enough about myself in all conscience.

Believe me your affectionate brother,

CHARLES LYELL.

To His Sister.

Frankfort : June 17, 1861.

My dearest Eleanor,—We were talking of you yesterday when we stopped at the railway station at Kreutznach, and wishing you were there again, and in as fine weather as we have been having ever since we left England.   Leonard and I have been making out the geology as we flew along over the Low Countries, and then spent two days among the limestone caves of the Meuse at Dinant above Namur, and then through the forest of Ardennes, which still deserves that name between Namur and Luxembourg.   I went from the last place to Saarbruck to see the finest collection of plants of the coal which perhaps exists in Germany, and the finest of insects of the same early period, chiefly fossil *blattæ* and *termites* (cockroaches and white ants), of which the Professor has collected more than thirty individuals.   I liked the look of Kreutznach.   Did any of you ever take a trip to Oberstein to see the agates there?   I should like to have stopped a day there.   We are to go for at least a three weeks' stay to Kissingen.

With love, ever your affectionate brother,

CHARLES LYELL.

To the Duke of Argyll.

53 Harley Street, London : July 9, 1861.

My dear Duke of Argyll,—I write to say that I have invited Professor Heer of Zurich, the best botanist in Europe for fossil tertiary plants, to come here in the autumn— August and September—to determine the Bovey coal-plants which have been collected by Mr. Pengelly at the expense of Miss Burdett-Coutts.

Heer is particularly desirous of examining the Mull plants, because he thinks they will form a link between the Miocene plants of Bovey Tracey, Switzerland, and the fossils of the surtebrand of Iceland, which Heer has lately worked upon. Among other wonderful discoveries they find the tulip-tree (*Liriodendron*) in that now cold and treeless island, and Heer thinks that Scotland in the Miocene period may have supported a similar flora.

If you can do anything, whether by new researches or by placing at Heer's disposal any stores already accumulated of the plant-bearing beds, it is a fine opportunity of having them thoroughly examined by a first-rate authority.

He will detect species where they might evade the search of any other naturalist, having so practised an eye. His work now finished on the 'Flora Tertiaria Helvetica' is a beautiful and most satisfactory one.

Believe me most truly yours,
CHARLES LYELL.

To LEONARD LYELL.

Folkestone: August 7, 1861.

My dear Leonard,[2]—If you visit Clova, or make out some other expedition, I hope you will write me an account of it. I was glad to hear you had found *Ancylus fluviatilis*. I met with it lately fossil in France, in great numbers, in gravel in which two species of extinct elephants and a rhinoceros had been met with. The gravel was laid open in a railway cutting on the banks of the river Oise.

Naturalists used to wonder how this *Ancylus* got spread over the country in separated lakes and streams, till some one found a young *Ancylus* adhering to the elytra of one of those large boat-beetles, *Dytiscus marginalis*, which you will see in the collection at Drumkilbo,[3] and which fly about at night from pond to pond, and may sometimes carry the *Ancylus* with them, if, like the *Patella* which you saw high and dry on the rocks here, he can manage to do without water for an hour or two, as most probably he can.

Tell mamma we gathered samphire in flower at the

[2] His young nephew of ten years old, who was in Scotland.
[3] The country house of Sir Charles's sisters.

Lyddon Spout yesterday, and many other marine plants, and wished she and you had been with us, as Mr. Bentham named all the plants for us, and gave us much botanical information. But we were so busy with the plants and with five miles of walking, that we had no time for fossils, and when we reached the gault it was almost dark. The tide was lower than when you were there with me.

Tell Uncle Tom that Mr. Powrie has been to Arbroath, and has written to say that he finds that he was mistaken, and not I, about the position of certain beds of conglomerate and purplish shale which he, Mr. Powrie, said in a paper read to the Geological Society, were not as I had represented them during a visit I made to the spot forty years ago. It is an interesting place, and I hope some day to examine it with you, but I must not wait another forty years, unless I intend to rival the Countess of Desmond,

> Who lived to the age of a hundred and ten,
> And was killed by a fall from a cherry-tree then.

I believe Aunt Mary told mamma about a letter of Mr. Darwin's, wanting Mr. Bentham to tell him how to get from Tenby in Wales a variety of *Orchis pyramidalis*, which you remember gathering here, 'without spurs,' which mamma one day pointed out to me as so characteristic of the flowers of that plant. The first specimen which Mr. Bentham gathered in the Warren when talking with me of C. Darwin's letter, was this very variety and monstrosity. But though we have picked two or three dozen since, not one of them departed from the usual type, and I fear they are all gone off now, which is a pity, as Darwin has written to me for another without spurs. But I shall try again if the rain will but stop. Yesterday we had a splendid day, the French coast very clear.

Give my love to Frank and Arthur and Rosamond.

Believe me, my dear Leonard,

Your affectionate uncle,

CHARLES LYELL.

To Sir Charles Bunbury.

London : August 26, 1861.

My dear Bunbury,—I wrote some time ago to Heer of Zürich, asking him, if he had any copies by him of his 'Fossil Flora of Switzerland,' to bring one with him to London in September, for that you wished to buy it. I begged him at the same time to bring, if possible, a copy for both you and me, of a beautiful French translation which I know Gaudin was making of the grand Essay on the 'Swiss Tertiaries' as compared to those of other countries, and on climate, all treated from a botanical point of view. In this translation Heer himself has added some valuable pages here and there on Aix-la-Chapelle plants of Debey, on Œninghen insects, &c. It would make a good 8vo. volume. All the rest of the book, the description of the plants, &c., does very well in German, but this theoretical portion, which is very well done, it is a luxury to have in French. Gaudin has also added some chapters to which his name is appended.

Heer has sent me a report on the Bovey coal-plants. They had only made out about fifteen species, and now that Pengelly has, at my suggestion, sent over the collection to Zürich, Heer finds forty-five species. I see he has named a *Lastræa, L. Bunburii.* It is Lower Miocene. A great or rather abundant *Sequoia*, allied to the Californian mammoth tree, forms the bulk of the coal, cones and seeds in plenty. *Sequoia Couttsiæ* is named in honour of the lady at whose expense the collection was made. There are palms (palmacites); cinnamons, two species; *Nymphœa, Proteaceæ, Quercus*, two vines; *Ficus*, two species; *Gardenia.* I quote from memory, having had to send on the letter to Pengelly the same day I got it. Fifteen species identical with continental Lower Miocene; many new species. I expect Heer here in a week, and he will go to Bovey Tracey, and I daresay find more species.

Ever affectionately yours,

Charles Lyell.

To PROFESSOR HEER.[4]

53 Harley Street, London: August 26, 1861.

My dear Professor Heer,—The day I received your letter
I sent it to Mr. Pengelly, and I have this morning heard
from him. He is as much delighted at the importance of
your results as I am, and he is more surprised, for I always
felt sure that you would find twice or thrice as many species
as they had dreamt of.

I am extremely gratified by the compliment which you
and M. Gaudin have had the kindness to pay me in your
dedication to me of the translation of your splendid essay,
which I am reading in French with renewed pleasure and
admiration. When I first received the copy, I did not see
that the dedication was there, and so acknowledged the
parcel in my letter to M. Gaudin without any special allusion
to this very flattering testimony of your esteem. I hope
the new edition will be approved of by you. I have gone
much farther than before in favour of a progressive development, and have also endeavoured to show the importance of
botany in geological classification.

Believe me most truly yours,

CHARLES LYELL.

To LEONARD HORNER, ESQ.

53 Harley Street London: November 5, 1861.

My dear Horner,—We had a good meeting of the
Philosophical Club, eighteen present. Sabine in the chair
in great force. Grove, Williamson, Miller of King's College,
Sharpey, Carpenter, Daubeny, Falconer, Frankland, Gassiot,
Hooker, Huxley, Bence Jones, Partridge, Sykes, Tyndall,
Wheatstone. Tyndall gave an account of an attempt he had
made to measure the heat of the moon's beams by an instrument on the roof of the Royal Institution, and he found in
a clear sky, that the moon when the instrument was turned
to it produced a chill. Melloni had found heat, but he

---

[4] Professor Oswald Heer, of the Technological Institute of Zürich, eminent
fossil botanist, author of the *Urwelt der Schweitz*, and other important
works.

found cold; he supposed there was watery vapour in the air, acted on by the heat of the moon. Grove and Miller discussed it, but could make little of the fact theoretically. At last Sharpey said that in controverting Tyndall's hypothesis, they seemed to forget that his having ascertained the fact was a great point, and they owed him thanks, although his own expectations had been

Quenched in the chaste beams of the watery moon.

With this happy quotation, which produced quite a burst of applause, the matter ended, and Huxley gave an account of the Neander-thal cranium, and asked what he should call it. Some one suggested ' an anthropoid,' which he has adopted since, in talking with me about it. I was asked to give some idea of the age of said skull, and of Lartet's Aurignac cave, which I did.

Ever affectionately yours,

CHARLES LYELL.

To GEORGE TICKNOR, ESQ.

53 Harley Street, London : November 10, 1861.

My dear Ticknor,—Milman was much pleased, when I called one day, to show me a couple of handsome volumes, the last of the series of eight of his ' Latin Christianity,' which had come out in the United States (New York, I think), since the war began. When people express surprise at such things, I remind them that everything went on here much as usual when we were in the worst of the Crimean War. I hope you read the Duke of Argyll on the ' American War,' the best speech I think of all, in or out of the Cabinet.

I like what I see of Adams, the new Minister here of the United States. But he must have much to put up with, if it were only the ignorance of American affairs shown by the questions put to him. If, says Mr. Adams, the questioners were not politicians it would be more excusable. He, the Minister, cannot pay visits, as the only time he went out of town for two days, he was telegraphed back again, so constant are the arrivals of news, and so numerous the points which he has to give answers about. The forfeiture of

merchant vessels because they have a one-sixteenth of their
cargo belonging to Southern States, was the last affair I
heard of, many Americans not daring to leave London and
Liverpool, and wanting to learn from Adams what is the
force and construction of some late Act of Congress.

I suppose you have read the memoir of De Tocqueville.
I never met him except that morning at Lord Stanhope's
when you were last here, when I thought him most agreeable
and conversationally eloquent.

Remember me to all your circle.

Ever affectionately yours,   CHARLES LYELL.

To LEONARD HORNER, ESQ.

Barton : December 26, 1861.

My dear Horner,—You were mourning like us the loss
we have sustained by Prince Albert's death.[5]   It was many
days before I could fully realise it.   I went down the day of
the evening he died, and at four o'clock, just as the illness
took its fatal turn, I found a bulletin at the Palace which led
us to hope that the danger was nearly over.   I hardly think
I ever saw a family where there was more domestic happiness
and good understanding between husband and wife, and
father and children.   As to the reserve with which he was
charged, I can only say that to me, and I am sure to a
multitude of others, he was remarkably open, and talked of
all subjects, even the most serious, without fear of committing himself.   Though I have spoken with him often since
then, the last time I had a full hour's talk, *tête-à-tête*, I owe
to Susan's asking me to arrange the affair of Mrs. Jameson's
bust.   That was soon settled, and he went into Educational,
Exhibitional, British Museum, and other subjects, and did
not mind speaking of his disappointment when he had been
overruled, and listened when I did not quite agree with him
as to the British Museum with patience, and like one who
wished to look at the matter on all sides.   I know of no public
man in England who was so serious on religious matters, and
so unfettered by that formalism and political churchism and
conventionalism which rules in our upper classes.

[5] Mr. Horner was spending the winter in Florence with his family.

The only point I ever remember distinctly differing from him, so that the subject was afterwards avoided, was Hungary. As to Germany he went quite as far as I do in his liberalism, but he could not help looking upon Hungary from a German point of view, though I believe he thought of the folly of Austria, with her Concordat, &c., much as I do.

Having seen his marriage, and the eldest children when they were hardly out of the nursery at Balmoral, and then the *fêtes* at the marriage of his daughter, and now his death as a grandfather, and having felt myself so far advanced in my own career when I first knew him, I cannot think of his being gone without seeming all at once much aged myself. The regular meeting of the Geological Society was put off because of this death.

As to my new book,[6] I get on very steadily and with pleasure to myself. I am now treating of the relations of the earlier history of man to the glacial period—a difficult subject, especially as I have to connect it with some new views respecting the glacial hypothesis, both as relates to Scandinavia, Scotland, and the Alps. I have deferred what I am to say on Darwin to the last or eighteenth chapter, knowing that were I once to begin there is no chance of coming out this year, but if all is ready but that, I may have courage to abridge it into one short chapter.

Believe me yours affectionately,
CHARLES LYELL.

To JOHN MURRAY, ESQ., *Albemarle Street.*

January 2, 1862.

My dear Sir,—I was at Sir Charles Bunbury's in the country, or I should have replied sooner.

The late Prince Consort certainly deserves the most eloquent *éloge* that one of your best writers can give him in the 'Quarterly.' He was always thinking of what could be done to improve the nation morally and socially, and in the fine arts, and how popular education, and that of the higher classes, could be advanced.

[6] On the *Antiquity of Man.*

I believe it was a common idea that he was too reserved, but he was very much the reverse. I am sure that not a few can affirm that he talked most freely and without restraint on a great variety of subjects on which many public men would have been somewhat afraid of committing themselves. He was certainly very careful, as became a man in his position, to avoid personalities, but he spoke out his opinions fearlessly, often on the most speculative as well as on practical subjects, and always listened patiently to what one had to say on the other side.

The quantity of work he got through, in spite of innumerable interruptions, was immense. His foreign correspondence alone, which the public here knew nothing of, would have been thought sufficient occupation for one who had nothing else to do. I remember on two occasions when he was called upon to write addresses, one as President of the British Association, and one again for the Social Science meeting, that he told me they taxed him rather too much, as his other engagements were so numerous, and many of them requiring much thought. He told me there was an unusual pressure upon him when he had to compose his British Association Address for Aberdeen, and that he felt strongly he had not done justice to it. Once when I requested him to attend a meeting of the Geological Society, he looked at his note-book, and I was astonished at the number of Wednesdays in advance that he already had engagements for, and on my expressing surprise, he showed me that it would have been the same had we met on any other day of the week. A very large proportion of these had reference to some useful object, or were connected with the duties of his station.

He had no special acquaintance with geology or mineralogy, or so far as I know with any branch of natural history, but he knew enough of all to be interested in them, and to understand what their cultivators were about. How much he desired to encourage them, and how much he wished that the elements of these branches of knowledge should enter more largely into the system of education in this country, is well known.

When I first knew him, about fifteen years ago, I found

that he had read my 'First Travels in North America' with some attention, and he referred more than once to what I had said on educational matters. He was then very sanguine of the progress that might be made in his own time in the diffusion of popular education. Twelve or thirteen years later he was equally zealous in the cause, but expressed his disappointment that 'all that we of the present generation in this country can hope, is to teach those who will educate the generation that is to follow.' Alluding to my citation of what Liebig had said to Faraday on the different relative value set by the Germans and English on the practical in science, and the purely theoretical, or discovery of new truths or principles ('First Travels,' vol. i. p. 309), he said he believed Liebig had justly estimated the character of the two nations, and that both might in this respect learn from the other.

He often alluded to the want of cultivation of the German language amongst the higher class, and the number of diplomatists who had been sent to German Courts, and even important embassies, who knew little or nothing of the language of the people. It is well known how much he tried to remedy this by giving prizes at Eton, &c. When first the excellence and originality of his speeches and addresses on various occasions attracted attention, it was very commonly asked, who wrote them for him? When I declared to some who put this question, that I was convinced he got no help from anyone in the way of ideas or opinions, only now and then some passages were put out of the German into the English idiom by friends (and this only in early days), I found people very incredulous. It seemed to me that for years he was underrated, at least that his great talents were not duly appreciated, and that his character was not understood.

I think the best tribute to his memory would be to collect all his speeches, and to make in your Review a judicious set of extracts. There would be much life in them, and some of the best will be found in newspapers, and not separately printed. There was a very good one I am told at a Royal Academy dinner which I missed.

I never made any notes of conversations which I had

with the Prince, but I have a vivid recollection of many, which I have never repeated to anyone, thinking it would be a breach of confidence.

I believe it was Sir James Mackintosh who said that a biography to be worth anything ought never to have been written, and I feel much the same in regard to such a letter as would really answer the object you have in view.

CHARLES LYELL.

To LEONARD HORNER, ESQ.

53 Harley Street, London : February 23, 1862.

My dear Horner,—I must send you a few words on the anniversary, which went off very well. Murchison read your letter, which was well received, and he then delivered an appropriate complimentary address to Godwin Austen on presenting the medals. Austen, who was remarkably gratified by the honour, replied at length, and said he should work much harder in future. The chairman then gave a biographical account of the late Dr. Fitton, which he had got up with much pains, and which was a just tribute to one who had taken so active a part in the Society as well as in our science. After which, Huxley delivered a brilliant critical discourse on what paleontology has and has not done, and proved the value of negative evidence, how much the progressive development system has been pushed too far, how little can be said in favour of Owen's more generalised types when we go back to the vertebrata and invertebrata of remote ages, the persistency of many forms high and low throughout time, how little we know of the beginning of life upon the earth, how often events called contemporaneous in Geology are applied to things which instead of coinciding in time, may have happened ten million of years apart, &c. &c., and a masterly sketch comparing the past and present and almost every class in zoology, and something of botany cited from Hooker, which he said he had done because it was useful to look into the cellars and see how much gold there was there, and whether the quantity of bullion justified such an enormous circulation of paper. I never remember an address listened to with such interest or received with such applause, though there were many private protests against some of his bold opinions.

The dinner at Willis's was well attended ; I should think eighty or more present. The Duke of Argyll made an excellent speech on proposing Ramsay's health. Monckton Milnes made a happy and humorous speech in reply to ' Members of the House of Commons.' I was requested to give the ' Universities,' which I coupled with Dr. Williams, Principal of Jesus College, with whom we stayed at Oxford, who spoke fluently in reply. Lord Ducie, Sir Philip Egerton, Sir H. James, most of the Council, and a full representation of Jermyn Street were there. The Duke of Argyll having talked of Scotland as a specific centre from which so many geologists had come, Warrington Smyth stood up for other centres of creation south of the Tweed, and late in the evening Huxley made them merry by a sort of mock-modest speech. I sat between Charles Bunbury and Dr. Williams, and had a pleasant time of it, and was pleased to think how much life there is coming on in the Society, when all of us who are above sixty are added to the extinct organisms.

Ever affectionately yours,
CHARLES LYELL.

To HIS SISTER.

53 Harley Street : March 7, 1862.

My dearest Caroline,—I do not wonder that you thought Curtis was dead, for really when I called after more than a year and a half, I wondered whether I should find that he was still surviving the mischief which the three cabs, all of which had passed over different parts of his body, had done him. I was shown in, and though quite blind found him looking well, wearing a handsome grey beard. . . . I found Curtis up to all that is going on in entomology, and envying the luck of a friend who lives near the great Lowestoft lighthouse, where the moths come by thousands so as to darken the light, and require to be swept away, some of them crossing from the Continent, and not being British species. Enough indeed to make an old blind collector's mouth water. Then he told me how the ' Glory of Kent,' *Bombyx versicolor,* which once cost 2*l.* 2*s.,* could now be had for 3*s.* 6*d.,* and other gossip of that sort. By the way, some day as I pass a dealer opposite the British Museum I will send you the said ' Glory.' So much for getting blind when near seventy, and having three cabs go over one.

Your affectionate brother,

CHARLES LYELL.

To CHARLES DARWIN, ESQ.

Freshwater Gate, Isle of Wight : August 20, 1862.

My dear Darwin,—Mr. Jamieson of Ellon has been again to Lochaber, and confirms his former theory of the glacier lakes. The chief new point is a supposed rise at the rate of a foot per mile of the shelves as we proceed from the sea inland. It seems to me to require many more measurements, before we can rely on it. He found some splendid moraines opposite the mouth of Glen Trieg. He found some shells of Arctic character in the forty feet high raised beach of the Argyllshire coast, and has asked me to learn about one of them, of which he sends a drawing.

I fell in yesterday in my walk with Mr. A. G. More, whom you cite in your orchid book. He considers you the most profound of reasoners, to which I made no objection, only being amused at remembering that, such being the case, you had performed a singular feat, as the Bishop of Oxford assured me, of producing ' the most illogical book ever written.'

We shall be here for a week longer. I have been with my nephew Leonard to Alum and Compton Bays.

Ever most truly yours,

CHARLES LYELL.

To LADY BUNBURY.

53 Harley Street : November 16, 1862.

My dear Frances,—The so-called gorilla which is going the rounds of the newspapers is a fine male chimpanzee, still at Liverpool. They told me to-day at the Zoological Gardens that some years ago there was a real live gorilla at

Liverpool which lived eight months and was mistaken for a chimpanzee, but they stuffed him and the mistake was afterwards detected. So the present hoax is a sort of compensation.

The Marquis d'Azeglio told me this morning that the ibex or bouquetin of which, as well as of the chamois, the King of Sardinia has sent a pair to the Gardens, is only a hybrid between the ibex and goat. It had been spoken of as of pure breed.

Mary and I saw Bishop Colenso yesterday, who is a very gentlemanlike and intellectual style of man.

I have to give evidence to-morrow before the Public School Commission—Lord Clarendon in the chair—recommending two hours a week on science and natural history, as an encroachment on Latin verses and translating Addison into Greek prose, to which the other forty hours must be devoted.

November 20.—I missed the Geological yesterday, and shall not go to Owen's paper on the ' Paleornis ' to-night at the Royal, nor to a dinner at the Rich's, nor to one which Katharine has invited me to, but in spite of every denial of the kind, I move slowly on. This last week an interesting examination by the Commissioners of Public Schools, in which they wished to have my opinion as to the feasibility of introducing the elements of natural history and physical science (two hours a week only) into Eton, Winchester, Harrow, &c., stirred me up a good deal and did not benefit me in my progress in my book.

Ever affectionately yours,
CHARLES LYELL.

To GEORGE TICKNOR, ESQ.

53 Harley Street, London : December 19, 1862.

My dear Ticknor,—I was glad to hear that you had at last determined to bring out your ' Life ' or ' Memoir of Prescott ' without waiting for the end of the war, which may possibly be deferred, not to the Greek kalends, for I don't think the South will hold out two years, but for a time so uncertain that it is better not to delay it. I should much have liked

it to have come out when people were less excited by battles
and such vital questions as the emancipation of the slaves.
People are beginning, I think, to estimate the financial
resources of the North and the real wealth of the country
better than they were. If the result of the struggle could
be the abolition of slavery by the year 1900, it would be
worth a heavy debt and many lives, at any rate when one
thinks of what most wars are waged for, not but that the
Union alone is worth a long fighting for. The distress in
Lancashire is increasing, and I suppose we must have
parliamentary aid. The cotton we get from India so raises
the price that they (the Indians) suffer severely, and their
native manufactures are knocked up.

The Milmans are well. He getting out a new edition of
his ' History of the Jews,' and will reply to George Cornewall
Lewis, who makes out the Egyptians so modern, and says
they were never a conquering people. Surely he never
could have seen the processions of captives and the sieges on
the temples.

Bishop Colenso on the ' Pentateuch ' is making as much
noise this year as did the ' Essays and Reviews ' last year. If
people had read what Norton, or the Germans, or our
William Greg in his ' Creed of Christendom,' have said so well,
there could be no sensation created by such a book. But
the policy is well sustained, never to reply to any lay attack,
as it only draws it into notice. But if a churchman enters
the lists, Convocation and meetings of the clergy, and the
' Record ' and other intolerant papers, set to work advertising
the delinquent publication, as if they were bribed by Longman, who sold 10,000 copies of Colenso the first day (price
6s.) and is going on since pretty steadily. The decision on
the case of the ' Essays and Reviews has left churchmen very
free on most points which they were afraid to venture on
for fear of legal penalties. One thing affirmed by Colenso
is universally admitted, that the strictness of the ordination
vows is preventing the young men of both Universities of
most talent and the finest moral sense from entering the
Church.

Believe me ever affectionately yours,
CHARLES LYELL.

CHAPTER XXXVI.

MARCH 1863–AUGUST 1863.

ON TRANSMUTATION OF SPECIES—LAMARCK—VISIT TO OSBORNE—MUSEUM OF THE YOUNG PRINCES — INTERESTING CONVERSATION WITH THE QUEEN—'ANTIQUITY OF MAN'—WELWITSCHIA—EXPEDITION TO WALES —MOEL TRYFAEN.

CORRESPONDENCE.

To DR. JOSEPH HOOKER.

London : March 9, 1863.

My dear Hooker,—Darwin has sent me a useful set of corrigenda and criticisms for the new edition I am busy in preparing.[1] He seems much disappointed that I do not go farther with him, or do not speak out more. I can only say that I have spoken out to the full extent of my present convictions, and even beyond my state of *feeling* as to man's unbroken descent from the brutes, and I find I am half converting not a few who were in arms against Darwin, and are even now against Huxley.

I feel that Darwin and Huxley deify secondary causes too much. They think they have got farther into the domain of the 'unknowable' than they have by the aid of variation and natural selection.

Asa Gray says that Lyell's doctrine is ' that the thing that is, is the thing that has been, and shall be.' Now if the thing that is, in the case of a man of genius born of ordinary parents and with ordinary brethren of the same parentage imply a slight leap, I do not see why Darwin should complain of my leap, given only as a speculation, from the highest unprogressive to the lowest progressive.

However, I plead guilty to going farther in my reasoning

[1] *Antiquity of Man.*

towards transmutation than in my sentiments and imagination, and perhaps for that very reason I shall lead more people on to Darwin and you, than one who, being born later, like Lubbock, has comparatively little to abandon of old and long cherished ideas, which constituted the charm to me of the theoretical part of the science in my earlier days, when I believed with Pascal in the theory, as Hallam terms it, of ' the archangel ruined.'

Monday evening.—On my return home to dinner I find your letter. I have not time to reply, but thank you much.

As the glacial chapters are of course not the most popular, I am the more pleased that you and Darwin like them.

I see you coincide with Darwin, and not with Crawfurd and others, who tell me they are so glad ' I did not lay down transmutation dogmatically as proved, though I have evidently come nearly quite round to it.'

I don't care what people have been expecting as to the extent to which I may go with Darwin, but certainly I do not wish to be inconsistent with myself. Though, as I have been gradually changing my opinion, I do not want to insist on others going round at once. When I read again certain chapters of the ' Principles,' I am always in danger of shaking some of my confidence in the new doctrine, but am brought back again on reconsidering such essays as Darwin's, Wallace's, and yours. I see too many difficulties to be in the danger of many new converts who outrun their teacher in faith.

I have not had time to profit fully by your valuable letter, but shall do so, and beg you to write freely if you have not said all in the way of criticism. I have heaps of approving letters, but few are able and willing to help one by such comments as yours and Darwin's.

Believe me ever sincerely yours,

CHARLES LYELL.

To CHARLES DARWIN, ESQ.

53 Harley Street: March 11, 1863.

My dear Darwin,—I see the ' Saturday Review ' calls my book ' Lyell's Trilogy on the Antiquity of Man, Ice, and Darwin.'

As to my having the authority you suppose to lead a public who up to this time have regarded me as the advocate of the other side (as in the 'Principles') you much over-rate my influence. In the new 'Year Book of Facts' for 1863, of Timbs, you will see my portrait, and a sketch of my career, and how I am the champion of anti-transmutation. I find myself after reasoning through a whole chapter in favour of man's coming from the animals, relapsing to my old views whenever I read again a few pages of the 'Principles,' or yearn for fossil types of intermediate grade. Truly 1 ought to be charitable to Sedgwick and others. Hundreds who have bought my book in the hope that I should demolish heresy, will be awfully confounded and disappointed. As it is, they will at best say with Crawfurd, who still stands out, 'You have put the case with such moderation that one cannot complain.' But when he read Huxley, he was up in arms again.

My feelings, however, more than any thought about policy or expediency, prevent me from dogmatising as to the descent of man from the brutes, which, though I am prepared to accept it, takes away much of the charm from my speculations on the past relating to such matters.

I cannot admit that my leap at p. 505,[2] which makes you 'groan,' is more than a legitimate deduction from 'the thing that is' applied to 'the thing that has been,' as Asa Gray would say, and I have only put it moderately, and as a speculation.

I cannot go Huxley's length in thinking that natural selection and variation account for so much, and not so far as you, if I take some passages of your book separately.

I think the old 'creation' is almost as much required as ever, but of course it takes a new form if Lamarck's views improved by yours are adopted.

What I am anxious to effect is to avoid positive inconsistencies in different parts of my book, owing probably to the old trains of thought, the old ruts, interfering with the new course.

But you ought to be satisfied, as I shall bring hundreds

---

[2] See *Antiquity of Man*, first edition, p. 505.

towards you, who if I treated the matter more dogmatically would have rebelled.

I have spoken out to the utmost extent of my tether, so far as my reason goes, and farther than my imagination and sentiment can follow, which I suppose has caused occasional incongruities.

Woodward is the best arguer I have met with against natural selection and variation. He puts conchological difficulties against it very forcibly. He is at the same time an out-and-out progressionist.

I am glad that both you and Hooker like the 'ice' part of the Trilogy. You are the first to allude to my remarks on Ramsay, who says ' I shall come round to his views in good time.'

Falconer, whom I referred to oftener than to any other author, says I have not done justice to the part he took in resuscitating the cave question, and says he shall come out with a separate paper to prove this. I offered to alter anything in the new edition, but this he declined. Pray write any criticism that occurs to you ; you cannot put them too strongly or plainly.

Ever yours sincerely,
CHARLES LYELL.

To CHARLES DARWIN, ESQ.

53 Harley Street : March 15, 1863.

My dear Darwin,—Your letter will be very useful. I wish to get such passages so far in the Darwinian direction as not to be inconsistent with my general tone, and what Hooker calls some of my original arguments in favour of natural selection. At the same time I am struck by the number of compliments, both in reviews and in conversation with the half-converted, which I receive, because I have left them to draw their own inferences, and have not told them dogmatically that they must turn round with me. Hooker admits that in science people do not like to be told too plainly that they must believe, though in religion they wish to have it laid down for them. Yet he may be wrong, for if the 'Times' were to write for the next fortnight against

the Southern States, and against the Poles, nine-tenths of
good society would whirl round, and the middle class which
would stand firm would be able to do so partly because they
read cheaper papers which are not interested in following
the lead of the ' Times.' . . . .

I wish I deserved what you say about taking criticism
kindly. I often think I should be as touchy as anyone if
the success of my works did not give me a constant opportunity of profiting immediately by every suggestion as to style
and moral tone, and above all as to facts and logic. Besides
the increased responsibility which I incur by the trusting
public, who before they had read a word induced the trade
to bid for 3,850 copies, I have the prospect, if I improve
my knowledge and my teaching, of future success in new
editions with comparatively little labour.

As to Lamarck I find that Grove, who has been reading
him, is wonderfully struck with his book. I remember that
it was the conclusion he came to about man that fortified
me thirty years ago against the great impression which his
arguments at first made on my mind, all the greater
because Constant Prévost, a pupil of Cuvier's forty years ago,
told me his conviction ' that Cuvier thought species not real,
but that science could not advance without assuming that
they were so.' When I came to the conclusion that after
all Lamarck was going to be shown to be right, that we
must 'go the whole orang,' I re-read his book, and remembering when it was written, I felt I had done him injustice.

Even as to man's gradual acquisition of more and more
ideas, and then of speech slowly as the ideas multiplied, and
then his persecution of the beings most nearly allied and
competing with him—all this is very Darwinian.

The substitution of the variety-making power for ' volition,' 'muscular action,' &c. (and in plants even volition was
not called in) is in some respects only a change of names.
Call a new variety a new creation, one may say of the former
as of the latter, what you say when you observe that the
creationist explains nothing, and only affirms ' it is so because
it is so.'

Lamarck's belief in the slow changes in the organic and
inorganic world in the year 1800, was surely above the

standard of his times, and he was right about progression in the main, though you have vastly advanced that doctrine. As to Owen in his Aye Aye paper, he seems to me a disciple of Pouchet, who converted him at Rouen to 'spontaneous generation.'

Have I not at p. 412 put the vast distinction between you and Lamarck as to 'necessary progression' strongly enough?

Huxley's second thousand [3] is going off well. If he had leisure like you and me;—and the vigour and logic of the lectures, and his address to the Geological Society, and half a dozen other recent works (letters to the 'Times' on Darwin, &c.), been all in one book, what a position he would occupy! I entreated him not to undertake the 'Natural History Review' before it began. The responsibility all falls on the man of chief energy and talent; it is a quarterly mischief, and will end in knocking him up.

I am sorry you have to go to Malvern. The good of the water-cure is abstinence from work; a tour abroad would do it, I am persuaded, as effectually and more profitably.

I hope my long letter will not task you too much; when I sit down to write to you, I can never stop. Hooker, not having heard from you, is growing anxious, and hopes it is because you are corresponding with me and not because of serious ill-health.

Ever affectionately yours,
CHARLES LYELL.

To LADY LYELL.

Osborne: May 6, 1863.

My dearest Mary,—At Southampton I found a Queen's Messenger, and on the way to the docks fell in with Lord Stanley of Alderley, who had come down by the same train, bound for Osborne. At the docks in a room there, Sir James Clark was waiting, having only arrived ten minutes before me. He had hired a nice open carriage, and given me the offer to go on in the 'Elfin' steam yacht with Lord Stanley to Osborne, or with him to Netley. I chose the latter, and

---

[3] *Lectures to Working Men.*

after a pleasant drive of some five or six miles after crossing the Itchen ferry in a steamboat, we saw the Abbey, and then went over the Military Hospital, an immense building for about 1,000 invalid soldiers, with four or five Professors of Medicine and Surgery, Museum, &c. Most of the patients are soldiers who have served in the East Indies, and it must be good practice for the young medical students from twenty to twenty-four years old to attend here on those who have Indian complaints of the liver, &c., just what they will find when they get out to India. A lecture was being delivered by one Professor, another showed me a collection of skulls for ethnology presented by army surgeons, another pointed out to me numerous bones with gunshot wounds, &c. After two hours there, and seeing geological specimens and Bracklesham fossils from an artesian well (*Cardita planecostata, Fusus longevus*), 180 feet deep, which they have dug, we left.

It is a beautiful site near the sea, this Netley Hospital, on a deep and extensive bed of gravel, and all the newspaper stories of its being in a bog were inventions. The colonel who commands assured me they are never annoyed at low tide by the smell of the mud.

We had expected the ' Elfin ' to return from Osborne to take us up at the Hospital, but Prince Leopold's tutor and Prince Alfred's had got hold of her to go on board the great ironclad and ram ship of war the 'Resistance,' and we received a message to take the Southampton packet. So we started in a boat with good sailors sent from the ' Elfin,' and glad I was of my cloak, for we were for an hour tossing about off Calshot Castle. Both the Southampton boat and the ' Elfin ' came in sight from opposite quarters at last, and it was twenty minutes before we knew which we should take. At last we got into the ' Elfin ' which took us to the 'Resistance,' but the Prince's party had not done their inspection, and the ' Elfin ' took us to the pier at Osborne, and Sir James and I landed and walked with our small bags through the pleasant grounds to the house, which we reached at half-past five o'clock, and we got a cup of tea. Dr. Lyon Playfair has been here for a day, but has left. I believe from what Sir James tells me, that I shall not get back till Saturday, but in good time.

It has been a heavenly day, or the toss in the boat, which
I by no means suffered from, might have been very disagreeable.   Had there been mist, it would have been an un-
pleasant adventure.   We saw the outside of the great ram,
the ' Resistance,' famously.

My love to all.   Believe me ever yours affectionately,

CHARLES LYELL.

To LADY LYELL.

Osborne : May 7, 1863.

My dearest Mary,—After I wrote to you Becker came to
my sitting-room, which opens into the bed-room, commanding a pleasant view of the park with the trees in their fresh
green foliage, as forward or more so than in London.   He
took me down to the dining-room, where I was introduced
by Sir T. Biddulph to his wife and to Lady Mount-Edgcumbe,
now in waiting, and her young daughter of eighteen on a visit.
Colonel Cowell, whom I talked with, you remember, on a
former occasion about his visit to the Dead Sea, was also in
the drawing-room, and we sat together at the dining-table,
and he told me of a late visit to St. Vincent in the West Indies
—the pitch lake of Trinidad, &c.   Some fifteen in all sat
down to dinner, and were very merry.

Latish in the evening, when cards were playing at the
table, Lady Augusta Bruce came in, and asked me after you,
and had a long talk about American affairs, on which she is
very enlightened, though leaning to the opinion that for the
sake of the world (*i.e.* England), a separation might be
better ; but I think I modified her views.

The Queen's open carriage and four, with herself and
Princess Alice, is just driving past my window for Cowes,
where she is to visit the Prince of Leiningen.

The secretary, Mr. Ruland, was at breakfast this morning,
and the same party as at dinner yesterday, except the Post-
master-General.   Sir C. Phipps talked to me with great
admiration of Arthur Stanley, and showed me an excellent
photograph of him.   I am going to take a quiet read, and
then walk for an hour before luncheon with Sir James Clark,
if not stopped by a thunderstorm which is threatening.   I

have just finished Milman's preface, which is excellent, and have lent it to Sir J. Clark.

Three o'clock.—I only got a short walk in the garden with Sir James Clark when the Prince of Hesse sent for me, and as I entered the Princess Alice received me, and said she remembered me since Balmoral days, and introduced me to her husband. Her manners are very charming, and she talked most freely on all subjects. He has really read me as far as the end of glacial chapters with attention. They had been discussing the time it would take for all the existing races to have come from one original pair. Arthur Stanley is evidently a very great favourite with the Princess. I had about half an hour's talk with them, and then retired to luncheon, having promised at four o'clock to go to the Swiss Cottage with Prince Louis to see the collection, which consists partly of objects brought home by his brother-in-law Prince Alfred, who he said wished to join us.

Ruland the secretary is here, so the Germans muster strong. When the Prince of Hesse wanted a word in English, he asked his wife in German to help him.

Becker is reading one of Mudie's copies of my book, lent him by Colonel Elphinstone. The Princess Alice said I should give her a copy of the second edition, as I had done to her sister. I believe I am to see the Queen before we go to the Swiss Cottage.

Geikie's book on the ' Glacial Period in Scotland ' is very well done, and may enable me to make the ' Elements ' a little different from the ' Antiquity,' in which last, however, there is nothing wrong, or discordant with Geikie's facts and arguments touching glacial matters.

Seven o'clock.—The Queen sent for me before four o'clock, and talked with me alone for an hour and a quarter. Mostly about Prince Albert, leading me also to talk of him. Arthur Stanley recommended her to read my ' Antiquity.' She asked me a good deal about the Darwinian theory as well as antiquity of man. She has a clear understanding, and thinks quite fearlessly for herself, and yet very modestly. Nothing could be more natural or touching than her admiration for the Prince. She said that for one who had so much enjoyment in the present. which he found wherever he was, it was remarkable that he was cheerful whenever he had to change place or business. If they were at Balmoral or Osborne, and were called to Windsor, he not only went, but never allowed himself to be put out. As soon as we had done talking over books, &c., I went to the Swiss Cottage, where the Prince of Hesse and Prince Alfred were waiting. The last showed me all over the museum—silicified woods from Antigua brought home by himself, a very nice collection ; a collection of Portland stone fossils, tertiary shells, a few of which he said I gave him when last at Osborne, stuffed birds which he or the Prince of Wales had shot in Canada and elsewhere, &c.

The Princess Helena and her next sister joined us at the museum. I then started with Becker and Ruland on a geological walk along the seaside, and in our way, fell in first with the Queen driving Princess Alice in a pony car. They went alone without any servant to the seaside, called a boat, and took a row afterwards. Prince Alfred took a boat and rowed the Prince of Hesse out to the Queen's boat. He had no sailor to help him. In our walk we fell in with Miss Hilliard and Princess Beatrice, who asked us to come to a small miniature fort constructed in the grounds by Prince Alfred, with a moat and rampart and drawbridge. Here she played all sorts of pranks with Becker, who barked as a dog, and got into a small miniature barrack, and then shut her up in it, and so on. She has wonderful spirits.

The weather is charming, but the gardeners and farmers are in despair at the drought. They fear there will be no hay.

I hope our aunts are arrived. My visit here has been a very agreeable one thus far. I am made to feel so very welcome.

I hope to hear from you this evening or to-morrow morning, my dearest love.

Princess Alice asked me after you and where I had left you.   Ever your affectionate husband,

CHARLES LYELL.

To LADY LYELL.

Osborne : May 8, 1863.

My dearest Mary,—Sir James and I leave this at ten o'clock to-morrow morning.   Yesterday we had Mr. Elliot to dinner, who says the Greeks are settling down, and getting reconciled with the new scheme.   He describes the enthusiasm for Prince Arthur as having been very great, ' the son of the widow,' as they called him.   Only 100 voted for a republic, but their doing so freely was an advantage, as showing that they were at liberty to choose.

I have been with Becker this morning over Prince Alfred's museum, and find there is less disorder than I thought—some rubbish to be thrown away of course, but most of it is very fairly grouped.   We went over the kitchen in the Swiss Cottage, in which the Royal children cook all sorts of things, quite a large *batterie de cuisine*, and they invite Becker and others to come and eat the products.

When the Swiss Cottage, in part of which the gardener lives, was built, the boys used to work two or three hours a day in earnest with the labourers, and got certificates of work done from the foreman, and sent in a regular bill, which the Prince Consort paid exactly according to the then rate of wages, to give them an idea of such things.

Then we went over the tool-house, the initials of every child on each watering-pot, wheelbarrow, &c. ; then over the separate garden-plots belonging to each, from the Princess Royal's down to Princess Beatrice's, each of equal size.   No great variety, because; if one of the elder ones chose to have a row of potatoes, and of strawberries, and of currants, &c., each of the others imitated.   The flowers do not take up a fourth of the whole.   I came in and read for some time, and then set out on a second walk with Sir James Clark, to a kind of steward's house, older than the Palace, once a monastery, with a great variety of shrubs, and a holly with a larger trunk than I ever remember seeing.   The *Erica arborea*, now in flower, flourishes exceedingly in the garden.   *Weigelia rosea*, I think, is the name of a splendid flowering shrub which Fortune brought twelve years ago from Shanghai in China. It is in great beauty here, standing the drought well.

30th.—The Queen has just started with Lady Mount-Edgcumbe and Sir James Clark to see Netley Hospital. It is almost the first time she has proposed anything of the kind, and they are all glad she is going. I take for granted that Princess Alice has also gone, but I did not hear.

The Queen has always dined privately with her own family, and will I suppose do so to-day.

A most beautiful small aneroid, not bigger than a very large pocket watch, has come down from Negretti, for the Queen to measure heights with at Balmoral. They have a fine telescope for star-gazing, and the night I arrived, saw Jupiter and his satellites, which I only heard about to-day. At every turn one meets the hand of the Prince Consort.

Six o'clock.—I have been walking with Becker and Ruland to the meteorological observatory, where there is a regular clerk of the works. For twelve years self-registering instruments for measuring the force and direction of the wind (to-day it is NNE.), and the fall of rain, and the temperature, degree of moisture, quantity of ozone in the air, barometric observations, &c., all printed monthly. These we inspected, and then out along the wooded walks near the sea; the wild furze and broom, hyacinths, some primroses in beauty.

Seven o'clock.—Sir James has just returned from Netley, where the Queen walked over the spacious Hospital till poor Lady Mount-Edgcumbe was knocked up. The Queen is actually taking a drive in the grounds after all the inspection and yachting. I received your letter this morning. I was glad to hear of the arrival of our aunts, and hope to see them soon after you get this.

Dearest love, your affectionate husband,

CHARLES LYELL.

Sir James has his *congé,* and we go to-morrow at ten o'clock. The Queen has sent word that she is to see me to-night to take leave.

To His Sister.

53 Harley Street, London: May 12, 1863.

My dearest Marianne,—I am in your, and I believe
everybody's debt in the way of letters, Caroline I believe for
one, whose account of the comparison of *Lycæna dispar* and
its foreign variety I was glad to have. The English entomologists declare that no one can deceive them by trying to
pass off a foreign specimen as British upon them. I have
got a fine specimen of the American variety of the common
Admiral, or *Vanessa Atalanta*, which I mention in my book,
and which I will send when you are at Drumkilbo again, for
at present none of you would have time or heart to enter on
a comparison of the European and American races, which
look so like and yet are always distinguishable.

I am busy to-day thinking what I am to say to-morrow
at a great meeting of the Literary Fund, Lord Stanhope in
the chair; having this morning got positive intelligence that
I shall have to reply to a toast on the ' Writers on Science.'
There is always a great muster of authors, and a good many
ladies. I have paid ten guineas as my subscription towards
the fund for distressed authors, besides a guinea for the
dinner, as I am one of the stewards. I have felt it right as
a ' successful author ' to contribute to the unsuccessful, and
believe it is really very well managed.

I daresay Mary and Katharine will have told you my Osborne
news. It was a great satisfaction to have a good long talk quite
alone with the Queen for an hour and a quarter, and about
one for whom I had such a regard, and for whom I felt, though
it would not be etiquette to say so, such real friendship, as I
did for the late Prince Albert. I do not think she has given
way more than is perfectly natural—all necessary duties she
has performed. The quantity of work thrown on her now is
great. I told her that when the Social Science people
pressed the Prince to be their President, he told me the
anniversary address would be a severe addition to his work.
No one of his speeches was more difficult, or, I think, better
done, especially on the connection of science and religion, so
difficult a question for a public man to deal with. She said

this address, and the thinking out the whole subject
thoroughly, as he always did, was one of the things which
overtasked him. She said he was always cheerful and
determined to think everything for the best, a short life,
among other things, of which he had sometimes a slight
presentiment, in spite of his good health. She has of course
been reading many serious books, and I asked her if she had
read what Sir Benjamin Brodie has said about death. She
said she had, and was much struck with his observing
that if we knew what those we had loved were doing in
another world, or if we even knew the exact time of our own
death, it would alter the whole complexion of our lives,
and probably make us perform our duties less well in this
life.

I had a talk with Princess Alice and her husband separately in their room on another occasion, and another evening
a long conversation with the Queen, Prince and Princess of
Hesse, and Prince Alfred—very cheerful, about books and
things in general. One morning I had a walk with the Prince
of Hesse and Prince Alfred, and no one else, and Prince Alfred
showed me over the museum in the garden, in which are all
the birds stuffed which he and the Prince of Wales had shot
in different countries. A fine set of half-polished silicified
woods of various kinds—palms, exogenous wood from
Antigua, fossil shells from Portland, Isle of Wight, &c. He
showed me a few Isle of Wight fossils I had given him,
named years ago, which I had forgotten. I was introduced
to Princess Helena by Prince Alfred, when she looked in at
the museum.

The dinner party included none of the Royal family, and
were most of them well known to me ; Sir C. Phipps, Sir T.
Biddulph, Colonel Cowell, Sir J. Clark, Lady Augusta Bruce,
two German ladies in-waiting on Princess Alice whom I had
not seen before, Lady Mount-Edgcumbe and a very agreeable daughter, and some others—a very merry party.

I have only as yet seen the Princess of Wales at a
distance, but on the 19th inst. Mary and I are asked to a
reception at St. James's Palace, which they are to have for
the Queen. The Queen asked me about the success of my
new book, about which I find Canon Stanley had spoken to

her, which had led her to request a copy from me. I had taken down a copy of the second edition, thinking that it might be useful, and Princess Alice asked me for one, and told me I should dedicate it to her. Her husband had actually read steadily through the Queen's copy as far as near the end of the glacial chapters.

I got some time for reading when I was at Osborne, and went over the meteorological observatory, which is first-rate, and the records of wind, rain, electricity, &c. &c., all kept by self-registering instruments, and well kept, and published monthly.

I went to Osborne on Wednesday and returned Saturday. The ' Fairy ' brought us back, Sir James Clark and me, in grand style from the pier at Osborne between ten and eleven o'clock, and then by fast train we went to town.

The Royal Academy dinner this year was a brilliant affair, and I was glad to have a good talk with Kinglake, author of ' Crimean War,' who sat next me. Lord Palmerston's speech also was very entertaining and lively. There is a pleasant French article on my book in the ' Revue des deux Mondes,' by Laugel.

The Queen told me that her sons had asked her if the Colenso whose Arithmetic they had studied was the Bishop, and had remarked ' Then he must be very clever.' I told her that my nephew Arthur had said, ' I don't like Colenso ; he gives me hard sums to do.' She laughed, and asking his age said, ' All mine were older.'

Believe me ever your affectionate brother,

CHARLES LYELL.

To THOMAS S. SPEDDING, ESQ.

53 Harley Street : May 19, 1863.

My dear Spedding,[4]—I was very glad to hear from you, and to know that you had been reading my book, which has met with great success, having as Mudie told Murray a few days ago, divided the reading world, so far as his library is a test, with Kinglake. We have sold nearly 5,000 copies. I wonder I have been let off with so little serious antagonism ;

[4] T. S. Spedding, Esq., of Mirehouse, Keswick.

only a few indignant remonstrances on the part of the
' Record ' and some of the Church reviews for ignoring the
Bible, and writing just as if I had never heard of such a
book, and could take for granted that the scientific readers
were as indifferent as myself at the irreconcilability of my
pretended facts and reasonings with Scriptural truths.

The question of the origin of species gave much to think
of, and you may well believe that it cost me a struggle to
renounce my old creed.   One of Darwin's reviewers put the
alternative strongly by asking ' whether we are to believe
that man is modified mud or modified monkey.'   The mud
is a great come-down from the ' archangel ruined.'   Even in
ten years I expect, if I live, to hear of great progress made
in regard to ' fossil man.'

I am in hopes that the struggle in America will rid the
country in the course of twenty years of that great curse to
the whites, slave labour, and if so, it may be worth all it
will cost in blood and treasure.   My New England friends
do not despair, though indignant at the mismanagement at
head-quarters.

Believe me ever most truly yours,
CHARLES LYELL.

To the REV. W. S. SYMONDS.

Barton Hall, Bury St. Edmunds : June 23, 1863.

My dear Symonds,—Since I wrote to you our plans have
been somewhat more matured (and I write again, as I forget
what day I told you we should probably leave town and
journey towards South Wales).   I propose to go on the 14th
straight to the inn near Gower Point, from which I see by
Murray's book that tourists see the caves of the Gower
peninsula, those caves which Dr. Falconer and Colonel Wood
have lately been examining, and shall try and form an
opinion as to the relation of the extinct animals to the raised
beach and to the glacial drift of South Wales.

My next point is the Cefn caves near St. Asaph, in the
north-east of Wales, examined by Trimmer, Ramsay, and
Falconer.   I have a letter from Ramsay about them.   They
are best calculated in North Wales to throw light on the

relative age of the extinct mammalia and the glacial drift, and in their neighbourhood Captain Thomas showed Ramsay drift with shells.

My other point is Moel Tryfaen, and the shells found at great heights by Trimmer and Ramsay near the Menai Straits. Not that I may succeed in seeing the shells, for they are very rare, but Ramsay has given me the name of a guide who can show me the drift and region where they have been found.

As it is only an hour or two by rail to Holyhead, I may go there. I hope to see a peat moss out of which Mr. Stanley, M.P. has dug two specimens of mammoth sent to the British Museum. He has invited me to see the spot.

You were good enough to suggest some months ago that, if the time suited, you might be able to accompany me on my tour. If you could, whether for a part or the whole of the three weeks' run, I should be very glad.

You said in one of your letters that there will be many who will go soon to Wales to see whether the enormous changes of level in Post Pliocene times at present inferred from the glacial and other phenomena are legitimate speculations. Even a brief glance of some of the ground will I am sure enable me better to judge than a great deal of reading, or will at least qualify me to read critically what has been said by others.

Believe me ever sincerely yours,
CHARLES LYELL.

To SIR CHARLES BUNBURY.

Pendock Rectory : July 19, 1863.

My dear Bunbury, —We are enjoying beautiful weather and a splendid view of the Malvern Hills from our windows, having yesterday had a fine drive through the vale of Evesham to Tewkesbury, and then here. I examined Strickland's Cropthorn beds, and found at the level of the elephants and *Cyrena fluminialis* what *may* be a 'core' from which flint knives were struck off, but perhaps Evans may say of it, as of certain prismatic flints of the elephant bed of Icklingham, that they are natural and not artificial productions.

He (Evans) has promised to go to Icklingham in the summer, and to call on Mr. Prigg and try and make out the question of the old trenches which contain tools.

My visit to Charbes and Saint-Prest and the Paris museums has satisfied me of ice-action in the time of *Elephas meridionalis,* and as to Desnoyer's proofs of 'pre-glacial man,' and horns and bones cut and scraped and broken by man, at the period of the Cromer forest bed. They are certainly very curious, not to be pooh-poohed, and yet tantalising because one wants more evidence. *À propos* to the same Cromer bed, King has just written to me to say that he has found the rhizomas of *Osmunda regalis.* I should like to know your opinion of the kind of evidence on which he relies so confidently. He says 'the rhizomas are so large that they must have risen like tree-ferns two or three feet out of the ground.' Mr. Symonds has got a specimen containing, in one small fragment of rock, two or three seeds representing the oldest known plants with the oldest yet known Silurian fish, which he tells me he is to give you for your museum.

With love to Frances, ever affectionately yours,

CHARLES LYELL.

To DR. JOSEPH HOOKER.

London : July 31, 1863.

My dear Hooker,—I have been reading the paper on 'Welwitschia' with as much pleasure and profit as one so ignorant of botanical details of structure can do. I am glad you threw out a few hints on its bearing on the development theory. It is a splendid anomaly, and had it been carboniferous instead of a living plant, would have afforded Agassiz a fine illustration of his favourite theme, that in the earliest periods adult organisms were what afterwards were only exemplified in the embryonic stages of more highly organised creatures, animal and vegetable.

I have been with my wife and nephew Leonard geologising for two days between London and Rochester. We found primroses in flower.

The Welwitschia would be enough for one year's work,

though a mere episode in yours. The illustrations most
telling.   Ever sincerely yours,

CHARLES LYELL.

To LEONARD LYELL.

53 Harley Street : August 23, 1863.

My dear Leonard,—I have been a long time answering
your letter, and have now to report that I went yesterday to
Mr. King's, whose shop is much improved, and bought three-
pennyworth of *Anacharis,* as you wished, which Aunt Mary
has put into the aquarium, when she counted the fish and
other creatures mentioned in your letter, and found the
number agree with your list, besides the small fry.

I have been driving to-day all round Battersea Park,
which is now beautified with flowers. Miles of Tom Thumb
geraniums, blue lobelias, yellow and other calceolarias, and
a background of dahlias. Some day I hope to take you to
see it.

Aunt Mary and I went over Victoria Park the other day,
where there is a greater extent of ground, with an equally
splendid display of flowers, and with a long piece of water
on which were boats let out to hire and much used.

I suppose you heard that Aunt Mary and I had a very
pleasant tour in Wales. On a hill called Moel Tryfaen, at a
height of 1,300 feet above the sea, I found twenty species of
fossil shells, all of living species, in sand and gravel fifty feet
thick. You would have known most of them familiarly,
for there was the common cockle, whelk, eatable mussel,
Mya truncata, the common turritella, and others, but with
them *Tellina proxima, Natica clausa, Mangelia pyramidella,*
and other northern species. On the whole an arctic fauna
like Spitzbergen. I have mentioned this hill in my last
book, and the shells, but only twelve were known before—
found some thirty-five years ago, and as Charles Darwin
could find none, the fact was disputed by some. Luckily a
new mining company wanting to get roofing slates, had
spent 150*l.* in laying open this section just in time for me
to see it. Next year probably it will all be closed up again,
but not I hope till forty species have been found.

These shells show that Snowdon and all the highest hills which are in the neighbourhood of Moel Tryfaen were mere islands in the sea at a comparatively late period, or when these living European mollusks were flourishing.

I had a good view of Colonel H. Bunbury's place, Abergwynant, the day I rode up to the top of the mountain called Cader Idris. July is often a rainy month in Wales, but this year we had constant fine weather. The sportsmen who came to fish complained of the want of rain, as the salmon could not get water enough to ascend the rivers, and were jumping in the sea in swarms off the coast of Aberystwith.

With my love to Frank, Arthur, Rosamond, and papa and mamma, believe me, my dear Leonard,

Ever your affectionate uncle,

CHARLES LYELL.

CHAPTER XXXVII.

FEBRUARY 1864–NOVEMBER 1865.

SCHEME FOR CAVE-EXPLORATIONS IN BORNEO—DUKE OF ARGYLL ON 'VARIATION OF SPECIES'—BERLIN—THE CROWN PRINCESS—CROLL ON 'CHANGE OF CLIMATE DURING GEOLOGICAL EPOCHS'—EARTH-PYRAMIDS AT BOTZEN—HEER'S 'URWELT DER SCHWEIZ'—LETTER TO MR. SPEDDING ON THE 'AMERICAN WAR'—SIR JOHN HERSCHEL'S DRAWINGS OF THE EARTH-PILLARS.

[In the beginning of 1864 his father-in-law, Mr. Leonard Horner, died, in his eightieth year. In April Sir Charles and Lady Lyell made an excursion to Midhurst (where he had been at school) and went on to Salisbury to see its valuable Archæological Museum.

He was elected President of the British Association, which held its thirty-fourth anniversary at Bath, where he dwelt in his Address at length on the thermal springs of that place, and having spoken on the phenomena of glaciers, he alluded to the antiquity of man, a subject which was beginning to attract general attention.

This year he was made a Baronet.

He spent the Christmas of 1864–5 in Berlin, with his brother-in-law Chevalier Pertz, and in the summer of 1865 he and Lady Lyell, accompanied by their nephew Leonard Lyell, revisited the Alps, and went to the earth-pyramids of Botzen and the Glacier Lake of Aletsch.

Through the greater part of his life, he had suffered from weak sight, and therefore was frequently read to, and he dictated much of what he wrote. During his last ten years, he had the benefit of a very efficient secretary, a lady [1] gifted with a rare intellectual power. From her daily intercourse with one who never failed to inspire all those who were with him with a love of his science, she acquired an extensive acquaintance with the subject, which has since enabled her to popularise the study of natural science among young people.]

[1] Miss Arabella Buckley, author of a *Short History of Natural Science ;* the *Fairy Land of Science ;* and *Life and her Children.*

CORRESPONDENCE.

To W. PENGELLY, ESQ.

53 Harley Street: February 12, 1864.

My dear Sir,—Mr. Alfred Wallace, whom you know by name, told me the other day that there are limestone caves in Borneo, within reach of Rajah Brooke's jurisdiction, which deserve more than any in the whole world to be explored, as he feels sure they must contain the bones of extinct species of anthropomorphous apes most nearly allied to man, just as the Australian caves afford us fossil species of extinct kangaroos and other marsupials. He proposed to me to get the Royal Society to make a grant for the exploration of some one of these caves, and asked if I knew anyone prepared to undertake it, and could be sure that Sir James Brooke would encourage the exploration, if I would do my best. Do you not sometimes meet Sir James at Miss Coutts's, or is he not somewhere at Torquay ? No one could better explain than you the peculiar interest of such an inquiry, and if I could get Wallace into correspondence with him, we might perhaps find some adventurer with competent knowledge to undertake the enterprise and get the scheme into such shape as to enable us to appeal to the Royal and Geographical Societies, and perhaps get up a private subscription in aid of the object. It is precisely the kind of investigation which no surveyor sent out to find gold or tin or coal, or other products of economical value, would be justified in following up, or even giving to it a small portion of his time, and you well know how much time and patience it would require.

Believe me, my dear sir,

Ever most truly yours,

CHARLES LYELL.

To GEORGE TICKNOR, ESQ.

53 Harley Street, London: April 28, 1864.

My dear Ticknor,—When I last wrote to you I had only read the first half of your ' Life of Prescott,' and if I admired and enjoyed it exceedingly, you may well believe that the last half interested me still more, for while I think its merit as a biography is equally great, it had the additional charm to me of embracing the time when I knew Prescott personally, and so many of his friends. It was like living over again those years which we spent in America, which we always look back upon as among the happiest of our lives.

I was very much struck with the skill you have shown in making Prescott tell his own story, and yet omitting those parts of his memoranda and letters which would have diluted the tale. As it stands, it is very racy and pithy.

I am to-day going to meet Sir James Brooke, the Rajah of Sarawak, and Mr. Ricketts, the first consul appointed by the Government here to that new region, which has so suddenly started into commercial importance, for the sake of organising a scheme for exploring some caverns of lime-stone in the Rajah's dominions, from which they have hitherto got nothing but eatable birds' nests. But I feel persuaded that on examining the floor, deeply covered with guano or the dung of countless bats or vampires, they will find fossil bones; and as we have obtained extinct kangaroos and other marsupials from the Australian caves, and extinct forms of armadillo, sloths, and American monkeys from the caves of Brazil, so I hope to get extinct ourangs, if not the missing link itself, from these Borneo explorations. Miss Coutts is a great friend of the Rajah, and we are all to meet at her house. She thoroughly enters into the spirit of the undertaking. If we can only get a few bones by a preliminary *reconnaissance*, by aid of the Rajah's officers and dyaks, we should easily raise funds here for larger excavations.

Ever affectionately yours,

Charles Lyell.

To Charles Darwin, Esq.

53 Harley Street, London : November 4, 1864.

My dear Darwin,—I was delighted to hear yesterday at the Athenæum that the Council had decided that you were to have the Copley medal, for when it was not awarded to you last year I felt that its value had been much lowered, and in my indignation at the want of courage implied in their hesitation, I sympathised with a friend who has long held that these medals do more harm than good, which, however, I have always been unwilling to believe.[2]

In the present instance it is of more than usual importance, not in a purely scientific point of view, for your reputation cannot be the least raised by it in the minds of those whose opinions you care for, or who are capable of judging for themselves as to the merits of such a book as the ' Origin,' but because an honour openly conferred by an old chartered institution acts on the outsiders and helps to increase that stock of moral courage which is so small still, though it has grown sensibly in the last few years. Huxley alarmed me by telling me a few days ago that some of the older members of the Council were afraid of crowning anything so unorthodox as the ' Origin.' But if they were so, they had the good sense to draw in their horns.

Believe me ever affectionately yours,

CHARLES LYELL.

To CHARLES DARWIN, ESQ.

Magdeburg : January 16, 1865.

My dear Darwin,—I was so busy with the last chapters of my new edition of the ' Elements ' before I left town a month ago, that I did not reply to your kind letter about my after-dinner speech on your Copley medal at the Royal Society anniversary. I have some notes of it, and hope one day to run over it with you, especially as it was somewhat of a confession of faith as to the ' Origin.' I said I had been forced to give up my old faith without thoroughly seeing my way to a new one. But I think you would have been satisfied with the length I went. The Duke of Argyll expresses in his address to the Edinburgh Royal Society very much what I have done (' Antiquity of Man,' p. 469), that variation or natural selection cannot be confounded with the creational law without such a deification of them as exaggerates their influence. He seems to me to have put the difficulty pretty clearly, but on the other hand he has not brought out as

---

[2] [It will be observed this does not apply to the late awards of the medals.]

fully as I should have liked him to have done, the great body of evidence so admirably brought to bear in your work, in proof of the bond of mutual descent, and the manner in which species and genera branched from common ancestors. He did not entertain this idea till he had read your book, and he is now evidently impressed with it, as I am; and he would, I think, go the whole length, were it not for the necessity of admitting, in order to be consistent, that man and the quadrumana came from a common stock. He does, indeed, in defiance of consistency, admit for the humming-birds what he will not admit for the *primates*, and Guizot's theology is introduced to support him; but the address is a great step towards your views—far greater, I believe, than it seems when read merely with reference to criticisms and objections. The reasoning about materialism appears to me admirably put, and his definition of the various senses in which we use the term 'law'; though, having only read the speech once, I am not yet able to judge critically on all these points. He assumes far too confidently that the colours of the humming-birds are for mere ornament and beauty. I can conceive a meaning in your sense for the advantage of the creature, or of its friends and enemies, in every coloured ray of light reflected from the plumes. We must indeed know far more than we do before we can dogmatise on the irrelevancy of particular colours to the well-being of a species. He ought also to define beauty, and tell us whether it is in reference to man or bird. I have no objection to the idea of beauty or variety for its own sake, but to assume it so positively is unphilosophical.

We have been about three weeks at Berlin, and I had some good geological talk with Ferdinand Roemer, Beyrich, Von Kœnen, Gustav Rose, Ewald, Dr. Roth, and Dove the meteorologist, besides Ehrenberg, Magnus, Lepsius, and Du Bois-Reymond, and an animated conversation on Darwinism with the Princess Royal, who is a worthy daughter of her father, in the reading of good books and thinking of what she reads. She was very much *au fait* at the 'Origin' and Huxley's book, the 'Antiquity,' &c. &c., and with the Pfahlbauten Museums which she lately saw in Switzerland. She said after twice reading you she could not see her way as to the origin of four things; namely the world, species, man, or the black and white races. Did one of the latter come from the other, or both from some common stock? And she asked me what I was doing, and I explained that in recasting the 'Principles' I had to give up the independent creation of each species. She said she fully understood my difficulty, for after your book 'the old opinions had received a shake from which they never would recover.' I shall be very glad to hear what you think of the Duke of Argyll's comments on the 'Origin.' I think that your book is a vast step towards showing the methods which have been followed in creation, which is as much as science can ever reach, and the Duke, I think, has not fully appreciated the advance which has been made, even in his own mind.

I had hoped that a copy of the 'Elements' would have been sent to you while I was still at Berlin. You will find much that is new, and nothing, I think, clashing with the 'Origin.' Please read my description of the Atlantis theory. I fear I shall return and find the book still unborn, which is too bad of the printer. Please let me know how your health has been during the last four weeks.

Ever most truly yours,
CHARLES LYELL.

P.S. In an article in the Berlin 'Punch' on the Pope's encyclical, in which all the innovations which trouble his Holiness are enumerated, 'Die Darwinische Lehre die uns alle Affen macht' was not forgotten.

Dover: January 19.

To SIR JOHN HERSCHEL.

53 Harley Street, London, W.: January 31, 1865.

My dear Sir John Herschel,—I was very glad to see your handwriting, and was on the point of writing to you on two subjects, to which I shall presently allude. The Irish siliceous rhomboids are remarkably regular examples of analogous results which I have seen in various countries where siliceous schists were traversed by divisional planes of cleavage, usually referred to expansion and contraction by heat after the original consolidation of the mass. But as to the real causes, I have never been able to satisfy myself. Sometimes I have seen very large fragments of shapes similar to that which you send. I will show it at our next meeting to some of those who have made a study of such appearances, and then write to you again.

The first question I wished to ask you, was whether you have read a memoir by Mr. J. Croll on the ' Physical Causes of Change of Climate during Geological Epochs,' in the August number 1864 of the ' Philosophical Magazine.'

It is well worth reading, but I am specially called upon to allude to it because I am now re-editing the ' Principles of Geology,' the ninth edition of which has been out of print for several years, and which my publisher says ought without delay to be reprinted ; a more arduous task than he is aware of, since already eleven years have elapsed since the last version saw the light, and Darwin, among others, has done much to shake our old opinions since that time.

In my eighth chapter, I repeated at p. 126 what I had said on your authority in preceding editions, as to the effect of the varying eccentricity of the earth's orbit. In your paper in the ' Geological Transactions,' vol. iii., you stated that the necessary calculations had not yet been made, but Croll now endeavours to found, on Leverrier's data, a difference of one-fifth in the winter temperature when the extreme of eccentricity occurs. If you have not read his paper, I should like to send it to you by post, and should be glad to know how far you think I may rely on his facts and reasons. Of their applicability to geology, I may, perhaps, form an independent opinion. His reasoning as to the direction of currents and the assumption that the physical geography of the globe in the glacial period resembled that now established, is entirely at variance with what we know, as is the assumed periodicity of hotter and colder periods, from the Cambrian to the recent era.

I feel more than ever convinced that changes in the position of land and sea have been the principal cause of past variations in climate, but astronomical causes must of course have had their influence, and the question is, to what extent have they operated ? So far as we at present know, the glacial period of the southern hemisphere coincided with that of the northern, which would not be the case if it were referable to the eccentricity of the orbit referred to by Croll. As I must allude to his paper, I shall be very glad to have your opinion on its merits. I daresay he sent you an author's copy. The other point on which I am desirous of consulting you relates to the earth-pyramids of Botzen in the Tyrol. More than forty years ago, at Dr. Fitton's I think, you showed me a drawing you had made of those columns. I have never seen anything published on the subject, and meant to visit Botzen, and may, perhaps, be able to go there this summer. But in 1857 I saw some remarkable exhibitions of the same phenomenon in some tributary valleys of the Upper Rhone, on my way to the glacier of Zermatt. I ascertained that the earth-pillars of Stalden in the valley of the Visp, a stream which runs down from Zermatt, belonged to an old glacial moraine, which forms a terrace all along the right bank of the Visp, and consists as usual of unstratified solid mud, with boulders rounded and angular, stuck irregularly in the muddy matrix, and where large blocks occur to form a capping, the rain has washed away some of the loose earth, and left pillars. An earthquake had occurred a few weeks before, which had thrown down chimneys and rent walls, the cracks being still open in the town of Brieg on the left bank of the Rhone, a few miles above its junction with the Visp, and these same shocks had thrown down some of the earth-pillars at Stalden.

I therefore saw, with no small wonder, two isolated earth-pillars, one of which is called the Dwarf's Tower, or in their patois, 'Zwergle-Thurm,' a few miles from Brieg in a valley on the right bank of the Rhone. They are of exactly the same moraine-like structure as those of Stalden, the principal one thirty-six feet high, seven feet diameter, with an angular block at the summit about seven feet diameter, and various other boulders sticking out from the sides.

The slope of the hill-side is very steep, at an angle of about 43°, a deep narrow glen, cut in mica-schist. If the two isolated columns mark the former position of a moraine of an extinct glacier, the entire removal of the rest of the moraine, and the subsequent excavation of several hundred

feet of a narrow glen, are remarkable, and it is strange
that an earthquake never threw down these columns; but I
suppose they are the last of hundreds that have been shaken
down. If the pine-trees were cut down, they would have
a most singular appearance.

I could not find any rubbed or glaciated pebbles in the
Dwarf's Tower, but I found some in the earth-pillars of
Stalden. One block of gneiss, twenty-two feet long and ten
high, caps two pillars at Stalden quite in the Stonehenge style.
The glaciation of some of the pebbles proves beyond a doubt,
in the Stalden case, that the pillars are part of the moraine
of one of the old extinct glaciers of gigantic size. I should
like to see your drawing, or a copy of it. Have you ever
published it? I should like to give a woodcut of it, as
contributed by you to my new edition, giving the date, for I
suppose in forty years the rain has done something to alter
it. I made some drawings of the Stalden and Dwarf Tower
pillars which may be of some use as diagrams, but yours
gave a much better idea of the appearance, and I have never
seen any published engraving, though so much has been
written about the old moraines.

With our kind remembrances to Lady Herschel and your
daughters, believe me ever most truly yours,

CHARLES LYELL.

To PROFESSOR HEER.

53 Harley Street, London: February 4, 1865.

My dear Heer,—I was very glad to receive your kind
letter, and your friendly acknowledgment of the receipt of
my 'Elements' makes me feel that I ought long ago to
have written to thank you for your 'Urwelt der Schweitz.'
At Berlin I began the reading of your 'Urwelt' with great
care, and soon found how much the botanical and entomological parts of my 'Elements' had lost by my not having
sooner gone to school in the pages of your excellent treatise,
the merits of which I discussed with Professors Beyrich and
Ferdinand Roemer, both of whom were already familiar with
the 'Urwelt,' and fully appreciated its value and originality.

After reading the first hundred pages I was interrupted, but shall recommence very soon.

When I say that the Fontainebleau sand shells are all specifically different from the rich falunian of the neighbouring Touraine, I speak after going into the question with Deshayes, Edward Forbes, Bosquet, and others. Your argument founded on the corals of the Upper Miocene of the West Indies and European Faluns, showing that the ocean of the Miocene period could not have been very deep, or could only have had a moderate depth, is well worthy of consideration. The affinity of the corals seems to me against the existence of a continent, but it does require at least a great many islands like the coral region of the Pacific (or Oceanica).

I long to have your report on the Miocene plants of the Amurland, a country where the living mammalia are now almost all of the same species as those living in Western Europe.

When I have read more of your book I will send you more comments, and hope you will do the same in regard to mine. Your criticisms are the more valuable because I am now re-editing the ' Principles of Geology,' and am therefore going over much of the same ground, and have often no opportunity of putting myself right.

We are quite well, and my wife joins me in kind regards to you and your family. The sudden death of Dr. Falconer has shocked us much. He is a great loss to science, and much knowledge dies with him.

Believe me ever most truly yours,

CHARLES LYELL.

To SIR JOHN HERSCHEL.

53 Harley Street, London, W. : February 21, 1865.

My dear Sir John,—The drawings[3] are most valuable, and it is tantalising not to be able to use all of them, or to have only to extract telling bits of them. The Stalden case is particularly good, because a portion of the moraine not yet divided into columns has two big stones lying at the top. But the beauty of the groups of the Botzen columns

---

[3] Of the Botzen earth-columns.

is far greater. They are most elegant, and I cannot believe that the large undivided masses which you allude to, in the lithographed sketch of which unfortunately the date is not given, can be taken from the same point of view; at least, that would imply a more rapid rate of waste than I suspect takes place.

I hope to get to Botzen this year, and shall take your drawing and try to find your point of view, so as to judge of the amount of change in forty-four years; if it is not very considerable, then the splitting up of the unbroken masses to which you allude cannot have taken place between the execution of the print and your visit. The Stalden one is of more interest to me at present, because I ascertained it to be the work of one of the old extinct glaciers, containing glaciated fragments of rock, so that ever since the era of extreme cold there has been no flood in this valley to interfere with the gradual formation of these columns by pluvial action. On this account it is valuable, because the columns go down so low, the base of some of them being only slightly above the river, which would not have been the case if a flood had ever cleared away the lower part of the moraine.

As you have been looking into my new edition, I should like you to read Chapter XV., which is new, as it will give you some idea of the new part which fossil botany is playing, and the arguments for a Miocene Atlantis will, I think, interest you. I see that Professor Phillips has been reading a paper to the Royal Society on the appearance of the planet Mars at different seasons, and the quantity of snow in different years. Would it be possible to say whether at present there is a preponderance of land in polar as compared to equatorial latitudes in that planet? Phillips has always ignored my cause—the varying distribution of land and sea as influential in accounting for former changes of climate, though he has never, I think, given any good reasons for slighting it. If in Mars, as well as in the earth, land and sea are always shifting places, it may sometimes happen that the position of land and sea in the planet nearest the sun may be so favourable to cold as to counterbalance the effects of greater proximity to the great source of heat.

Ever most truly yours,

CHARLES LYELL.

To Thomas S. Spedding, Esq.[4]

53 Harley Street : March 12, 1865.

My dear Spedding,—It is always a great pleasure for me to hear from you, and so rare for me to differ from you in opinion that, had I not been four times across the Atlantic, and seen both North and South of the United States, I should doubt my own faith when I found you questioned its soundness.

I must begin by stating that I entirely differ on the first point you lay down, namely, the right of Secession. I think it was about fourteen years from the first outbreak of the war with Great Britain, that the thirteen States tried to get on with such a constitution as might have justified any one of the independent States in seceding. But they found it would not do, and they then made a constitution which is as much violated by Secession as would our constitution be if Ireland, Scotland, Wales, or even Yorkshire should declare themselves independent because they were dissatisfied with a new Parliament, and complained that they were not fairly represented.

Pray understand me, that I admit that every people have the right of rebellion or revolution whenever they are oppressed, or suffer such grievances as renders such an extreme measure natural.

But you have only to read Vice-President Stephens' earnest address to the legislature of his own State (Georgia) *before* the outbreak, when he tried to point out to them the folly of the proposed separation, to see that so far from having any just ground of rebellion, the South had been dominant to the last in foreign and domestic politics, had always had the lion's share in the choice of Presidents and other civil appointments, and in officering the army and navy. In short they rebelled simply because Lincoln's election showed them that the Free-Soil or Republican party were at last determined to resist the extension of slavery into new territories, although they would still permit them to retain the institution at home, because they were pledged to do so by the constitution. The Morrill Tariff was never

4 Of Mire House, Cumberland.

put forth by the Southerners as a ground of complaint, although it was afterwards by Spence, who, as he was a paid advocate of the South, ought not to be compared to Goldwin Smith.

The South had become very aggressive in regard to slavery when the North, who had too long yielded to them, determined to resist. No doubt the South were right in foreseeing that the non-extension of slavery into new regions would in the long run be tantamount to its extinction, or, at any rate, to their rapidly diminishing importance in the political scale. But this did not justify the rebellion.

I differ from you as to our having ever had any proof of the unanimity of the South in favour of Secession. No one of the Southern States took the steps, which by their con- stitutions they were bound to take, in order to obtain a sanction for an organic change in their constitution, even for one of much less moment than Secession from the Union. I believe that the slave-owning oligarchy did not dare call a Convention and put the question to a popular vote. I doubt whether they would have carried it even in South Carolina, and feel sure that they would have been beaten in Georgia and Tennessee by large majorities ; the Negroes going for nothing, and the white population alone voting. Sherman's unopposed march through Georgia and South Carolina does not surprise me. You will grant that if the population in those parts of our island where the artisan and working class are most intelligent, and the Baptists and some other sectarians very numerous, had been polled, whether in the first or fourth year of the war, they would have given a majority of votes for the North; and yet if our aristocracy had chosen to drag us into a war in favour of the South, they might have raised regiments out of the very class which felt most strongly for the North and against the slave-owners, and they would have fought with spirit when once excited, and fairly embarked in the feud. I have my fears as to the result in this country of a wide extension of the suffrage ; but if the opinions of such a class as the arti- sans of Lancashire, Birmingham, Newcastle, &c., were better represented in Parliament, the Government would have been more vigilant in preserving neutrality, and fewer ' Alabamas '

would have been built and allowed to steal out, and we
should be in less danger now of a war, or at least of having
to support additional taxes in order to be prepared for the
worst after four years of insolent and malignant writing by
such papers as the 'Times' and 'Saturday Review,' which
has estranged from us and embittered that party in the
United States which had the greatest feeling for England,
while it has done nothing to soften the hatred which millions
of Irish birth, with the democratic party which side with
them, cherish against us.

There has been a great combination of motives which
have led the aristocracy to take part against the North.
The most influential, I think, has been a John Bullish desire
to see the dismemberment of a power which if it does not
divide will soon interfere with British supremacy. Next in
order comes a sincere wish, for the good of mankind as they
think, for the failure of democracy. Thirdly, I think the
Liverpool merchants, shipbuilders, and others were anxious
for the destruction of the shipping of their rivals. Fourthly,
the general ignorance (in great part wilful), of the ruling
classes, of the real state of things in America, cannot be
exaggerated, and if such men as Gladstone and Earl Russell
had been only six weeks in the United States, they would
never have said what they did. If you had lived among the
Americans you would not have wondered at the loyalty they
have shown for the Union, which outside of the New
England States has had more to do with the spirit they
have shown than the feeling against slavery. The two
motives taken together—the integrity of the empire, and the
non-extension, and for the last two years the extinction of
slavery—constitute to my mind better grounds for a pro-
tracted struggle than those for which any war in our time,
perhaps in all history, has been waged. Although I think
the Irish might have rather a better case than the Southerners for repeal, yet I would fight for any number of years
rather than let that island be independent of us; and there
is no Englishman who, if he had settled in one of the North-
Western States of the Union, would not have the same
feeling in regard to allowing the States at the mouth of the
Mississippi to belong to some other power. How could the

loss of Bengal be compared to this, and yet how many
thousands of lives and millions of pounds should we not
have expended, rather than have allowed even India to
secede? If you visited America you would soon see that the
vast majority have such a feeling of security of property,
equality of rights and of religious sects, as makes them feel
toward the Government as the rich and upper classes do
towards our own.

Lincoln and his colleagues are not the sort of men that
you and I would put into a Cabinet, so far as their conventional manners are concerned, but you must recollect that
this is not the feeling of ninety-nine in a hundred of the
electors on the other side of the Atlantic. But after all, are
Lords Palmerston, Clarendon, and some others, men of higher
principle than Lincoln, or as high? I am intimate with
men equal to any here in literary attainments and in polish
of manners, and of independent fortune, in the United
States, whom I used to wish to see in power instead of the
coarser class into whose hands the reins of Government
have been placed. But these men and the majority of
capitalists would, I am sure, have knocked under to the
South, and the slave-owner would have made a compromise
by which his institution would have been more rampant
than ever. If slavery, which was more injurious to the
white man than to the Negro, and which to a certain extent
poisoned the political institutions of the North, as well as
keeping the South in ignorance, is got rid of, it will be
owing to a very extended suffrage among a class which
had had much instruction, for working men, but to whom
the aristocracy of wealth and refinement were not prepared to make great sacrifices for such an object. Nevertheless, when they were once embarked in the contest a
very large proportion of young men of the highest attainments and fortune went to the war, many of them serving
in the ranks, and this year after year, when the enormous
sacrifices and danger which they incurred were thoroughly
realised.

The 'Times,' whose leaders against the United States
were always copied into their newspapers, did much in one
way to help the North. It made them feel that if they fell

to pieces (and the Secession of the South once yielded would
have been followed by other dismemberments), they would
be treated by foreign nations, at least by us, with that contempt to which the weak alone are exposed. The writers
of the money articles of the ' Times ' hoped to damage them
financially, and they must have succeeded so far as to cause
the English to sell out, instead of taking any of the new
loans, about one-tenth of which has, I am told, been purchased
by the Germans and other continentals. But the result has
only shown how ignorant the ' Times ' was of the wealth of
the Union, as they have already borrowed some 300 million
from themselves, and a fourth hundred million is now forthcoming, almost all from the same source. It is surprising that
year after year the prophecies of the ' Times ' correspondent as
to the impending bankruptcy of the North, the impossibility
of their obtaining any more men, and their disunion had
been falsified, while the opposite statements of the correspondent of the ' Daily News ' on the same subject were as
regularly verified by the events, including the Presidential
election, that the ' Times ' should have tolerated the continuance in office of writers who must have associated with
Confederate agents at New York. You do not seem to think
that people in good society here have been ' Times '-ridden,
but I assure you I could always tell what people would think
on each separate question and event throughout the war,
if, before going to a party, I glanced at the leading journal ;
and if, as often happened, incidents of great importance were
under discussion, such as the treatment of prisoners by the
Southerners, as proved before a commission, or the fact that
only eleven per cent. of foreigners entered into the army, as
officially shown, few regular readers of the ' Times,' when told
of the suppression of such documents, cared to know the truth,
because it was disagreeable to have a chord struck out of harmony with the antipathies which had been excited by their
daily reading. They would sometimes tell you that life was
not long enough to read more than one paper. The contrast
in the feelings of the Germans, when last Christmas we mixed
with the reactionary Tory party, as well as with the Liberals
at Berlin, with the sentiment in good society here, was most
striking, and Judd, the American Minister there, remarked

to me how enviable was his position as compared to that
of Mr. Adams in England. But had people read the 'Daily
News,' 'Weekly Times,' and 'Star,' in this country, as well
as the 'Times,' a state of feeling among the literary and
scientific men more like that which I found in Prussia would
have prevailed.

By the way, the ability, literary and political, with which
the leaders of the 'Weekly Times' have been written for
the last four or five months, during which I have taken it,
on European as well as American affairs, has struck me as
marvellous for a penny paper.

You cannot say too much of the spirit, courage, and
military skill which the Southerners have displayed. They
have certainly shown the power of an aristocracy to command and direct the energies of the millions, and as by far
the largest true Anglo-Saxon army that was ever organised
in history, we may feel proud of their prowess, for there was
not that large mixture of Celtic and German blood which
was on the Northern side. I also respect men like Lee, who,
after Secession was determined on, and after long and painful
hesitation, decided that Virginia with its State sovereignty,
and not the Federal Union, was his country. This doctrine had
been carefully inculcated into the minds of the Southerners,
a minority only of whom consisted of the mean whites, while
the majority, such as formed the bulk of the soldiers of
Stonewall Jackson, were good honest farmers, largely drawn
from the upland country, having few, and sometimes no
slaves, but very ignorant compared to the Northerners of the
same class, even unable to read and write, because the planters
wished them to be so, and they could seldom get a school
within forty miles' distance, and then it was too expensive
a one and meant only for the planters' children. Although
a notion prevails here that refinement was the comparative
privilege of the Southerners, they exhibited in truth a low
form of civilisation. They have produced no historians,
poets, great preachers, novelists, nor any literary men;
orators and politicians alone, for to those departments their
aristocracy devoted themselves. The rebellion was chiefly
brought about by a small number of ambitious men, many
of whom while holding office under the Federal Constitution

treacherously prepared for Secession by filling the Southern
forts with cannon and ammunition, sending away ships of war
to distant ports, &c. &c. These men, for whom so much sympathy has been shown, formed a party which has always given
the most trouble to England, and if they are subjugated, it
will be far more easy for the Northerners, who are industrious,
and who have a distaste for military service, to keep the
peace, provided we can stave off an outbreak for a few years.
They will be bound over by heavy recognisances to keep the
peace, for at first the interest of their debt at 6 or 7 per
cent. will be about equal to our own, and a terrible burden.
But I am very hopeful, for there will be a great tide of free
labour pouring into the South, where slave labour has
hitherto had a monopoly. The resources of a magnificent
territory will be for the first time developed, and having,
since I first visited America, witnessed what has been
done in Texas and California, which did not belong to
the Union when we first went there in 1841, I feel sure
that people here will be astonished at the rapidity with
which the wounds will be healed when once the contest is
over.

My late friend M'Ilvaine, once secretary of the United
States Bank, before the days of repudiation, used to assure
me that the ignorance of the Washington Legislature of
political economy, and their want of sound free-trade notions,
was extreme. They are showing this now, and have aggra-
vated their debt by not being able to check the over issue of
paper by the State banks, to say nothing of enormous corruption of contractors and others. If one did not recollect our
history in the war with France, and the Protectionists of
thirty years ago, one might despair. But some adversity
was wanting to improve and chasten the republic, and they
will have it now.

It is strange that Canada is so anti-free-trade, and makes
me grudge expenditure in their defence. Australia, I fear,
is equally desirous of 'protecting native industry.' But we
must have patience. I expect the military discipline to
teach even civilians in the United States habits of subordina-
tion to central authority, which they needed, and that the
large debt will strengthen the Federal power which formerly

could not control the States from buccaneering and other
enterprises.

Had the States been dismembered, there would have been
endless wars, more activity than ever in breeding slaves in
America, and a renewal of the African slave trade, and the
future course of civilisation retarded on that continent in a
degree which would not, in my judgment, be counterbalanced
by any adequate advantage which Europe would gain by the
United States becoming relatively less strong.

You must recollect that from the first I had a deep conviction that the North would be staunch for the Union, and
for non-extension of slavery, and that they must prevail,
having such vast odds, not only of wealth and numbers, but
of knowledge and intelligence on their side.

I believe that if a small number of our statesmen had
seen what I had seen of America, they would not have
allowed their wishes for dismemberment to have biassed their
judgment of the issue so much. Certainly I never reckoned
on the South being able to make so splendid a stand, which
will be memorable in history; but then you must recollect
that I took for granted that Lincoln would never attempt
general emancipation. When abolition was exchanged for
non-extension, we could not reasonably have expected that
to be carried without four or five years of war, or what was
equivalent to a conquest of the South. The result, I think,
will be worth all this dreadful loss of blood and treasure.
As to the internal happiness of the States, so long as they
are not divided by hostile tariffs, and are not forced like the
Europeans to keep up ruinous standing armies during peace.
I am sure that had you travelled in the United States you
would feel, as I do, that their political organisation, with all
its faults, does not prevent their rapid and successful development. Whatever it may be for the rich, who have shown,
however, their loyalty to the constitution in this war, I
certainly think that for the millions it is the happiest
country in the world.

As for their external policy, it has hitherto been mainly
guided, not by a democracy, but by a slave-oligarchy. But
when that is got under, there is another curse, which like
slavery they owe to this country, and that is the millions of

Irish, who though their condition is wonderfully improved by the schools and good wages which they obtain there, form a class to whom the suffrage ought never to have been extended.

As to the foreign policy of the United States, could it be more selfish and aggrandising than that of Bismarck, or of France when they appropriate Savoy to themselves; and have we much to boast of? Italy was not helped by us to make the step she has made. But all this you will grant, and you will say that it only shows the danger of too much power wielded by any one country. But as I always felt sure of the success of the North, I have been vexed at what I consider the folly of this country doing so much to irritate the better class of Americans. We have, at the cost of several millions a year of taxes for the rest of our lives, indulged in four years of vituperation of a people who were doing exactly what we should have done, and it will be unnecessary now for the Americans to make war with us ; they have simply to bide their time, and whenever we get into a quarrel with any power, they have only as neutrals to imitate us, and the whole of our commerce will be swept off the seas. I hold that it is by the ignorance even more than to the prejudices of our ruling class, that this state of things has been brought about, and it has been partly owing to the monopoly of influence by one single journal, which I think tells far more in an affair of this kind than the Cabinet or even the Legislature.

Believe me, my dear Spedding, ever most truly yours,

CHARLES LYELL.

P.S. That there were never more than eleven per cent. of foreigners in the Federal army, I have from the American Minister, Mr. Adams. But this does not of course include born citizens of the United States of Irish and German parents. It only disproves the absurd charge of the Federals having fought with a hireling force, which people in *good* society still believe here.

To Professor Heer.

53 Harley Street: March 16, 1865.

My dear Heer,—I have been frequently citing your very instructive ' Urwelt der Schweitz,' in the new edition of my ' Principles of Geology,' and I am now referring to what you say of the huge blocks found at many points in the Flysch. This ' Flysch,' which I studied at Vienna, Salzburg, Genoa, and elsewhere, has always perplexed me much. You say truly that to suppose glaciers when the climate as indicated by the organic remains was so hot, would be a violent hypothesis. The absence of organic remains, and suspension of life in the seas during a glacial episode, might be conceived, and I should be inclined to imagine this, if I could rely on the facts which you give as to the large blocks of lias, oolite, granite, gneiss, talcose rock, &c., observed in the conglomerates of the Flysch.

I should be very glad if you could assure me that there is no doubt of large-sized angular blocks having been seen fairly embedded in the Flysch. I am well disposed to believe it after having seen them in the Italian Miocene.

You will be glad to hear that our friend Dr. Joseph Hooker has succeeded to his father at Kew, and is now fairly installed as the head of that department, which is becoming every day more and more creditable to this country as a scientific institution.

A good dissecting-room and a regular paid anatomist has been added to the Zoological Gardens, which will now become a good school of comparative anatomy, as so many rare animals die there. I am very busy at present, trying to make up my mind how much former changes in the eccentricity of the earth's orbit may have exaggerated the winter's cold, and helped to cause a glacial period.

Did I tell you, when I last wrote, that Agassiz has found *roches moutonnées* in the Organ Mountains near Rio Janeiro ? If there were glaciers in that latitude, how did the existing tropical plants survive the chill of the glacial period ? It seems improbable, according to Darwinian or any other principle, that the peculiar and distinct tropical vegetations of Australia, South America, and Africa should be all of post-glacial origin. There has not been time for such an amount of development of creational power, unless we admit the doctrines of the paroxysmalists.

Believe me, my dear Heer, ever most truly yours,

CHARLES LYELL.

To MR. PENGELLY.

Munich : July 26, 1865.

My dear Mr. Pengelly,—I hear from Mr. John Moore, that you have written an important paper about the Trade-winds, Gulf-stream, Sahara, &c. As you may not have had separate copies to give me one, I shall depend on your telling me in what Journal I can buy it, for it is a subject which I am working at. Is it not singular that the hot water of the Tropical Indian Ocean should, as we are told, flow round the Cape of Good Hope, across the Atlantic to Brazil, while the water of the Gulf of Mexico flows NW. towards the polar regions? If the Equatorial Sea, north of the line, sends its hot water towards the Arctic regions, the Equatorial Sea, south of the line, ought to send its warm waters towards the Antarctic latitudes. I suppose it is the geographical conformation of the land, and bottom of the sea, that causes this different behaviour of the southern and northern currents. If you have been studying the trade-winds, you can perhaps enlighten me. I presume that if there were no such winds, the heated tropical waters expanding and becoming lighter, would flow towards the north and towards the south from the equator, while cold under-currents would be flowing in opposite directions.

Believe me ever truly yours,

CHARLES LYELL.

To SIR CHARLES BUNBURY.

Dover : August 31, 1865.

My dear Bunbury,—I had expected to write to you from Calais, but we found it answer best to come over and sleep here. We have had rather a rough passage, but are all pretty well recovered, and waiting for our luggage from the Custom House. The Kissingen waters have done me as yet

neither harm nor good. Our Alpine tour answered very
well. The weather upon the whole favourable, and I
accomplished the only two geological points I was bent on
clearing up. First the pyramids of Botzen, and the similar
earth-pillars of several places in the Vallais, or rather in the
ravines of small tributaries to the Upper Rhone; and
secondly, to see the Glacier Lake of Aletsch, and whether it
would throw any light on the parallel roads of Glen Roy,
which it does most decidedly ; a large terrace being formed
exactly on a level with the Col which separates the valley of
the said Aletsch Lake from another valley, that of the Viesch
Glacier. Some persons have shrunk from the idea that the
drainage of the old Glen Roy lakes was effected in a direction opposite to that by which the valleys are drained where
the parallel roads occur. This is just what actually happens
when the Aletsch Lake, or Märjelen See, as they call it, is
blocked up by a barrier of ice, and gets quite full. It then
flows over a Col, or rather the stream issuing from it passes
over the Col on a level with the terrace bounding the lake.
I luckily saw the lake forty feet lower than usual, and was
able to examine the structure of the great shelf which is
level with the Col.

As to the earth-pillars, I had made up my mind before,
that those of Stalden, near Zermatt, were remnants of a
moraine of one of the great extinct glaciers, and I can prove
that every other case agrees as to origin in glacial time.
Some of the columns are from sixty to eighty feet in height.

Such a thickness of coherent unstratified mud, with the
requisite accompaniment of erratic or capping stones, could
scarcely be found in any but a glacial formation.

Leonard has been a most agreeable companion. He has
got on in German as well as in geology.

We shall start for the Birmingham Meeting on Tuesday
next, and return to town on Saturday, where I shall much
enjoy being stationary after so much wandering.

Ever affectionately yours,
CHARLES LYELL.

To Sir John Herschel.

53 Harley Street: November 28, 1865.

My dear Sir John Herschel,—I have often been on the point of writing to tell you the result of my comparison on the spot of your excellent drawings of the Stalden and Ritten earth-pillars and pyramids, with the scenes which they now present after nearly half a century. I shall now dictate a few words on the subject to my amanuensis, having found it necessary this last year to abstain as much as I can from using my own eyes, whether in reading or writing, by which means I get on well, and seem to have acquired a new lease for my eyes, and for ordinary purposes they feel much as they used to do.

I took my eldest nephew, aged fifteen, with me on my last expedition to Switzerland and the Tyrol, which added much to our pleasure, for he is quite an enthusiast in geology, conchology, chemistry, and mineralogy.

Except occasional attacks of lumbago, I am battling well with sixty-eight years, but am obliged to be very careful of myself. Remember us kindly to Lady Herschel and your daughters.

Believe me, dear Sir John, ever most truly yours,

Charles Lyell.

I found the Stalden columns when I first saw them in 1858 had suffered greatly from the earthquake of 1855, which had rent the walls of houses in the town of Visp, some of them not repaired when I was there. The principal capping stone had been thrown down, and many others probably loosened, for great havoc had been made in the seven years between 1858 and 1865. I suspect that but for this shock I should have found as many pillars as you saw.

I convinced myself that the unstratified mass in the valley of the Visp, and another on a still grander scale in the valley of the Borgne, which leads to Evolena near Sion, in the Vallais, are the moraines of old extinct glaciers. It is not merely the absence of stratification and of all organic remains, but decidedly glaciated stones, pebbles, and erratics which bear out this opinion. The same holds true with the

Botzen red mud, where I found all the same proofs of a morainic origin.

I could get no good photograph of these last, and no engravings in the least degree comparable to your drawings. I am therefore very desirous of availing myself of your permission to use your drawing of the Ritten pillars, perhaps as a frontispiece to my new edition. I took much pains to try and ascertain the exact point of view from which you took your sketch. I often thought I had discovered it, and it was not till the last day that it occurred to me that you must have been looking up the ravine, whereas I had made up my mind you were looking down, in consequence of the view which I got from the bridge the first day answering so nearly to your representation, making allowance for such slight differences in the columns as forty years might well bring about. Have you sufficient recollection of the place to be able to tell me, first, whether you are looking up the stream; secondly, whether on ascending the valley one would pass between *a* and *b*; thirdly, is not the torrent entirely hidden from sight by the wood? I conceive that it would go to the bottom between the principal groups of pillars in the line from *a* to *c*, then from *c* to *d*, and then in the line *d, e, f*; but of this last I am not sure, for there hardly seems space between the foreground on which the tree *e* stands and the opposite rock *g*. As I presume that the size of the copy you sent me agrees with that of your original drawing, the tracing which I enclose will suffice to show you what I mean, as my letters would in that case be exactly opposite the points in your original to which I allude.

I suspect that the scene has not altered materially, not even in any of its details, since you were there.

CHAPTER XXXVIII.

JANUARY 1866–APRIL 1868.

ON GEOLOGICAL CHANGES OF CLIMATE—TICKNOR'S 'SPANISH LITERATURE'—
THE INQUISITION BEARING ON NATURAL SELECTION—DUKE OF ARGYLL'S
'REIGN OF LAW'—MR. DARWIN ON 'VARIATION OF ANIMALS AND PLANTS
UNDER DOMESTICATION'—REMINISCENCES OF FARADAY.

[This year, 1866, he made excursions to the coast of Suffolk, to
Weymouth, and later on to Forfarshire, returning south by Whitby.
He received the Wollaston Medal at the Geological Society for
eminence in his science.

In 1867 the Exhibition was held in Paris, and after going there,
he went later on to Scotland, and attended the British Association at
Dundee. The tenth edition of his 'Principles of Geology,' much
enlarged, was brought out this year in two volumes.

In the year 1868 he made expeditions in various parts of
England, and was at the Norwich British Association, and afterwards
Sir Charles and Lady Lyell spent September at Tenby and its neigh-
bourhood with his brother's family.]

CORRESPONDENCE.

To PROFESSOR HEER.

53 Harley Street, London : January 21, 1866.

My dear Heer,—I am much obliged to you for your letter
to me announcing the death of our friend Gaudin, for whom
I had a sincere regard, and who was one to whom I always
felt drawn so much, that had I seen as much of him as you
did, his departure would have made an irreparable void in
the circle of my friends. He was indeed a most loveable
person, and he would have occupied a very leading place in
science if he had had more leisure to devote to it, and
had his health and physical strength been more on a par
with his talents.

I should have written sooner to thank you for your valuable pamphlets, but I have been very much absorbed with a new theory of changes of climate dependent on variations in the eccentricity of the earth's orbit. I am called upon to give an opinion as to the influence of this cause in a new edition which I am preparing of the 'Principles of Geology.'

I by no means abandon my old doctrine, that the principal cause of former fluctuations in temperature, and of warm and glacial periods, has been the ever-varying position of the sea and land; but it is now ascertained that the extreme distance by which the earth sometimes deviates from a circular orbit is such as to carry it away 11,000,000 of miles from the sun in aphelion, instead of 3,000,000 as at present, and Mr. Croll of Glasgow has pointed out that the cold of winter in aphelion would be very intense during such maximum eccentricity.

I ought to have said that this 11,000,000 of miles occurred 230,000 years ago, and was not the extreme possible maximum, which I believe would be 14,000,000 of miles. The proximity of the earth to the sun in perihelion would be proportionally great; but Mr. Croll maintains, and with some reason, that most of this excessive heat would be expended in melting part of the accumulated ice of winter, and this process would cause so much fog and cloud as to prevent the intensity of summer heat from overcoming all the effects of winter's cold. When I have worked more upon the tables which the Astronomer Royal is having calculated for me, I will let you know more about it.

I received very lately your beautifully illustrated paper on the 'Pflanzen der Pfahlbauten.' Your opening address of last year to the Zürich Natural History Society interested me greatly. Among other things your remarks on the flora of the interglacial period. I am very glad to hear of your forthcoming work on the 'Plants of the Chalk,' and I hope you will be able to devote a portion of it to the fossils collected by Dr. Debey at Aix-la-Chapelle, for I begin to despair that we shall ever have any account of them unless you undertake the task.

My friend Professor Huxley is preparing a paper on

the five new genera of reptiles discovered in the ancient coal
of the county of Kilkenny, in Ireland; there are eight or
nine species, all of these new, some of them four or five feet
long, of various forms, one of them with small fore feet like
a siren, and abortive hind feet, having the shape of an eel.

All the genera are of the labyrinthodont family, and
therefore, though they have well-ossified vertebral columns,
their low place in the scale accords well enough with the
theory of progressive development. At the same time, when
you recollect that ten or twelve years ago no single reptile
was known from a rock so ancient as the coal in any part of
the globe, one cannot reflect on the rapidly increasing number of these carboniferous amphibia without being prepared
for the discovery of some more highly organised vertebrata
of the like antiquity.

Believe me, my dear Heer, ever very faithfully yours,

CHARLES LYELL.

To CHARLES DARWIN, ESQ.

53 Harley Street: March 10, 1866.

My dear Darwin,—Your precious MS. has arrived safe.
I will return it registered in a few days. I am much obliged
to you for the privilege of reading it; and in regard to the
notes prepared for the new edition, I am amused to find
how many of the topics are the same as those treated of in
the letters of yourself, Hooker, and Bunbury, in commenting
on the observations by Agassiz of marks of glaciation in the
Organ Mountains. By the way, you allude to Hooker's discovery of moraines in the Sikhim Mountains, which I believe are only about 7° farther from the equator than the
Organ Mountains. It is very interesting to read Hooker's
letter dated 1856, and to see the impression which the MS.
made on him, causing him to feel, as he says, 'shaky as to
species' so long before the 'Origin' was published. We
certainly ran no small risk of that work never seeing the
light, until Wallace and others would have anticipated it in
some measure. But it was only by the whole body of doctrine being brought together, systematised, and launched at

once upon the public, that so great an effect could have been wrought in the public mind.

I have been doing my best to do justice to the astronomical causes of former changes of climate, as I think you will see in my new edition, but I am more than ever convinced that the geographical changes are, as I always maintained, the principal and not the subsidiary ones. If you snub them, it will be peculiarly ungrateful in you, if you want to have so much general refrigeration at a former period. In my winter of the great year, I gave you in 1830 cold enough to annihilate every living being. The ice now prevailing at both poles is owing to an abnormal excess of land, as I shall show by calculation. Variations in eccentricity have no doubt intensified the cold when certain geographical combinations favoured them, but only in exceptional cases, such as ought to have occurred very rarely, as paleontology proves to have been the case.

Ever most truly yours,
CHARLES LYELL.

To SIR CHARLES BUNBURY.

53 Harley Street : September 3, 1866.

My dear Bunbury,—I have been so absorbed in my proofs and new climate chapters, that I have been writing to no one, but must tell you a piece of botanical news which has interested me much. In recasting my chapter on ' Progressive Development,' I have stated that there was no unquestionable case of an Angiosperm so old as the carboniferous period. Mr. Carruthers of the British Museum, on reading my proof, told me that he had just seen in Edinburgh a specimen from the coal-shale of Granton, which is an undoubted Monocotyledon, and which has been figured some years ago in the ' Transactions of the Edinburgh Botanical Society,' under the name of *Pothocites Grantonii,* Dr. Paterson having named it with the approbation of Greville and other authorities of Auld Reekie. Mr. Kippist, Librarian of the Linnæan Society, showed me coloured plates of various species of Pothos, and of allied genera of the Aroideæ, to prove that sometimes the spathe is some way

from the bottom of the flower, and therefore that the
broken-off part may, as Paterson supposes, represent the
spathe. After this I shall not despair of your Antholite
being angiospermous. But how rare are such specimens!
about one in a million. Agassiz has written an interesting
paper on the ' Geology of the Amazons,' but I regret to say
he has gone wild about glaciers, and has actually announced
his opinion that the whole of the great valley, down to its
mouth in lat. 0, was filled by ice, and dammed up by a
moraine since destroyed by the sea, by which means he
accounts for freshwater deposits such as we find in the
valleys of the Mississippi and Rhine. He does not pretend
to have met with a single glaciated pebble or polished and
striated rock *in situ,* and only two or three far-transported
blocks, and those not glaciated. As to the annihilation
during the cold of all tropical and extra-tropical plants and
animals, that would give no trouble to one who can create
without scruple not only any number of species at once, but
all the separate individuals of a species capable of being
supported at one time in their allotted geographical province.

Ever affectionately yours,

CHARLES LYELL.

To DR. JOSEPH HOOKER.

73 Harley Street : October 20, 1866.

My dear Hooker,—I am sure there is some screw loose
in all the climate theories now afloat. When I send you my
thirteenth, or astronomical chapter, you will better understand one of the principal difficulties, arising, I think, from
the estimates made by Herschel and Poisson of the heat
received by the earth from the sun, as distinct from the
assumed heat of space. If they are right, the difference
between the sun's heat in the northern hemisphere and in
the southern ought to be very great, even with the present
eccentricity, and five times greater, as Croll says, with a
maximum eccentricity, and the winter's cold ought to be in
proportion. But if it were so, the monkeys would never
have lived through the cold of the last period of maximum

eccentricity at the tropics. But I suspect that the vapour to which you allude, and on which Tyndall has written so much, may equalise the heat 'and cold caused by greater proximity to, and distance from, the sun.

I hope you will have patience to read my thirteenth chapter, as I know you will tell me frankly what you think of it. It has cost me a great deal of time, for I had to go to school in astronomy, and moreover many of the things it was necessary to learn were not to be found in the ordinary elementary books, and I had to write to Airy and Herschel.

Without John Moore's assistance, I should never have got on.

> With many thanks,
> Ever affectionately yours,
> CHARLES LYELL.

To GEORGE TICKNOR, ESQ.

73 Harley Street, London : February 8, 1867.

My dear Ticknor,—I have been looking through your 'Spanish Literature' in search of some remarks which I had fancied you had made on the effects of the Inquisition in sacrificing the lives of so many men of superior mental power, and the moral and intellectual deterioration of the Spanish population which it produced.

But although I have found passages relating to the mischief done by the systematic persecution persevered in for centuries, it strikes me that I have missed the place I was in quest of. Possibly it was in some one of Prescott's works. At any rate I will tell you in what manner I was desirous of applying the exterminating action of the Inquisition to the subject I have now in hand.

You have read Darwin on the 'Origin of Species,' and know what he means by natural selection. If it were possible so to frame the institutions of a country, that the population should be chiefly kept up by breeding from those individuals, male and female, which are the best morally and intellectually; this selection would tend, according to Darwinian principles, to improve the race, and if an opposite

system were persevered in an opposite result might be expected. Perhaps the best experiment hitherto made is that
which we owe to the Inquisitors, and the result in regard to
the deterioration of the race is perhaps as satisfactory as
Darwin could desire. Yet even in Spain there have been
so many disturbing causes that the experiment has been
very incomplete. Many aspiring and powerful intellects
naturally went into the Church, and is it not true that
their vows of celibacy were sufficiently set at naught to
allow the children of such ecclesiastics to constitute no
insignificant part of the population? Can you give me any
facts as to children of eminent ecclesiastics who were acknowledged by their parents as such? Have you any idea
of the numbers who perished by the hands of the Inquisitors,
and for how many centuries this went on, and what may
have been in round numbers the population of Spain at the
period of the most severe persecution? Of course a large
number of those who were sacrificed would not have great
claims to superiority, but it would be enough for my purpose
if as a whole they constituted that portion of the people who
were superior as independent thinkers and inquirers, and had
more moral courage and character than the ordinary herd.

The argument would turn in a great degree on statistics,
and some of the data are not obtainable, especially the
extent to which each generation was recruited from the
clergy. It may be said that although the Spaniards are
doing so little now in literature, and nothing in science,
and producing no great statesmen or men of commanding
genius, yet we cannot be sure that the race has suffered
serious degradation, or that if like Italy they could be set
free, they might not be found as capable as ever of yielding
a crop of great men. But Italy has been kept down by
foreign domination, whereas Spain has had fair play, and
very little meddling from without. It seems to me therefore
that selection has really done its work, and that a race has
been produced as incapable of keeping pace with the progress
of the age, or boldly originating new ideas or questioning
the truth of old ones, as orthodoxy could desire.

Ever affectionately yours,
CHARLES LYELL.

Extract of Letter to A. WALLACE, ESQ.

May 2, 1867.

I forgot to ask you last night about an ornithological point which I have been discussing with the Duke of Argyll. In Chapter V. of his ' Reign of Law ' he treats of humming-birds, saying that Gould has made out about four hundred species, every one of them very distinct from the other, and only one instance, in Ecuador, of a species which varies in its tail-feathers in such a way as to make it doubtful whether it ought to rank as a species—an opinion to which Gould inclines—or only as a variety or incipient species, as the Duke thinks. For the Duke is willing to go so far towards the transmutation theory as to allow that different humming-birds may have had a common ancestral stock, provided it be admitted that a new and marked variety appears at once with the full distinctness of sex so remarkable in that genus. According to his notion the new male variety and the female must both appear at once, and this new race or species must be regarded as an ' extraordinary birth.' My reason for troubling you is merely to learn, since you have studied the birds of South America, and I hope collected some humming-birds, whether Gould is right in saying that there are so many hundred very distinct species without instances of marked varieties and transitional forms. If this be the case, would it not present us with an exception to the rule laid down by Darwin and Hooker, that when a genus is largely represented in a continuous tract of land, the species of that genus tend to vary? In regard to shells I have always found that dealers have a positive prejudice against inter-mediate forms, and one of the most philosophical of them— now no more—once confessed to me, that it was very much against his trade interest to give any honest opinion that certain varieties were not real species, or that certain forms, made distinct genera by some conchologists, ought not so to rank. Nine-tenths of his customers, if told that it was not a good genus or good species, would say what they wanted was names, not things. Of course there are genera in which the species are much better defined than in others, but you would explain this, as Darwin and Hooker do, by the greater

length of time during which they have existed, or the
greater activity of changes, organic or inorganic, which have
taken place in the region inhabited by the generic or family
type in question. The manufactory of new species has
ceased or nearly so, and in that case I suppose a variety is
more likely to be one of the transitional links which has not
yet been extinguished than the first step towards a new permanent race or allied species. Sclater gave me *Euplocamus
melanotis* and *E. lineatus* as an example of two distinct
species—Tenasserim and Pegu—which pass through every
intermediate variety in the intervening country of Arracan ;
but when I asked if it was proved that the two extremes of
the series would not intercross and produce fertile hybrids, he
said the experiment had not been fairly tried. I suppose
we have a good deal to do before we get all the facts which
we shall one day probably have in confirmation of the theory
of transmutation, or of such a divergence as will be accompanied by the sterility of hybrids. It may require several
hundred thousand years. M'Andrew told me last night
that the littoral shells of the Azores being European, or
rather African, is in favour of a former continental extension ;
but I suspect that the floating of sea-weed containing their
eggs may dispense with the hypothesis of the submersion of
1,200 miles of land once intervening. I want naturalists
carefully to examine floating sea-weed and pumice met with
at sea. There should be a microscopic examination of both
these means of transport.

CHARLES LYELL.

To SIR CHARLES J. T. BUNBURY.

73 Harley Street, London : July 9, 1867.

My dear Bunbury,—I have been reading your essay on
' South American Vegetation ' in ' Fraser ' with great interest
and profit, and am very glad that Kingsley has succeeded
in persuading you to take up your pen again, and let the
public benefit by your store of scientific knowledge. It was
no easy task to render popular the botany of a country, of
which so few even of the genera are known even to a great
many who are acquainted with European plants. I re-

member Hooker experiencing the same difficulty in his Himalayan book. But you have managed to make the most of the araucarias and calceolarias and fuchsias, and by describing the Desfontanea in such a way that one can see it.

You have put life into the paper by what you have said of the wide range of some plants, and the discussion whether they are varieties or species, and by the geographical and geological speculation, especially the migrations of plants along the Andes during the glacial period.

I should have liked you to have gone a little more in one direction into the great question of whether species are derivative, or are primordial creations. I was in hopes you would have pointed the moral, when you spoke of two features in South American botany to which you properly call attention. First, the manner in which the vegetation as it goes south from the tropical region of Brazil still continues to be Brazilian, instead of new genera and species fit for temperate latitudes making their appearance; and secondly, the invasion of European immigrants into the Pampas—a wonderful fact, and which would not, I think, have happened if plants were specially created for each country instead of being modified forms of types already there, and dating from an older geological period. It is marvellous how exotic genera thrive best as a rule, both animals and plants, and Darwin has, I think, well explained why. But perhaps this would have led you into too wide a digression, even if you had been as thoroughly imbued with the transmutation creed as I am beginning to be by writing my second volume.

Believe me, my dear Bunbury,
Ever affectionately yours,
CHARLES LYELL.

To CHARLES DARWIN, ESQ.

73 Harley Street : August 4, 1867.

My dear Darwin,—I must write a word before starting to-morrow morning for Paris, to thank you for your last letter, and to say what a privilege I feel it to be allowed to read your sheets[1] in advance. They go far beyond my anticipations, both as to the quantity of original observation, and the materials brought together from such a variety of sources, and the bearing of which the readers of the ' Origin ' will now comprehend in a manner they would not have done had this book come out first. The illustrations of the pigeons are beautiful, and most wonderful and telling for you, and the comparison of the groups with natural families difficult to divide will be most persuasive to real naturalists. The rabbits are famously worked out, osteology and all. The reason I have not got on faster is, that I have been correcting the press of my recast of Mount Etna, which I have reviewed twice since my former edition of fourteen years ago, also the Santorin eruption of 1866, and my grand New Zealand earthquake, which produced more permanent change than any other yet known. I have also had to rewrite my chapters on the ' Causes of Volcanic Heat,' the ' Interior of the Earth,' &c. But all this is in the printers' hands, and I can now give myself to variation and selection.

Believe me, my dear Darwin, ever affectionately yours,

CHARLES LYELL.

To the REV. W. SYMONDS.

73 Harley Street, London : October 20, 1867.

My dear Symonds,—I think I told you last year that Huxley received a specimen of his genus Hyperodapodon (first founded on an Elgin fossil contemporary with Telerpeton and Stagonolepis) from the New Red of the neighbourhood of Warwick. The discovery was made just in time to enable Sir Roderick to cancel and recast some pages of his new edition of ' Siluria,' in which he had defended his old position, that the higher sandstones of Elgin were Old Red. I was pleasantly reminded yesterday of the expedition we made together to the Elgin district by a discovery which Huxley had just made, that among some fossils sent from central India this same Hyperodapodon had made its appearance in those beds studied by Hislop, and which had already

———
[1] *The Variation of Animals and Plants under Domestication.*

been supposed to be Triassic, because they had afforded us several Dicynodons and Labyrinthodons. You remember that the first Dicynodons found at the Cape of Good Hope by Bain were conjectured by him to be Triassic, but this was thought a mere guess until the same genus was found in India associated with Labyrinthodon. Is it not curious that the united evidence got from Elgin, Warwick, the Cape, and central India should leave so little doubt as to the true age of the beds in all these regions? Charles Moore has found land shells about the age of what they call on the Continent Infra-Lias. The evidence, now that his paper is being printed, seems strengthening. You of course heard that my friend Dawson, requested by me to search in that carboniferous stratum containing *Pupa vetusta* in abundance, has found in it a *Helix* of the sub-genus Zonites. Think what persistent types Pupa and Helix are thus proved to be, and how little we know of air-breathers older than the tertiary which these revelations must convince us existed, though we continue to be so marvellously ignorant of them.

Believe me, my dear Symonds, ever affectionately yours,

CHARLES LYELL.

To DR. BENCE JONES.

April, 1868.

My dear Dr. Bence Jones,—I will endeavour to comply with your request, to give you in writing my reminiscences of Faraday during a short time when I was associated with him on a commission of inquiry. We had been requested by the Government to attend a coroner's inquest, and report on an unusually fatal colliery accident, which occurred in September 1844, at Haswell, about seven miles east of Durham.

Instructions were given us by Sir James Graham, then Secretary of State for the Home Department, and we were to inquire into the causes of the explosion, and if possible to suggest the means of preventing the occurrence of similar catastrophes in future. You will see that it is twenty-four years since the event happened, and although I have occasionally spoken to you and others on the subject, you will make allowance for any shortcomings in my recollections,

when I tell you that I have no notes to refer to, nor any
letters written at the time. Various incidents, however,
occurred, while we were engaged on this mission, well
calculated to draw out Faraday's character, and they made
a lively impression on my mind, the more so as I had never
had an opportunity of knowing him except in his laboratory,
and because I was not prepared to see him play his part
with so much spirit and self-reliance, when suddenly
launched into a new sphere of action. I had been acquainted with him for many years, and yielded to no one in
my admiration of his talents and my appreciation of his
scientific eminence ; but I had always looked upon him as a
singularly modest and retiring person, and one who would
shrink from the stir and responsibility of such an undertaking as that in which we were about to enter. He had in
truth undertaken the charge with much reluctance, but no
sooner had he accepted it, than he seemed to be quite at
home in his new vocation. In a few hours after he had
agreed to accompany me, we were carried by a fast railway
train to the scene of the catastrophe, and were immediately
introduced to the coroner and his jury, who were in the
midst of their inquest.

Faraday began, after a few minutes, being seated next
the coroner, to cross-examine the witnesses with as much
tact, skill, and self-possession as if he had been an old
practitioner at the Bar. He seemed in no way put out or
surprised when the most contradictory statements as to fact
and opinion were given in evidence, according to the leaning
which the different witnesses had, whether from a desire to
screen or exculpate the proprietors, or be regarded as
champions of the pitmen. The chief question was whether
any undue considerations of economy had induced the owners
of the mine to neglect such precautions as are customary
and indispensable to the safety of the men employed in the
work.

We spent a large part of two days in exploring the
subterranean galleries where the chief loss of life had been
incurred, that we might satisfy ourselves whether the means
of ventilation had been duly attended to. We conversed
freely with the workmen and overseers, and Faraday especi-

ally questioned them about the use of the Davy-lamp. We had both of us been much struck with the uneducated condition of the men examined on the inquest, and suspected that they would not so fully appreciate the dangers to which they were exposed as workmen who were better instructed. Few of them could write, and one even of those who had been promoted to the place of 'master wasteman,' had been unable the day before to sign his name as witness.

Among other questions, Faraday asked in what way they measured the rate at which the current of air flowed in the mine. They said they would show us. Accordingly one of them took a small pinch of gunpowder out of a box, as he might have taken a pinch of snuff, and, holding it between his finger and thumb, allowed it to fall gradually through the flame of a candle which he held in the other hand. A little cloud of smoke began immediately to traverse the space between two of the loose wooden partitions by which the gallery we were in was divided into separate compartments. When the smoke reached the end of the nearest compartment, his companion, who had been looking at his watch, told us how many seconds had transpired, and they then calculated the number of miles per hour at which the current of air moved. Faraday admitted that this plan was sufficiently accurate for their purpose, but, observing the somewhat careless manner in which they handled their powder, he asked where they kept their store of that article. They said they kept it in a bag, the neck of which was tied up tight. 'But where,' said he, 'do you keep the bag?' 'You are sitting on it,' was the reply; for they had given this soft and yielding seat as the most comfortable one at hand to the commissioner. He sprang upon his feet, and in a most animated and impressive style, expostulated with them for their carelessness, which, as he said, was especially discreditable to those who should be setting an example of vigilance and caution to others who were hourly exposed to the dangers of explosions.

As it fell to my lot to examine into the geological structure of the rocks in which the mine was worked, I had engaged several of the men to collect for me the more abundant fossils. When they brought these to me, chiefly plants, and were very inquisitive about them, I told them what I knew of their characters, and how far they were allied to families of living plants. I entered on this the more fully when I found that my fellow commissioner was no less desirous than the men to be informed on the subject. As I was going away, after having paid the men for their trouble, I was about to throw away all the specimens which had been collected. Faraday, perceiving this, begged me to wait till the men were out of sight, as he said their feelings might be hurt if they saw that nothing had proved to be worth keeping. He had been much pleased with the intelligent curiosity they had shown about the nature of the fossils, and equally struck with their entire want of information respecting them, although it was evident that they had thought and speculated much on the frequency and meaning of such remains in the rocks, and had not been satisfied with the only hypothesis which had been suggested in explanation, namely, that they had been washed there at the time of the Deluge.

Hearing that a subscription had been opened for the widows and orphans of the men who had perished by the explosion, I spoke to Faraday on the subject, and found that he had already contributed largely without having said anything about it to me. He apologised for this, saying that he feared I might think it necessary to give as large a sum as himself, whereas I ought not to be guided by his notions on such matters, which would be generally regarded as peculiar, and which he shared with members of the religious community to which he belonged. In our conversation with the miners we had been often embarrassed by the number of local terms used by them, and by their conflicting statements as to the methods and precautions commonly adopted in the ventilation and management of coal mines. I therefore proposed that we should call in to our assistance an experienced surveyor of mines of my acquaintance, residing in a distant part of the country, one who is now no more, and who, in obedience to our summons, came to us at once. By his help we were rendered much more independent in forming a judgment on several points at issue between the proprietors and the men ; but unfortunately our

new assistant, when his services were ended, undertook without communication with us to draw up an elaborate report
to the Government, setting forth extensive schemes of his
own of legislative interference with the working and inspection of all the coal mines in Great Britain. This he would
have sent in to the Secretary of State, as if officially
authorised to do so, had we not heard of his intention in
time to interdict such a proceeding. Presuming on the
kindness with which Faraday had treated him, and the
marked mildness and modesty of his manner, he evidently
thought he might take the lead as if he had been at the
head of the commission, and he must have been surprised at
the firm and decisive manner in which his forwardness was
checked. On our return to town we found the Home Office
besieged with an extraordinary number of letters from individuals, and addresses from local committees formed in
the mining districts, each proposing a different plan for the
prevention of colliery accidents. Many of the schemes were
scientifically unsound, and most of the rest implied such an
outlay of money as if insisted upon by law would have led
to the shutting up of the mines.

It was natural that Sir James Graham, then fully occupied with his ordinary official business, should wish to refer
all these documents to us, while on the other hand we saw
clearly that unless we fixed on some limit to the time we
were willing to give to duties of this kind, we might be involved for an indefinite period in controversial correspondence
on such matters which were equally uncongenial to both of
us. Faraday had been engaged in experimental inquiries
in which he took a deep interest, when he had thought it
right to interrupt them by agreeing to take part in the
commission; but he explained to me that no pecuniary remuneration would compensate him for a further suspension of his investigations, and he believed that the improvement in the management of coal mines might be as well or
better promoted by others whom the Government might
employ. We therefore determined that our labours should
cease with the sending in of our report. When this was
finished Faraday asked me to sign first, because my name
stood first in the commission. I begged him not to insist

on this, as it would place me in a false position, seeing that the brunt of the work had unavoidably fallen to his share, his remarks on the ventilation of mines comprising what was really valuable and original in the report. In fact, my contributions had been restricted to a few words on the geology of the rocks, and some concluding recommendations that steps should be taken to instruct the workmen in the elements of those sciences, such as chemistry, pneumatics, and geology, which were intimately connected with their occupation. My fellow commissioner had agreed with me that the safety of the miners would be increased if they received such instruction, and with more knowledge those prejudices would disappear which always impede the adoption of any change of system. Faraday at last agreed to sign first. It was in fact by mere accident that my name stood before his in the commission, Sir Robert Peel having sent first to me to know if I would go to Haswell, and when I pointed out to him how much more the questions at issue would turn on points of chemistry and physics than on geology, he determined to ask Faraday to go also. Throughout the whole proceeding I found Faraday willing to take more than his share of the work and responsibility, and to be very indifferent as to the amount of credit assigned to him. I was often astonished at the number of hours in the day for which he was able to keep his mind on the full stretch, for I was aware that two or three years before he had been obliged to abstain entirely from mental exertion in consequence of having overtaxed his powers.

Our companionship on this commission left on my mind a lasting impression of the amiability of his temper and the versatility of his genius.

From that time I saw him as frequently as was consistent with his rules of secluding himself from general society, and up to the time of his death he always received me, when we met, on the footing of an intimate friend.

Very truly yours,

CHARLES LYELL.

CHAPTER XXXIX.

CORRESPONDENCE.

To the REV. W. S. SYMONDS.

Southend : April 29, 1868.

My dear Symonds,—I have been going over with much
pleasure the geology of the coast sections of Hampshire
between Southampton and Poole, and between Poole and
Swanage. One cannot too often contemplate the numerous
proofs of fresh and shallow water in the dense Purbeck and
Wealden strata, covered as they were afterwards by the deep
sea of the Chalk, and the splendid leaf beds of the Lower
Bagshot with their fan-palms, cinnamon laurels, and ferns,
overlaid by the purely marine Barton clay or Upper Bagshot.
The ferns of these Eocene pipeclays retain not only their
sori, but in each dot or fruit bag you can, under a powerful
microscope, count the spores as distinctly as in the fruit of a
living frond. The Purbeck beds are not only rippled, but the
calcareous slabs have preserved on their under side fine casts
of the sun-cracks of the clay beds below. I carefully
examined in the cliff that single dirt bed, or old soil, only six
inches thick, in Durlston Bay near Swanage, in which Brodie
and Beckles found more than a dozen species of mammalia.
There is no discovery which to my mind speaks so eloquently
of the fragmentary nature of the record, and of the know-

ledge that may one day be rescued even from the shreds which have been spared to us, than this thin, ancient layer at the base of the Middle Purbeck. Whenever you have an opportunity, visit the Blackmore Museum at Salisbury, and get Mr. Stevens, the curator, to show it to you. Even since I mentioned it, I found some splendid antlers of the reindeer at Fisherton, that grand repository of a paleolithic fauna. I am delighted at the idea of your being with Leonard and me at Tenby. I hope to attend the first half of the British Association, which will begin at Norwich August 19.

Sincerely yours,
CHARLES LYELL.

To PROFESSOR HEER.

73 Harley Street, London : June 7, 1868.

My dear Heer,—Yesterday we had our anniversary meeting at the Greenwich Observatory, and one of the astronomers remarked to me, that although the inclination of the earth's axis must have been constant for such periods of moderate duration, as for example, a few tens of thousands of years, yet if you come to millions, there is nothing in the facts at present observed in astronomy to preclude the supposition of a change in the axis. I told them not to go too fast, for although the flora arctica showed a luxuriant forest vegetation near to the pole, and vines in high latitudes, there were as yet no fan-palms, much less a Phœnix palm, as in the Miocene of Switzerland, so that if they could place the polar circle under or near the equator in Miocene times, they might create as many difficulties as they would remove.

What you told me in your letter of April 7 of your ' Flora Baltica ' was very interesting. I got so well acquainted with the deciduous cypress, or *Taxodium distichum*, in the swamps of the Lower Mississippi, that your finding that very species in the brown coal of the north of Europe makes me seem to understand the state of things which then prevailed. I am only afraid that a hundred species of these fossils will give you too much hard work. I am glad you are able to compare some of the Bovey Tracey species with those of other regions.

Believe me, my dear Heer, ever truly yours,

CHARLES LYELL.

P.S. I ought to have thanked you for your beautiful paper on the 'Chalk of Moletrin in Moravia,' and for the beautiful plates which illustrate it.

To PRINCIPAL DAWSON.

London : July 1, 1868.

The new edition of your 'Acadia' is a work of which you may well feel proud, and I am much pleased with the re-dedication of the book to me.   I have not yet had time to study the chapters on the 'Fossils of the Carboniferous and Devonian Rocks ; ' you did well to give a restoration of the flora of the latter in the frontispiece, not omitting the insects.   It marks the great step which has been made in our science in twelve years, and it is remarkable, as you say, in how many points Acadia is prominent in geology, even if, as I suspect, the *Eosaurus* may turn out to be *Labyrinthodont*.   The *Conulus* is also a grand addition, and a wonderful proof of a persistent type.

In regard to the antiquity of man, I have great faith in the judgment of the Danish naturalists as to the long duration of the Neolithic period in their country.   The extensive peat bogs have been well searched, and although they cannot draw a precise line between the accumulation of peat in the last eighteen centuries, and that of pre-historic date, they can yet form a tolerably correct notion of the relative importance of the accumulation of the two periods.   To suppose a rotation of crops in the forest trees to have been brought about by general conflagrations is too catastrophic to suit my notions, and would require some monuments of the work of fire.   The absence in all Scandinavia, where they have collected so many thousands of implements, of a single specimen of the paleolithic type of oval hatchets, or spear-shaped unpolished tools, is a very strong negative proof of man not having entered that country in the ancient period of the mammoth.   Farther south the evidence is steadily augmenting in favour of the great geographical changes which have taken place since England, France, and other parts of Europe were inhabited by the earliest fabricators of the paleolithic tools.

The Duke of Argyll has argued in an article in 'Good Words,' that as the stone period of the Esquimaux coincided in date with the iron period of London and Paris, so the ancient flint folk north of the Alps may have been contemporary with men of a more advanced civilisation farther south, and he wishes to make out that the stone age may be connected with degradation, and the driving out of a surplus population into regions having the worst climate. But when we go south of the Alps, and examine the drift containing the bones of mammoth and other extinct mammalia, as at Rome, or near Madrid, or in Sicily, we find stone implements of the ancient type instead of the monuments of an iron age. Lately, in the Presidency of Madras, our Indian surveyors have found on the coast a bed of Laterite at the height of 500 feet above the sea level, containing quartz hatchets marvellously similar in size and shape to the flint implements of Amiens. Certainly there are no signs in warm latitudes of a civilisation like that of old Nineveh or Egypt having flourished at such a remote date as to allow of great subsequent changes in climate, or in physical geography, or in the mammalian fauna.

What think you of the alleged discoveries going on in the American continent, or in the auriferous gravel of California, in which bones of Mastodon, according to Dr. Snell, together with some implements, were found in the deep places of Table Mountain? These old alluviums have in one respect a more ancient aspect than our higher level valley gravels, as they belong to a period when the drainage was more distinct in direction from that of the present rivers. This, however, would not prove a higher antiquity, because the overlying basaltic lava shows that great volcanic action was going on in the same country, which would greatly accelerate geographical changes. You have done well to point out that there was a stone and reindeer period in Nova Scotia, only three hundred years ago, in the latitude of central and southern France.

What you have written on the glacial period I like much,

and I am sometimes disposed to attribute the numerous lakes of your country to submarine current and ice action, but I do not see my way as yet. I should like to know what may be the velocity of these currents at the bottom.

CHARLES LYELL.

To LADY BUNBURY.

73 Harley Street: July 26, 1868.

Dearest Frances,—I have just been installed in my new chair, which they have adjusted beautifully to my height and other requirements, and when I am tired of reclining on the sofa which you gave me, I shall be glad to indulge in sitting in this new piece of furniture which you have so kindly added to my study. Many thanks, dear Frances, for thinking of such luxuries for me. I have also to thank your husband for having referred me to Buckle on the ' Conversion of Henri IV.' It is really curious to compare the parallel passages in Buckle and Motley. That in Buckle is exceedingly eloquent, and has of course some truth in it, but the tone of morality in Motley is higher. I only get on with the latter by snatches, but am very much interested. I agree with Charles in wondering whether Philip II. could possibly have deceived himself to such an extent as to believe he never injured any man knowingly. The historian is evidently much puzzled, and tries to make out that he thought himself a god, and there may be something in that. When Motley sums up all his misdeeds, I rather wondered that he omitted that glaring act of repudiation of all his heavy pecuniary obligations which he was guilty of a year or two before his death. The misery he must have caused by that was so great that it might have made him think he had injured some even among the most orthodox of his subjects. It helped to explain to me how the military power of Spain failed to crush the Dutch republic. The narrative of the polar voyages was very new to me. The siege of Ostend very well told; such a skilful suppression of uninteresting details. If you have not read it, do indulge in this history.

Believe me your affectionate brother,

CHARLES LYELL.

To Principal Dawson.

73 Harley Street : July 31, 1868.

My dear Dawson,—I hope that long before this you have got my letter, written in the beginning of this month, in which I spoke of the first part of your 'Acadia,' which I have since been reading steadily with increased pleasure and profit, and which I find Etheridge is reading with equal satisfaction. It is so full of original observation and sound theoretical views, that it must, I think, make its way, and will certainly be highly prized by the more advanced scientific readers.

I think Huxley's doctrine of Homotaxis [1] would never have been put forth in such strong terms had he been more of a field geologist. What we have to remember is, that our records are so fragmentary; that at several intervals all memorials of minor periods of vast duration, as measured by centuries, may be wanting; and on comparing the coal-period of New South Wales or China with that of Nova Scotia, it is conceivable that no one stratum or set of strata in one of these regions may be strictly contemporaneous with any one in the other region. But the whole may have been formed in all three areas between the close of the Devonian and the beginning of the Permian periods. The more our Indian surveyors advance, the more do they become persuaded that they have in the Himalaya counterparts of several of our European subdivisions of the Lias and Oolite, and on comparing the northern and southern hemispheres, the proofs of real contemporaneity of similar formations, as opposed to homotaxis, are growing stronger.

You have done well to give that wonderful section of the South Joggins in detail. I certainly think that a thickness of 16,000 feet implies the existence and waste of great continental areas and of large deltas. The evidence of subaerial conditions is overwhelming, and I always think with pleasure of the privilege of having seen in the Great Dismal

---

[1] 'Similarity of Arrangement.' See Anniversary Address to the Geological Society, 1862, by Professor Huxley.

Swamp such accumulations of pure vegetable matter going on, and hundreds of fallen and prostrate trees rotting away. I am also glad to have observed, near New Madrid, so many areas where hundreds of erect trees submerged ever since the earthquake of 1811, are still standing erect and leafless in the water.

What amount of subsidence must we imagine to allow of the superposition of strata 16,000 feet thick? Must we have a sinking to that amount? In all great deltas the tendency of conversion into dry land by the deposition of sediment must be counteracted in those areas where there is a suspension of such deposition by the condensation of mud into shale, loose vegetable matter into coal, and loose sand into more compact sand and sandstone. When, therefore, fifty generations of your Sigillaria have accumulated uninterruptedly, there will have been a subsidence owing to the materials below having been gradually pressed down and packed into a smaller space, quite independently of a general sinking down of the earth's crust in that quarter. This may help the coming on of those conditions which produced the lagoons in which the Naiadites and Serpula flourished.

I had forgot to say how surprised I was that Leslie, whom I have seen lately here, should have raised such strange objections to some of your conclusions. We know over how vast an area the Appalachian coal-measures extend. Suppose them to be 3,000 feet thick, and that the rate of subsidence was two and a half feet in a century, as soon as the river or rivers coming from where the Atlantic now is filled up a certain space, it would push the delta farther on; but if the rate of sinking was five feet instead of two and a half, the mass of sediment deposited would be thicker near the old land, instead of a larger area of sea being reclaimed. If the Nova Scotia coal-measures are thicker than those of other regions, it does not prove that the quantity of mud and sand was greater, but that the rate of sinking was such as to cause the deltas to be limited in horizontal extension. I cannot see what objection there can be to isolated basins; there may have been many independent rivers, like those which enter the Gulf of Mexico. If different coal-basins of the United States resemble each other, as I presume Leslie means to say, it ought to be a relief to him to find Nova Scotia is exceptional as to thickness; for a general uniformity would be the real puzzle, for it would imply that corresponding movements in the earth's crust extended simultaneously over a large part of the northern hemisphere, of which we have no counterpart in the present state of things.

Ever yours,
CHARLES LYELL.

To the REV. W. S. SYMONDS.

Tenby: September 10, 1868.

My dear Symonds,—As my sister, Mrs. Lyell, is writing to you, she will tell you about my brother's illness and the consequent uncertainty of our stay here. If he recovers I shall hope to carry out our scheme of seeing something of the geology of this part of the world with you, and you will give us as much notice beforehand as you can of your own intended movements. I should like much to know what you think of the relations of the strata here, and with those of Devonshire. The limestone on which Tenby stands was called by De la Bêche carboniferous, and the few fossils which I have seen from it, such as *Productus Scoticus,* seem to confirm that view. Then the true coal-measures overlie. The millstone grit I have not yet been able to make out, but there is a break in the section of the cliff here between the mountain limestone and the coal, in which there may be sandstone and grit which I may find in the interior. Below the mountain limestone at Manorbier, we saw yesterday some 500 or 600 feet of vertical red shale and red sandstone, with some beds of green shale, and a few whitish beds of quartzose grit; this is named by De la Bêche Old Red, which it resembles exactly, and though we saw hundreds of strata finely exposed, we could not see a single fossil in it. But at the junction of this formation and the carboniferous limestone at Skrinkle Haven, nearer to Tenby than Manorbier, there are several hundred feet of fossiliferous beds, in which limestone, red shale, quartzose grit, conglomerate, blackish shale, alternate again and again, many of the beds containing teeth and scales of fish, some of them Orthis and

Rhinconela, which we have not had time to collect in, and see whether the fossils are Carboniferous or Devonian. If they are not Devonian, are we to conclude that in going from Devonshire to Pembrokeshire all the Devonian beds are lost, and nothing but the true Old Red left between the Silurian and Carboniferous? I forgot to say that in the junction bed there are ripple-marked sand and shrinkage cracks, showing what shallow water preceded the deep-sea condition of the carboniferous limestone, in which last we have seen several corals, especially Caryophylia.

Mr. Smith of Gumfreston says the limestone here is divisible into three masses, one of them, qy. the uppermost, characterised by Encrinites, another by Corals, and among them *Lithostrotion basaltiforme.*

Leonard is working away both at the recent shells and older rocks. The first three days more than fifty species of the former picked up on the beach, and their number always increasing. They throw much light on the Crag. They who study tertiary geology, and who do not begin at this end, are like children taught Greek and Latin before they have learned a word of English.

I am summoned to join a party.

Believe me ever most truly yours,

CHARLES LYELL.

To the DUKE OF ARGYLL.

Tenby: September 19, 1868.

My dear Duke of Argyll,—I have just read with great interest your spirited and clearly written article in reply to Wallace on 'Nidification.' [2] If I did not feel sure that portions of it will be embodied in some of your future works, I should grudge its being placed in a periodical just struggling into existence, though it may perhaps be most usefully published in the same journal as the paper which it controverts. I objected in my ' Antiquity of Man ' to what I there called the deification of natural selection, which I consider as a law or force quite subordinate to that variety-making or

[2] ' A Theory of Birds' Nests,' by Alfred R. Wallace. *Journal of Travel and Natural History,* No. 2.

creative power to which all the wonders of the organic world must be referred. I cannot believe that Darwin or Wallace can mean to dispense with that mind of which you speak as directing the forces of nature. They in fact admit that we know nothing of the power which gives rise to variation in form, colour, structure, or instinct.

But before Darwin had written his first work, I used to maintain that there was a meaning and utility in every one of the marks and shades of colour in the butterfly's wing, and I still think so. It may be inherited after it has ceased to be useful or indispensable. The explanation offered by Darwin of the dull colour of most female birds appears to me a great step. The assumption that birds and mammalia of the same species vary in colour both in a wild state and under domestication is not imaginary, but founded on fact. A breeder might give rise to a race of white pheasants; but natural selection would not allow of such a race, because they would be too conspicuous, at least under ordinary conditions, and in the present state of the world. How far sexual selection is one of the causes of modification of colour is a point on which I never could form any opinion. I was surprised when I first saw the importance which Darwin attached to it. He seemed to me to exaggerate at least its probable influence. It does not strike me that Wallace's theory requires that there should be a tendency whether in cocks or hens to produce bright hues, but merely that in the course of time, and among the millions which will be born, individuals will be gifted with plumage of every tint from the brightest to the most dull, and that those varieties that are ill suited for the conditions prevailing at any given time or place will be killed off, while the fittest will survive. There is no necessity of supposing any consciousness on the part of the birds of the bad effects which certain conspicuous colours would produce.

That we should ever come to know so much of the system of nature as to be able to give a reason for each particular variation of colour is more than could be expected; but Darwin has been making vast progress in this direction, and the discovery that plants with bright and conspicuous flowers require the aid of insects to fertilise them, while those which are inconspicuous are fertilised by the wind, appears to me a grand generalisation. I believe he also finds that the odoriferous plants, and those which afford honey, require insects for their impregnation.

What you say of hereditary habit being in each individual as independent of experience or observation as are instincts, is most true; but if we could trace such a habit to a remote progenitor, and show that he acquired it under some new conditions in the physical world to which he was subjected, the insight thus gained into the history of the origin of certain instincts would be a great gain to the naturalist.

Although the advocates of natural selection often ascribe too much to that one cause, it has enabled them at any rate to assign a reason for many phenomena which would never have been brought to light by those who are satisfied with saying that all things were pre-ordained to be as they are. You are certainly not one of that school, and it appears to me that the reason which you assign why some of the smallest birds having sombre and neutral tints build dome-shaped or covered nests is very satisfactory, and so far as I know, original, and I shall be curious to know whether Wallace will not be forced to admit that in all these cases it is a better theory than the one which he had proposed.

In some of the passages which you have cited from Wallace you have shown that he is obscure and perhaps illogical. In others the discordance of opinion arises, I suspect, from the fundamental difference of the point from which each of you start. The assumption of special creation would make the origin of all things simple; but if once a naturalist taking all the geological evidence into account inclines to the opposite view, that of transmutation, as more probable, we are led to speculate on the manner in which the present instincts, habits, structure, and colour came to be gradually acquired by each species, living under the surrounding conditions which are not those that prevailed in the period immediately antecedent. The transmutationist may reasonably hope that we shall get to understand more and more the working of those forces by which nature brings about changes in the organic and inorganic world. The part played by natural selection, which was almost entirely overlooked by Lamarck, is evidently very important, and I now indulge a hope that the working of other secondary causes, hitherto not attended to, will help to explain many perplexing phenomena which these who are satisfied with special creation would never bring to light.

We are much relieved by hearing this morning from Arthur Milman a decidedly better account of the Dean of St. Paul's.

Believe me, my dear Duke, ever most truly yours,

CHARLES LYELL.

To the REV. W. S. SYMONDS.

Tenby : September 28, 1868.

My dear Symonds,—I have just returned with Leonard from a very successful expedition to St. David's, where Dr. Hicks devoted the best part of three days to show us some 2,000 feet of strata, most of them fossiliferous, lying below the original ' lingula flags ' of Sedgwick, which last I have always held to be the only rocks for which the Cambridge Professor had any right to claim his name of ' Cambrian,' as distinguished from Sir Roderick's ' Silurian.' Within the last two months Dr. Hicks has found in purple shales mineralogically like those of the Longmynd, organic remains (a new Lingulella and a bivalve entomostracan) 1,200 feet below the fossils previously discovered. Higher up in the Lower Cambrian series are twelve genera of Trilobites, &c., like those of Barrande's primordial, but one or two new forms and many new species ; one species of Paradoxides more than twenty inches in length ; all in the ancient azoic. I was much pleased with the sight of the raised beach at the NE. end of Caldy Island, with its pebbles and littoral shells some fifteen feet above high water mark. Leonard also took me to a good example of the same beach at Giltar Point, with littoral shells, and there are three layers separated by blown sand of full-grown eatable shells at the top of Giltar Point, apparently refuse heaps of the ancient molluscophagous inhabitants; these, I suppose, are nearly 200 feet high. He has also found a raised beach at the top of

Merlin's Cave, a cavern in the limestone in the cliff near the western suburbs of Tenby.

When there are raised beaches and submerged forests, it is difficult enough to restore in imagination the succession of geographical changes. The difficulty becomes greater when you have to make due allowance for the relative age of two other sets of phenomena, viz. the glacial drift and the lines of lofty cliffs bounding the present sea, and sometimes undermined by the waves.

I made two attempts to dig through the submerged forest of Amroth at low tide. The first day we went down five feet without going through the beds containing vegetable matter. The second day, nearer the cliff, Leonard got to what seemed a sea beach, after passing through the beds containing the stumps and roots of trees, at the depth of about four feet.

Dr. Hicks says that in the absence of marine shells we ought not to assume that a bed of clay or mud with sub-angular and rounded pebbles was an old beach like the present, because it may have been glacial ' till ' with boulders. In support of this view he says that the submerged forest of Whitesand Bay certainly grew on ' till ' containing glaciated stones, and I remember seeing ' till ' with true arctic shells at about the level of the sea at the Great Orme's Head, and there are submerged forests in that neighbourhood.

When you say that the old forest extended across the Bristol Channel, which I think highly probable, do you suppose that the lofty cliffs of Ilfracombe and of Tenby and its neighbourhood were not only in existence, but were even of pre-glacial date?

Believe me, my dear Symonds,

Ever most truly yours,

CHARLES LYELL.

To PROFESSOR HAECKEL.

London : November 23, 1868.

My dear Sir,—I have to thank you for your kindness in sending me a copy of your important work on the ' History of Creation,' and especially for the chapter entitled ' On

Lyell and Darwin.' Most of the zoologists forget that anything was written between the time of Lamarck and the publication of our friend's ' Origin of Species.'

I am therefore obliged to you for pointing out how clearly I advocated a law of continuity even in the organic world, so far as possible without adopting Lamarck's theory of transmutation. I believe that mine was the first work (published in January 1832) in which any attempt had been made to prove that while the causes now in action continue to produce unceasing variations in the climate and physical geography of the globe, and endless migration of species, there must be a perpetual dying out of animals and plants, not suddenly and by whole groups at once, but one after another. I contended that this succession of species was now going on, and always had been; that there was a constant struggle for existence, as De Candolle had pointed out, and that in the battle for life some were always increasing their numbers at the expense of others, some advancing, others becoming exterminated.

But while I taught that as often as certain forms of animals and plants disappeared, for reasons quite intelligible to us, others took their place by virtue of a causation which was beyond our comprehension; it remained for Darwin to accumulate proof that there is no break between the incoming and the outgoing species, that they are the work of evolution, and not of special creation.

It was natural, as you remark, that Cuvier's doctrine of sudden revolutions in the animate and the inanimate world should lead not only to the doctrine of catastrophes, such as Elie de Beaumont's sudden formation of mountain chains, but to a similar creed in regard to the organic world. A. D'Orbigny gave us twenty-seven stages or groups of living beings, all the species in each of which were so distinct that none of them passed from one to the other stage. Agassiz still inclined to the same notion, the sudden annihilation of one set of inhabitants of the globe, and the coming upon the stage in the next geological period of a perfectly distinct set. I had certainly prepared the way in this country, in six editions of my work before the ' Vestiges of Creation ' appeared in 1842, for the reception of Darwin's gradual and insensible evolution of species, and I am very glad that you noticed this, and also the influence of Cuvier's work, which in an English dress, translated by Professor Jamieson, went through almost as many editions in this country as in France, and exercised great authority long after my 'Principles' began to be popular. No part of your new work has produced such an effect as the pictures of the embryological state of man and the dog at corresponding ages, and figures of man, dog, tortoise, and fowl in the embryo. I am surprised to observe how much even very good naturalists are persuaded by the evidence of these woodcuts, as if although they previously had admitted the analogy, they had never thoroughly felt the force of it till they saw the facts as represented by you. Some of those to whom the ape origin of man is a very unwelcome doctrine, have tried to lessen the effect of these woodcuts by saying they must be a little flattered, like the physiognomies of men and apes given in your frontispiece, for some of these last are accused of being caricatures, the artist being supposed to have put a little too much of the ape into the human, and of the man into the Simian countenance and features.

For my own part I think I would rather have seen the embryological figures in the frontispiece. I think they would have induced a greater number to become readers, and would have inspired more faith, but perhaps the publisher would be of a different opinion. It is one of those cases in which one would like to have the real photographs of each countenance, as there could not then be any cavilling, however great the resemblance between the two portraits.

CHARLES LYELL.

CHAPTER XL.

APRIL 1869–FEBRUARY 1875.

GEOLOGY ON THE COAST OF NORFOLK—MR. WALLACE'S REVIEW ON NATURAL SELECTION—GLACIERS SCOOPING OUT LAKE-BASINS—FOSSIL FLORA OF ALASKA—'LIFE OF DANIEL WEBSTER'—DEATH OF HIS WIFE—FORFAR- SHIRE GEOLOGY—MR. JUDD ON 'GEOLOGY OF MULL'—HIS DEATH—DR. HOOKER'S LETTER—FUNERAL—WESTMINSTER ABBEY—TRIBUTE.

[He made excursions to Norfolk and Suffolk in 1869, and later on to Westmoreland on his way to Forfarshire, and thence he made an expedition with Lady Lyell and his nephew to Ross-shire by Inchna- damff and Ullapool, returning by the parallel roads of Glen Roy.

In 1870 he visited Dunster, Minehead, and Ilfracombe in Somer- setshire, and in August went to Scotland, to the Isle of Arran, and after spending some weeks at Ambleside joined the British Associa- tion at Liverpool.

In April 1871 he made a tour in Cornwall, to Tintagel, the Land's End, and Penzance, and in the following summer he spent some time in Yorkshire and Westmoreland.

In April 1872 Sir Charles and Lady Lyell went to the south of France, to the Caves of Aurignac. In these last journeys, when his sight and bodily powers had become more enfeebled, he enjoyed the advantage of the companionship of Mr. Hughes,[1] whose zeal for geology and whose powers of observation rendered him a most useful as well as agreeable fellow-traveller.

In the spring of 1873 Sir Charles had the affliction of losing his beloved wife, his companion and helpmate, with whom he had had forty years of unbroken happiness.

This severe blow fell heavily on his already broken health, but to his science alone he felt he must turn, to arouse him for the short time he had yet to live, and he determined to fulfil what he had planned with her; so calling on the friend who had promised again to accompany them, and with a sister (who devoted herself to him for

---

[1] T. M'Kenna Hughes, Esq., Woodwardian Professor of Geology at the University of Cambridge.

the rest of his days), he went abroad this autumn, and visited his friend Professor Heer of Zürich.

In June 1874 he went to Cambridge, to receive the honorary degree of LL.D., and that same month was admitted to the freedom of the Turners' Company in the City of London. He then spent some weeks in Forfarshire, and found pleasure in visiting some of his earliest geological haunts, and in finding that his theories of fifty years past still held good. He invited Mr. Judd [2] to join him, and went with him to several points of interest.

On November 5, the fiftieth anniversary of the Geological Society Club, of which he had been a member from its foundation, he attended the dinner, and spoke with a vigour which surprised his friends.

His failing eyesight and other infirmities now began to increase rapidly, and towards the close of the year he became very feeble. But his spirit was ever alive to his old beloved science, and his affectionate interest and thought for those about him never failed. He dined downstairs on Christmas Day with his brother's family, but shortly after that kept to his room.]

CORRESPONDENCE.

To Sir Charles Bunbury.

Cromer : April 6, 1869.

My dear Bunbury,—I will dictate a few lines to you on the geology which Leonard and I have been seeing since April, when we went to Aldborough, where we resumed the work which was interrupted last year by the British Association meeting at Norwich.

From Aldborough we went by Southwold to Lowestoft, seeing the coast at Kessingland and Pakefield by the way. It is a great satisfaction to see a continuous section for miles unbroken of such deposits as one only gets a peep of in isolated pits in your county, without any means of guessing their relative age. We first had this advantage in the cliff at Easton Bavent, where the yellow sands and gravel beds of Lord Stradbroke's park at Henham and Wangford are well displayed. But I wished much I could have had the advantage of walking with you along the Kessingland and Pakefield cliff, about fifty and sixty feet high, where at

---

[2] Now Professor of Geology at South Kensington School.

the base for more than a mile, in a bed of what I formerly called green till, a homogeneous unstratified clay, I found upright plants or shrubs, standing vertical with their roots in the same green soil (apparently tap-roots), also vertical and a foot or more long. In one place, near Pakefield, this lower stratum was laminated, and contained prostrate flattened trees, a foot or more in diameter. Over this green till, with plants *in situ* (of which I have kept a few specimens to show you), reposes stratified sand many yards thick, and over this drift, with plicated boulders of chalk, lias with fossils (*Avicula cygnipes*), numerous ammonites, belemnites, pieces of mica-schist, sandstone, greenstone, and other rocks. It is strange to see this glacial drift covering the bed for a mile and a half with trees which must have grown *in situ*, and must have sunk down so as to allow first the sand, and then the boulder clay to accumulate over it.

We could find no shells, freshwater or marine, with the plants. But in the Norfolk cliffs at Happisburgh, about sixty feet high, we found a different kind of drift, the matrix of which is a green clay much resembling in appearance, and in being generally without laminæ or stratification, the Pakefield plant-bed, but containing boulders of all kinds, with numerous pieces of chalk and chalk-flints, many of the rocks well glaciated. But although, when washed out of the till, they formed fine piles on the beach, they are so sparse as not to interfere with the general green colour of the mass. In a bed of gravel, two feet thick and several yards long, in the middle of this green till, we found a bed of pebbles like a fragment of an old beach, with a single valve of *Tellina solidula*, perfect, and some pieces of *Cardium edule*. In the same green boulder clay, on both sides of Happisburgh, we found small pieces of the same *Cardium*, and of what seemed to be *Cyprina Islandica*, and one perfect valve, but usually only small pieces of the Tellina, and no other shells whatever, in this glacial drift. I suppose therefore we must set it down as a marine formation; and underneath it, from Happisburgh to Cromer, comes the famous lignite bed and submarine forest, which must have sunk down to allow of the unquestionable glacial formation being everywhere superimposed.

We found the lignite during a walk along the cliffs this morning, and were shown the place where prostrate trees and some stumps of the forest were exposed by some recent high tides, though they are now buried again under shingle and sand. We also found ferruginous or reddish-coloured concretions with dicotyledonous leaves, which, though we did not find them *in situ*, must, I suppose, have come from the lignite or forest bed, from which I suppose Heer might in time construct a flora to add to the spruce fir, Meny-anthes, &c.

Leonard and I have just returned from Sherringham, where I found that the splendid old Hythe pinnacle of chalk in which the flints were vertical, between seventy and eighty feet high, the grandest erratic in the world, of which I gave a figure in the first edition of my ' Principles,' has totally disappeared. The sea has advanced on the lofty cliff so much in the last ten years, that it may well have carried away the whole pinnacle in the thirty years which have elapsed since our first visit.

We are going to-morrow to revisit Mundesley.

Believe me ever, with love to Frances,

Affectionately yours,

CHARLES LYELL.

To CHARLES DARWIN, ESQ.

May 5, 1869.

I am pleased at the impression which the historical part of Wallace's review [3] made on you. It reminds me of Cuvier's daughter, a charming and intelligent girl, telling me she had been reading my book (vol. i. of ' Principles ') to her father, and that they had been struck with the complete antagonism of my views to those which he had propounded in his ' Theory of the Earth.'

I was always made to feel myself a welcome guest at Cuvier's *soirées*, but he never alluded to my book, and but for Mademoiselle Cuvier's saying she had been reading it to him in their carriage as they drove out, I should never have known he had seen it.

[3] ' Geological Time and the Origin of Species.' *Quarterly Review.*

I quite agree with you that Wallace's sketch of natural selection is admirable. I wrote to tell him so after I had read the article, and in regard to the concluding theory, I reminded him that as to the origin of man's intellectual and moral nature I had allowed in my first edition that its introduction was a real innovation, interrupting the uniform course of the causation previously at work on the earth. I was therefore not opposed to his idea, that the Supreme Intelligence might possibly direct variation in a way analogous to that in which even the limited powers of man might guide it in selection, as in the case of the breeder and horticulturist. In other words, as I feel that progressive development or evolution cannot be entirely explained by natural selection, I rather hail Wallace's suggestion that there may be a Supreme Will and Power which may not abdicate its functions of interference, but may guide the forces and laws of Nature. This seems to me the more probable when I consider, not without wonder, that we should be permitted to give rise to a monstrosity like the pouter pigeon, and to cause it to breed true for an indefinite number of generations, certainly not to the advantage of the variety or species so created.

At the same time I told Wallace that I thought his arguments, as to the hand, the voice, the beauty and the symmetry, the naked skin, and other attributes of man, implying a preparation for his subsequent development, might easily be controverted; that a parrot endowed with the powers of Shakspeare might dictate the 'Midsummer Night's Dream,' and that Michael Angelo, if he had no better hand than belongs to some of the higher apes, might have executed the statue of Lorenzo de' Medici.

In reply to this and other analogous comments, Wallace said : ' It seems to me that if we once admit the necessity of *any action* beyond " natural selection " in developing man, we have no reason whatever for confining that action to his brain. On the mere doctrine of chances, it seems to me in the highest degree improbable that so many points of structure all tending to favour his mental development should concur in man, and in man alone of all animals. If the *erect posture,* the freedom of the *anterior* limbs for *purposes of locomotion,* the *powerful* and *opposable* thumb, the naked skin, and *the great symmetry of force*, the *perfect organs of speech*, and his mental faculties, *calculation of numbers, ideas of symmetry,* of *justice,* of *abstract reasoning,* of *the infinite,* of *a future state,* and many others, cannot be shown to be each and all *useful* to man in the very lowest state of civilisation, how are we to explain their co-existence in him alone of the whole series of organised beings ? Years ago I saw a Bushman boy and girl in London, and the girl played very nicely on the piano. Blind Tom, the idiot *Negro,* had a *musical ear or brain,* perhaps superior to that of any living man. Unless you can show me how this rudimentary or latent musical faculty in the lowest races can have been developed by survival of the fittest, can have been of *use* to the individual or the race, so as to cause those who possessed it to win in *the struggle for life,* I must believe that some other power caused that development, and so on with every other especially human characteristic. It seems to me that the *onus probandi* will lie with those who maintain that man, body and mind, could have been developed from a quadrumanous animal by natural selection.'

As to the scooping out of lake-basins by glaciers, I have had a long, amicable, but controversial correspondence with Wallace on that subject, and I cannot get over (as, indeed, I have admitted in print) an intimate connection between the number of lakes of modern date and the glaciation of the regions containing them. But as we do not know how ice can scoop out Lago Maggiore to a depth of 2,600 feet, of which all but 600 is below the level of the sea, getting rid of the rock supposed to be worn away as if it was salt that had melted, I feel that it is a dangerous causation to admit in explanation of every cavity which we have to account for, including Lake Superior. They who use it seem to me to have it always at hand, like the ' diluvial wave, or the wave of translation,' or the ' convulsion of nature or catastrophe ' of the old paroxysmists.

I have just got a letter from Professor Leslie, and an important paper by him in the American ' Philosophical Society ' for 1862, and another on a projected map, ' intended to illustrate five types of earth-surface in the United States,' published in 1866. He was formerly a catastrophist, but of

late years he seems to have anticipated Geikie and Croll in
regard to sub-aerial denudation, giving, like them, too little
to the sea. But he is a man intimately acquainted with the
Appalachians, and he gives his reasons for not believing that
the ice-sheet has had any hand in eroding the Appalachians.
It has polished the surface, and carried erratics so far as
mid-Pennsylvania, and no farther; but the surface erosion is
just as great in Southern Pennsylvania and Virginia, &c.,
which was not reached by the ice, and where there is not a
single glacial scratch or groove. He says that the large map
which he has planned will make the ice-scooping of lakes in
the United States appear as absurd as if applied to tropical
Africa or the Albert Nyanza Lake.

　　　　Believe me ever affectionately yours,

　　　　　　　　　　　　　CHARLES LYELL.

　　　　　　　　To PROFESSOR HEER.

　　　　　73 Harley Street: November 4, 1869.

　　My dear Heer,—I ought sooner to have acknowledged
the receipt of your beautifully illustrated paper from the
' Fossil Flora of Alaska,' which I have looked at with more
interest from what you told me in your letter of the light
which it throws on the isothermals of the Miocene period.

　　That this flora should have ranged with so little variation over 10° of latitude in the Miocene period, will have
a great bearing on those who are ready with so little
ceremony to shift the place of the earth's axis of rotation
when it suits their theoretical views. The late deep sea
dredging expeditions to the Faroe Islands have shown that
in our northern sea, cold and warm areas occur at the depths
of from 2,000 to 3,000 feet below a surface having a uniform
temperature, and from which such a variation of climate
would not have been expected. These different climates
occur in the same latitudes in areas sometimes fifty, sometimes less than twenty geographical miles apart. The
marine fauna consists of distinct species in the cold areas
where the current comes from the North or the Arctic regions,
and in the warm areas where the water is brought from the
South or Gulf-stream. It is only a small proportion of cos-

mopolitan species and genera which are common to the two
distinct climatal areas. Last year they came away with
an idea that the cold areas where the water is about 30°
Fahr. were very barren of organic remains, but a totally
new dredging apparatus has shown that this was a great
error. They have also visited a region 260 miles south of
the island of Ouessant in the Bay of Biscay, where the depth
was 15,000 feet, or within a very few feet of that depth, the
same as the height of Mont Blanc, and they have found
abundance of life of different classes at this enormous
depth.

As you will soon see more particulars in the ' Proceedings
of the Royal Society,' I will not attempt to give you more
of what I have learnt from conversation with the dredgers.

Believe me, my dear Heer, ever most truly yours,

CHARLES LYELL.

To GEORGE TICKNOR, ESQ.

May 31, 1870.

My dear Ticknor,—I have been reading with great
interest the eloquent passages which you pointed out in the
' Life of Daniel Webster ' by Mr. Curtis. Such eulogies
from two men of such weight, and addressed to a large
public during a man's lifetime, are a splendid monument to
his memory.

I go out as little as possible, as I like to go to bed at
nine o'clock, and it is scarce any use to start now for an
evening party till half-past ten o'clock. We were asked to
an evening at the Van de Weyers', to meet the King of the
Belgians, and it was no loss to me to learn that they were
obliged to put off the party because some previous balls had
caused a crack in the wall of their house, and a surveyor
said that an additional weight would bring the rooms down.
The Motleys gave a grand reception to the king in the
same street, which Mary went to.

But to return to Curtis's ' Life of Webster.' I have been
too busy with my own new edition of the ' Elements of
Geology,' to be able as yet to do more than turn over the
pages, but I was surprised to see in a note at p. 586, that

woman's suffrage was a matter already familiarly discussed in 1852 on your side of the water. We have heard a great deal of it lately on our side, and though I approve of much in J. S. Mill's ' Subjection of Women,' and have always thought that there was much to rectify in our legislature, I confess I should look with alarm if I thought a great extension of suffrage should take place, including women as well as men. We should never have passed the disestablishment of the Irish Church, if the vote of the English clergy had been strengthened by that of all the women whom they could have influenced to oppose it. It would be a formidable Tory measure, and might, I think, delay educational reforms. But it seems undeniable that property ought to be represented under a system like ours, whether in the hands of women or men.

With love to Mrs. Ticknor and Anne,

Yours very affectionately,

CHARLES LYELL.

To PROFESSOR HEER.

73 Harley Street, London : June 15, 1870.

My dear Heer,—A few days ago I sent you part of Mr. President Bentham's address to the Linnæan Society, in which he disputes all determinations of Proteaceæ by fossil leaves, and says that the few imperfect specimens of fossil fruits referred to the same order of plants are not sufficient to prove their former existence in Europe. He does not say to what other order we may suppose that all these leaves, so like those of Proteaceæ, may belong. I suspect from one expression in his speech, that he thinks we ought to require more proof because the present countries of the Proteaceæ (the Cape and Australia) are so distant, and so detached from Europe. But he does not reflect that the Falunian or Upper Miocene flora must have derived its species direct or by modification from the Lower Molasse or Older Miocene (Oligocene) period, and before this time there were Eocene Proteaceæ in Europe ; and still farther back, as at Aix-la-Chapelle, there was a rich flora in the time of the White Chalk, comprehending, if I mistake not, Proteaceous genera (?)

Now the land of the Cretaceous period differed so entirely in its geographical distribution from our present continents, that the species of plants that you have called Proteaceous, to whatever family they may have belonged, are as likely to have spread from Europe to the Cape or to Australia, as to have travelled from those countries into the European area. Mr. Bentham must admit that the plants which you and other botanists have conjectured to be Proteaceous, do not belong to any of our actual European trees. They once flourished here, and when they died out they may have found an habitation elsewhere. As we are certain that they lived in Europe at the period of the White Chalk, and went on diminishing in successive tertiary periods till they became extinct in Europe, it is safer to suppose that they went from Europe to Southern Africa and Australia, than to assume that they came from those countries into our area. For if we were to adopt the latter hypothesis, we must begin by assuming that there was land in South Africa and Australia before the White Chalk was formed, of which, however probable it may be, we have no positive proof, still less that such land was inhabited by trees which had a foliage like Proteaceæ.

In conversation I told Mr. Bentham that you had often begun by determining the leaves, and years afterwards discovered the fruit, as in the case of the chestnut found in the North.

I believe that these determined botanical sceptics do harm by undervaluing paleontological evidence, and exerting themselves to bring it into contempt. If they succeed in making geologists believe that fossil plants are of little or no value, this important branch of organic remains will not be cultivated with the zeal and scientific skill to which it is entitled. I wish therefore to make as good a fight as I can in a new edition of my 'Elements,' of which we are going to print 5,000 copies or more. Any assistance you can give me will be of real use to the cause.

Believe me, my dear Heer, ever faithfully yours,

CHARLES LYELL.

To Dr. Hooker.

73 Harley Street, London : February 14, 1871.

My dear Hooker,—I got your letter at Barton, and was much pleased to hear that you had been dipping into the 'Elements,' and were pleased with it and criticising it. I was afraid of alluding to the rarity of lakes on the south side of the Himalaya, though I can quite believe that there ought to be such lakes if the ice-scooping theory were true, to the extent of accounting for such basins as the Lago Maggiore, Geneva, &c. I saw yesterday James D. Hague, a mining engineer of the United States, returned from a survey of California, who says that the American geologists have convinced themselves that the great basin between the Rocky Mountains and the Sierra Nevada, in which the Utah Salt Lake, amongst other water-bearing depressions occurs, was, in comparatively modern tertiary times, a great fresh-water lake, 500 feet deep, and of much larger dimensions than Lake Superior. If, as they suppose, this lake was formed chiefly by anticlinal and synclinal folds, connected with the structure of the adjoining mountain chains, an analogous origin may be ascribed, in great part at least, to Lake Superior and many minor lakes.

I found that Charles Bunbury had been marking in your 'Student's Flora' all the species of flowering plants growing in his park, and he was full of praise of the manner in which your work was executed after going into much of the details of it. I am reading Mivart's 'Genesis of Species,' and am only half through; it improves greatly as I proceed. I thought his first objection, that so many other Ungulata ought to have long necks as well as the giraffe, a very poor one against natural selection. But the difficulty about the eyes of the cuttle-fish, dragon-flies, and man, is very well put.

Believe me ever most truly yours,

Charles Lyell.

P.S. I am very glad of your hint about ice-shoots or ava-lanches. I am convinced they would explain what nothing else can, the large boulders and erratics on the lower side of the Forfarshire tarns.

To Sir Charles Bunbury.

Crown Hotel, Penrith : August 18, 1871.

My dear Bunbury,—I have been enjoying very much my tour of inspection, avoiding any regular work, and trying to make it a tour of rest, which is difficult. After being in the more central region of Buxton, I crossed the watershed of Axe Edge to the county of Cheshire above Macclesfield, where Prestwich announced that he had seen marine shells in the drift at the height of upwards of 1,200 feet. I am satisfied that the Moel Tryfaen marine fauna is found in the stratified drift not only at that height, but nearly, if not quite, at as great an elevation as at Moel Tryfaen, but the newly-discovered locality, ascending to nearly 1,400 feet inland from Macclesfield, has not been trigonometrically measured—the point is about seventy-four English miles, as the crow flies, from Moel Tryfaen. Is not this a surprising fact? implying that such a great body of land has been uplifted in post-pliocene times, for the shells, more than fifty in number, are all of recent species, and by no means like those of Moel Tryfaen ; as a whole very glacial. The rarity of marine shells in intervening tracts is so puzzling. In the Lake district, though there are stratified drifts 1.200 feet high, I could not hear of a fragment of marine shells any more than in the Matlock and Buxton region. Yet if the country was now submerged 1,400 feet or more, I should think the sea would find its way to Chatsworth.

Ever affectionately yours,

Charles Lyell.

To Professor Heer.

73 Harley Street, London : March 16, 1873.

My dear Heer,—I have just seen Mr. Carruthers, who tells me that he now thinks that the cumulative evidence in favour of the existence of the order Proteaceæ in the Eocene period in Great Britain is overwhelmingly strong. Two or three years ago, when Bentham published his article denying that this could be proved by the evidence of any of the fossil leaves yet discovered, or by any of the cones in the Sheppey

Eocene clays believed by Bowerbank to be Proteaceous, Mr. Carruthers was afraid to adopt such opinions, so that I am very glad that the structure of some British tertiary wood, and a great abundance of Eocene leaves found at Bournemouth, in Hampshire, has converted him to opinions which you have long held. I was always persuaded that your opinion would turn out to be correct, but shall be glad to be able to cite the keeper of the botanical collections in the British Museum as being convinced that your views are well founded.

You kindly wish me to tell you of my health and that of Lady Lyell. As I am now half way through my seventy-fifth year, you will not be surprised to learn that my eyes, which have always been weak from boyhood, are beginning to fail me, so that I am obliged to depend on other people for writing from dictation all my letters to correspondents, and for reading all the books which I study; but I am able to walk, enjoy life and society in moderation, and if you could come to England when I am at home, I should be happy to show you hospitality. Your letters are always a great treat to me.

[My dear Mr. Heer, I am perfectly well, thank you, and should be very glad to see you again.

Very sincerely yours,[4]

MARY E. LYELL.]

Yours very faithfully,

CHARLES LYELL.

To PROFESSOR HEER.

73 Harley Street, London : July 7, 1873.

My dear Heer,—I have to thank you very much for your last two letters, which have been of extreme interest to me. Indeed, the determination which you have made of the cretaceous plants of Greenland and Spitzbergen, sent home by Professor Nordenskjold, appears to me one of the most important scientific discoveries which has been made for

---

[4] *Note by Mr. Heer* : 'Mad. Lyell starb an 25 April 1873 ; also 5 Wochen nachdem sie obigen geschrieben.'

some time. My object in writing to you now is to say that I am going to make a tour on the Continent with my sister, and hope about the second or third week in August to be in Zürich. It would give me so much pleasure to see you, if you are likely to be there, and if your health is good enough to enable you to accompany me in some short expeditions such as my own health will allow of. But if, as I fear from your last letter, you may not be equal even to so much of a geological excursion to Utznach, &c., as I hope to accomplish in company with my friend Hughes, now Professor of Geology at Cambridge, I shall still trust to my being able to obtain instructions from you respecting the position and nature of the lignite of Utznach, Dürnten, Wetzikon, &c.

I have not written to you since the sudden and unexpected death on April 24 of my dear wife, with whom you were so well acquainted, and you can, I am sure, appreciate the shock which this has given me. I endeavour by daily work at my favourite science to forget as far as possible the dreadful change which this has made in my existence. At my age of nearly seventy-six, the separation cannot be very long, but as she was twelve years younger, and youthful and vigorous for her age, I naturally never contemplated my surviving her, and could hardly believe it when the calamity happened. A feverish cold carried her off almost without pain or suffering.[5]

I should be very glad to hear from you what chance I have of finding you in Zürich or the neighbourhood.

It will be a great pleasure to see an old friend whom I knew in happier days.

Ever affectionately yours,
CHARLES LYELL.

To MISS F. P. COBBE.

73 Harley Street : July 20, 1873.

My dear Miss Cobbe,—I have been so taken up with my Geology (a new edition of the 'Student's Elements' having to be prepared), that I begin to be afraid that I shall not keep my promise of writing to you before I go abroad, if I

---

[5] See Appendix B.

delay any longer. Your articles on a 'Future State' in the
'Theological Review' have interested me much, but they
confirm my opinion that we are so much out of our depth
when we attempt to treat of this subject, that we gain little
but doubt in such speculation.

I have, however, been much struck with your answer to
those strange opinions thrown out by W. R. Greg in his
chapter on 'Elsewhere,' which is, I think, very original and
satisfactory; for when one looks back forty years, and feels
compunction for many things one has done, it is wonderful
what allowances one makes, because we feel that we are
judging ourselves, and regard our former self with great in-
dulgence, while at the same time it is like contemplating a
different individual.

I am told that the same philosophy which is opposed to
a belief in a future state, undertakes to prove that every one
of our acts and thoughts are the necessary result of antecedent events and conditions, and that there can be no such
thing as free-will in man. I am quite content that both
doctrines should stand on the same foundation, for as I
cannot help being convinced that I have the power of exerting free-will, however great a mystery the possibility of this
may be, so the continuance of a spiritual life may be true,
however inexplicable or incapable of proof.

But I will not weary you with more of my lucubrations,
which, as I am obliged to dictate them to an amanuensis,
may appear in a stiffer form than if I was able to use my
own pen.

I am told by some that if any of our traditionary beliefs
make us happier, and lead us to estimate humanity more
highly, we ought to be careful not to endeavour to establish
any scientific truths which would lessen and lower our estimate of man's place in nature; in short, we should do nothing to disturb any man's faith, if it be a delusion which
increases his happiness. But I hope and believe that the
discovery and propagation of every truth, and the dispelling
of every error, tends to improve and better the condition of
man, though the act of reforming old opinions and institutions causes so much pain and misery.

I expect to leave town for the Rhine before the end of

this week, and shall be in Switzerland in August. My sister
Katharine is on her way to Innsbruck with Arthur and
Rosamond.

Ever most faithfully yours,

CHARLES LYELL.

To the DUKE OF ARGYLL.

Shielhill, Kirriemuir: August 16, 1874.

My dear Duke of Argyll,—I was very glad to get your
letter, and much interested in the two raised beaches of the
Island of Jura. I have long been thinking that we may be
under a great delusion when we find proofs of upheaval in
ascribing to the movement an almost indefinite lateral extension, whereas all the evidence which we have in regard
to the modern effects of earthquakes runs quite in a contrary
direction.

I wish I could feel sure that you have in your library at
Inverary a copy of the eleventh edition of my ' Principles of
Geology,' but on second thoughts I have determined to send
you by book-post the second volume, which you can return
to me when you have quite done with it. I beg you will
read attentively my account of the earthquake of New Zealand in 1855.

I may as well tell you that a few months before I left
town a diploma was sent to me from a New Zealand Institute, given to me as the author of the most full and correct
account of what happened during that great convulsion, and
it was quite clear from letters then received, that the information I had given, derived from authorities peculiarly trust-
worthy, had been in no wise impaired by subsequent events
or criticisms.

Having this event in my mind, I thought your first
address to the Geological Society peculiarly telling, and I
do not see how any one who is willing to interpret the
former changes of the earth's surface by the light of those
now going on can pretend to refer any rock basin to ice-
scooping in preference to such movements as took place
north and south of Cook's Straits in 1855.

Mr. Judd, whose paper on the ' Geology of the Hebrides '

occupies worthily so large a portion of the August number
of the 'Quarterly Journal of the Geological Society,' has been
staying here at my sister's for the last month. Coming
fresh from Naples, Sicily, the Lipari, and Ponza Islands, and
having seen their examples of volcanic eruptions of every
kind mineralogically diversified, from the trachytic to the
basaltic extreme of composition, he finds here in this
county every variety of igneous rock, exhibiting exact
counterparts of modern lavas, from quartz, porphyry of Volcano, one of the Lipari group, to dolerite and basalt, such as
you have in Fingal's Cave and other parts of the Western
Islands. The great east and west dyke which traverses this
county, described by me nearly fifty years ago as containing
serpentine diallage, greenstone, quartz, porphyry, dolomite
hypersthene, &c., Judd has re-examined, and I am happy
to say he finds nothing to alter, and what he has observed
with me bearing on the section given in the 'Geological
Transactions' for 1826, he also finds quite correct. He regards our numerous east and west greenstone dikes cutting
through the old red sandstone, as lavas which filled up rents
which passed vertically through the rocks which they traverse, and which, but for the subsequent denudation of 2,000
feet, would end in Puys like those of the carboniferous strata
of Fife, or like Arthur's Seat.

Have you read in the newly issued volume of our
Society's Memoirs the paper by Jamieson of Ellon, on the
'Last Stage of the Glacial in North Britain'? I think it is
very suggestive, especially the intimate connection pointed
out between the present excess of rainfall on the western
side of Scotland as compared with that of the eastern being
analogous to the excess of glaciation on the corresponding
sides. It seems to show a connection between the present
meteorological agencies and those of the glacial period
which may help us to explain the great differences in the
glacial age of the excess of snow and ice in parts of the
northern hemisphere not very distant, and it may perhaps
help to explain how plants and animals of tropical forms
managed to escape being extinguished during the ice age.
I am greatly in hopes that the north and south axes of
movement, which Judd thinks he is going to establish for

the Miocene period, will explain the manner in which the valleys of Argyllshire and the west of Scotland were converted into sea lochs and fiords, so that we shall have to dispense with the hypothesis of ice action, and exchange it for that of fire.

I have a great deal more to say, and to ask you to observe, but shall wait till you have received my book.

With my compliments to the Duchess of Argyll, believe me ever truly yours,

CHARLES LYELL.

To CHARLES DARWIN, ESQ. *(dictated).*

Shielhill, Kirriemuir, Scotland : September 1, 1874.

My dear Darwin,—I have been intending from day to day to congratulate you on the Belfast meeting, on which occasion you and your theory of evolution may be fairly said to have had an ovation. Whatever criticisms may be made on Tyndall, it cannot be denied that it was a manly and fearless out-speaking of his opinions, and no one can wonder that the Belfast clergy of the Calvinistic school, three or four of them, as I suppose you saw, preached against such opinions on the Sabbath in the middle of the scientific week. It was principally, I believe, on the question of the efficacy of prayer, that objection was taken to the tone adopted by scientific writers of late, though I do not remember whether this was specially alluded to in the President's address; but Professor Jellet in one of the churches read what may be considered a regular argumentative paper on the efficacy and propriety of prayer, and I was glad to see that although part of his argument may have been special pleading, yet he fairly admitted that truth was the chief object to be kept in sight, and that unless prayer could be shown to be rational no Christian sanction and authority or Scriptural support ought to have any weight. I have been spending nine days, for the sake of change of air and sea breezes, on the coast between Arbroath and Montrose. Near the latter place I saw the Rev. H. Mitchell, who has contributed a very good paper on the Old Red Sandstone of this part of the world to our 'Quarterly Geological Journal.' He showed me his specimens of crustacean footprints, a long series of tracks, with the mark of the body trailing along, accompanying ripple-marks, and beautiful rain-drops. This seemed to bring *Pterygotus Anglicus* vividly before one, while the entire absence of marine shells in our Devonian beds, 10,000 feet thick, seems confirmatory of their freshwater origin. Perhaps there were lakes as large as Lake Superior.

Mr. Judd (whose important paper on the 'Five Great Volcanos of the Hebrides' you will have seen), has been staying with me here, and I should have much to tell you of what I have learnt of our geology. He quite confirmed what I have published about Forfarshire.

Ever affectionately yours,
CHARLES LYELL.

To *the* DUKE OF ARGYLL.

Shielhill, Kirriemuir : September 14, 1874.

My dear Duke of Argyll,—The reading of Judd's paper, and his estimate of the probable height of the five great volcanic cones of the Hebrides, has left a very grand impression on my mind of those mountains, and when he has fully worked up the paleontology and shown what a rich succession of life the primary, secondary, and tertiary strata of those islands exhibit, it will make a wonderfully grand addition to British, I may say to European, geology.

With regard to your question as to whether the movement giving rise to the formation of raised beaches (like those of the Island of Jura) may not sometimes be of a *local* character, I may refer to an interesting observation of Mr. Jamieson of Ellon, to the effect that the twenty-five-foot beach of the valley of the Forth gradually diminishes in height as we pass northwards, till in Aberdeenshire it is only eight feet above the present sea-level.

In reference to what you say on the outlines of the mountains in Skye and Rum, Mr. Judd tells me that the great *intrusive* masses in the Western Isles (whether composed of granite or gabbro), frequently assume a pseudo-stratified appearance when viewed from a distance. Having read his paper since you wrote your letter to me, you will

understand what he means by gabbro, of which the Coolin
Hills are constituted; and if you wish for any further explanations, he will be happy at any time to give them in
reference to the geology of the Hebrides.

Mr. Judd informs me that he was acquainted with the
existence, though he has not yet visited your interesting
little coal-field in Kintyre. The remarkable point about the
patch of coal strata in Morvern appears to him to be, that
not only is it considerably to the north of any other exposure
of carboniferous strata in the British Islands, but that it
lies on the north-west flank of the great Grampian axis
constituted by the series of granitic intrusions which extend
from Peterhead on the north-east to the Ross of Mull on
the south-west.

Since I sent you the representation by rubbing of a fossil
from the coal in Morvern, I have heard from Sir Charles
Bunbury, to whom I had sent the original specimen, that he
considers it to be *Lepidodendron aculeatum,* a common coal
species; he had supposed it to be a Sigillaria. The associated
fossil is, as Judd thought, a Calamite, so that there can be
no doubt of the geological age of the formation.

Judd quite coincides with you in your views as to the
existence everywhere in the Hebrides of evidences of enormous denudation, side by side with equally striking proofs of
grand subterranean movements; and like yourself, he regards
the contours of the existing surfaces as the product of the
continued working side by side of these two classes of forces.
He was pleased to hear that you had an opportunity of seeing the remarkable outliers of tertiary basalt resting on
Silurian gneiss (two of which were noticed by Macculloch,
and which have their exact counterparts in the north of
Ireland).   Ever truly yours,

CHARLES LYELL.

To CHARLES DARWIN, ESQ.

73 Harley Street, London : September 25, 1874.

My dear Darwin,—There is no subject to which Judd
oftener referred, and told me he had done so when discussing
volcanic questions with Scrope, than your subsidence of St.

Jago, as being a general law of volcanic regions. The sinking down referred to by me ('Principles' vol. ii. 458, 'Student's Elements,' 149–166), as occurring in New Zealand, and which has been confirmed by the New Zealand geologists, corroborates your St. Jago experience.

Just before I left Scotland Judd made an excursion to revisit the Hebrides, and during a ten days' absence discovered for the first time carboniferous strata preserved under some two thousand feet of tertiary basalt in the Island of Mull, and he brought to me what I guessed was a Sigillaria and the cast of a Calamite; but when I sent them to Charles Bunbury he pronounced them to be *Lepidodendron aculeatum* and a Calamite, ordinary carboniferous fossils. All this is explained by admitting that in the time of the Miocene volcanos there was a sinking down like that of St. Jago in the island, near the loftiest cone of that island, and one of the five great volcanos which he believes to have existed in the Hebrides, and which may have rivalled the Peak of Teneriffe in height, in the Miocene period. How much grandeur the scenery of the Hebrides must have presented in that same Miocene period, when Madeira was already in continued action, as well as Porto Santo and the Giant's Causeway, &c.! The Duke of Argyll has sent me word that he has found a fossil *Salisburia* (now a genus growing in Japan) among the fossils of Ardtun in Mull, where he formerly found *Asa platanoides*. I wonder whether the cones of the Hebrides, if they were as high as Etna or the Peak of Teneriffe before the sinking down which you observed in St. Jago had taken place, were covered with snow in the Miocene period, when the vegetation at their base, at Ardtun for example, was sub-tropical. As there is now a vast difference in the vegetation of the desert region of Etna usually covered with snow, so in Miocene time it does not follow that volcanos ten or twelve thousand feet high should not have had on their summits a flora very different from that at their base; but these you will say are idle speculations.

It is remarkable how perfectly a sinking in Miocene times, like that which you have supposed or proved for the Cape de Verde Islands, would in Judd's opinion give a satisfactory solution to the preservation in Mull and in the cliffs on both sides of the Sound of Mull, of those intercalations of poikilitic, triassic, liassic, oolitic, neocomian, cretaceous, and newer strata, of which a full account is to be given before the close of Judd's paper.

All the work which I have done with Judd in Forfarshire has confirmed me in the belief that the only difference between Paleozoic and recent volcanic rocks is no more than we must allow for, by the enormous time to which the products of the oldest volcanos have been subjected to chemical changes such as those which turn an olivine basalt into serpentine.

Ever affectionately yours,
CHARLES LYELL.

His death took place on February 22, 1875 (having had another family bereavement only a fortnight before, in the unexpected death of his brother Colonel Lyell, who had been almost daily with him, up to the time of his sudden and fatal illness).

Though expected by friends it was deeply felt. Dr. Hooker [6] wrote to Miss Lyell that same day :—

' I have just heard the distressing news, and can hardly yet say how much I feel it. My loved, my best friend, for well nigh forty years of my life. To me the blank is fearful, for it never will, never can be filled up. The most generous sharer of my own and my family's hopes, joys, and sorrows, whose affection for me was truly that of a father and brother combined. I deeply feel for you all; two such blows to you and your sisters, and to Mrs. Lyell, surely hardly ever came so rapidly, so remorselessly as it were.

' I have just headed a memorial to Dean Stanley, praying that he may be interred in Westminster Abbey, the Dean having volunteered his hearty assent, and every influence in his power to have it granted. Sir Edward Ryan first told me of it, and now joins with me in the earnest hope that you will allow this tribute to be paid to the most philosophical

---

[6] Now Sir Joseph Hooker, C.B., K.C.S.I.

and influential geologist that ever lived, and one of the very best of men.

'The memorial is being signed by Fellows of the Royal, Linnæan, and Geological Societies, and is most powerful.'

The following is the memorial in pursuance of which the body of Sir Charles Lyell was placed in Westminster Abbey:—

'We, the undersigned Fellows of the Royal, the Geological, and the Linnæan Societies, respectfully pray that the remains of Sir Charles Lyell may be interred in Westminster Abbey. For upwards of half a century he has exercised a most important influence on the progress of geological science, and for the last twenty-five years he has been the most prominent geologist in the world, equally eminent for the extent of his labours and the breadth of his philosophical views.

'JOSEPH HOOKER, President of the Royal Society.
JOHN EVANS, President of the Geological Society.
GEORGE J. ALLMAN, President of the Linnæan Society.
W. G. ADAMS, F.R.S.
J. M. ARNOTT, F.R.S.
GEORGE BUSK, F.R.S.
H. DEBUS, F.R.S.
JAS. FERGUSSON, F.R.S.
W. H. FLOWER, F.R.S.
AUGUSTUS W. FRANKS, F.R.S.
FRANCIS GALTON, F.R.S.
J. H. GLADSTONE, F.R.S.
J. A. GRANT, F.R.S.
W. R. GROVE, F.R.S.
J. HEYWOOD, F.R.S.
T. ARCHER HIRST, F.R.S.
W. HUGGINS, F.R.S.
T. MCK. HUGHES, F.R.S.

T. H. HUXLEY, F.R.S.
CHARLES MANBY, F.R.S.
J CARRICK MOORE, F.R.S.
E. A. PARKES, F.R.S.
JOHN PERCY, F.R.S.
JOSEPH PRESTWICH, F.R.S.
A. RAMSAY, F.R.S.
H. RAWLINSON, F.R.S., Pres. R.G.S.
G. OWEN REES, F.R.S.
EDWARD RYAN, F.R.S.
W. SHARPEY, F.R.S.
W. SPOTTISWOODE, Tres. R.S.
WARRINGTON SMYTH, F.R.S.
THOMAS THOMSON, F.R.S.
JOHN TYNDALL, F.R.S.
ALEX. WILLIAMSON, For. Sec. R.S.
E. H. BUNBURY, F.G.S.
L. L. DILLWYN, F.L.S., F.G.S.
A. GROTE, F.L.S., F.G.S.
JOSEPH WHITWORTH.'

The pall-bearers were the Duke of Argyll, Sir Edward Ryan, Mr. Justice Grove, Professor Huxley, Dr. Hooker, Mr. John Evans (President of the Geological Society), Mr. J. Carrick Moore, and the Rev. W. S. Symonds.

The Dean of Westminster conducted the service.

The following Sunday the Dean of Westminster preached

the funeral sermon in the Abbey, of which the following is
an extract:—

'Of him who is thus laid to rest, if of any one of our
time, it may be said that he followed truth with a zeal as
sanctified as ever fired the soul of a missionary, and with a
humility as child-like as ever subdued the mind of a simple
scholar. For discovering, confirming, rectifying his conclusions, there was no journey too distant to undertake.
Never did he think of his own fame or name in comparison
of the scientific results which he sought to establish. From
early youth to extreme old age it was to him a solemn
religious duty to be incessantly learning, constantly growing,
fearlessly correcting his own mistakes, always ready to
receive and reproduce from others that which he had not in
himself. Science and religion for him not only were not
divorced, but were one and indivisible.

'The instinct which impels us to seek for harmony between
the highest truths of science and the highest truths of the
Bible is an instinct far nobler and truer than that which
would seek to part them asunder. In this higher instinct,
he who has departed fully shared. The great religious
problems of our time were never absent from his mind.
The infinite possibilities of nature gave him fresh ground
for his unshaken hope in the unknown, immortal future.
His conviction of the peaceful, progressive combination of
natural causes towards the formation of our globe filled him
with a profound and ever profounder sense of "the wonder
and the glory of this marvellous universe." The generous
freedom allowed to religious inquiry in the National Church,
the cause of humanity in the world at large, were to him as
dear as though they were his own personal and peculiar
concern. With that one faithful, beloved, and beautiful
soul, who, till within the last two years of his life, shared
all his joys and all his sorrows, all his labours and all his
fame, he walked the lofty path, "which the vulture's eye
hath not seen, nor the lion's whelp trodden "—the pathway
of the just, "lightening ever more and more towards the
perfect day," in which we humbly trust that they are now
at last reunited in the presence of that light which they
both so sincerely sought.'

He was buried in the nave of the Abbey. A gravestone of fossil marble from Derbyshire, selected as an appropriate tribute, bears the following inscription :—

CHARLES LYELL
BARONET F.R.S.

AUTHOR OF
' THE PRINCIPLES OF GEOLOGY '
BORN AT KINNORDY IN FORFARSHIRE
NOVEMBER 14 1797
DIED IN LONDON
FEBRUARY 22 1875

THROUGHOUT A LONG AND LABORIOUS LIFE
HE SOUGHT THE MEANS OF DECIPHERING
THE FRAGMENTARY RECORDS
OF THE EARTH'S HISTORY
IN THE PATIENT INVESTIGATION
OF THE PRESENT ORDER OF NATURE
ENLARGING THE BOUNDARIES OF KNOWLEDGE
AND LEAVING ON SCIENTIFIC THOUGHT
AN ENDURING INFLUENCE
' O LORD HOW GREAT ARE THY WORKS
AND THY THOUGHTS ARE VERY DEEP '
PSALM XCII. 5.

A marble bust by Mr. Theed, after the original one by Gibson, is placed near the grave.

A few words from one [7] who had frequent opportunities of seeing him in his latter years may be appropriate here, as they are full of comprehension of the child-like simplicity, and the vigour of intellect combined with a deep earnestness, which were his chief characteristics :—

' The last of the elder generation of our great men of science, Sir Charles Lyell, leaves behind him the memory of a character almost ideally representing what such men should be; so free from egotism, vanity, or jealousy, so ready to be pleased with every innocent jest or amusement, so ready to listen patiently to the remarks of those infinitely below his intellectual calibre, and withal so affectionate and tender of heart, that no child could be more simple; and, on the other hand, so filled with reverent enthusiasm for the glory and grandeur of the universe to whose study

[7] Miss F. P. Cobbe.

he devoted himself, and so ready to open his mind to each new truth, that no man could better deserve the high title of a true philosopher. Nor did his philosophy, though it released him from some of the bonds of early prejudice, ever lead him to renounce those highest truths to which the lesser ones of science lead up. It was his frequent observation that religious sentiment deserved as much confidence as any other faculty of our nature, and in full faith and hope in God and immortality he passed calmly into the dark valley of age and death.' [8]

[8] See Appendix C.

APPENDICES.

APPENDIX A.

Vol. I. p. 376.

From CAPTAIN BASIL HALL, R.N., *to* LEONARD HORNER, Esq.

Geneva : September 7, 1833.

My dear Sir,—Upon the whole I am glad that our admirable friend Lyell has disentangled himself from King's College and the Royal Institution. He will do more for his science, and for himself, and for us who love both, by observing and writing, than he could by possibility do by lecturing alone. Could he do both, it would be better certaialy, but if the lectures are to interfere with his publications, and still more with his observations, and his journeys to see men as well as rocks, it is clearly the wisest plan to cut the cockneys, and address himself solely to men of science, or to those who, without that pretension, are capable of understanding his most excellent writings.

I think his book by many many degrees the best work, not only on his subject, but on any scientific subject with which I am acquainted. It reduces an intricate, obscure, and most enormously copious subject, to one which is almost mathematically arranged, clear and condensed. His generalisations are quite inimitable, except by the singular beauty and force of his detailed descriptions and his patient investigation of disputed points ; to which I may add the calm, dispassionate, gentlemanlike style in which he handles, not one, but every controversial subject which the subject requires should be discussed. And yet he does this with so much liveliness both of manner and of diction, and with such genuine earnestness, that I, for one, cannot help being swept away with the gentle but irresistible current of his persuasive eloquence. For, indeed, his book contains the essence of eloquence; right reason, extensive and exact knowledge, cultivated taste, and a disinterested and philosophical desire to state the matter in such a way that truth may be the result. There is, moreover, an elegance of fancy throughout, and a touch of humour, or rather of wit, which are in happy companionship with the simplicity and general elegance of the composition. It is already, in my apprehension, the first book of the day, and every time I read it, I am filled the more and more with respect for the author's talents and his knowledge, and feel more and more grateful to him for the pleasure he has given me. I trust he will be able soon to print a cheaper edition, for the book would soon be very extensively circulated, if its form and price were such as to enable the great body of readers to get at it.

It is the most abominable shame in the world Lyell's book not being translated into French. Their science is very much a matter of talk and pretension, and in geology they believe that beyond their eternal Paris Basin, there is nothing worth looking at or thinking about. Did you see in one of the five hundred Eloges on Cuvier, what a Paris *savant* said of him. 'It would almost appear,' said the modest eulogiser, 'as if the Almighty had placed this vast and wonderful collection of fossil remains for the express purpose of being discovered and descanted upon by our great countryman!' Poor Cuvier! how little did this ass know of the taste of the mighty man about whom he was braying, to an audience, who if they had possessed a grain more sense than the orator, ought to have made a 'fossil remain' of him forthwith, by pitching him over the quay into the silt of the Seine.

Ever most truly yours,

BASIL HALL.

APPENDIX B.

Vol. II. p. 451.

[The following Tribute to the Memory of Lady Lyell by George S. Hillard, Esq., is so just and true, that it is placed here in order that she, who was the constant and cherished companion of her husband for forty years, might be united with him in these volumes.]

From the Boston (United States) ' Daily Advertiser,'
May 19, 1873.

' DIED IN LONDON, AFTER A SHORT ILLNESS, APRIL 24, MARY ELIZABETH, WIFE OF THE CELEBRATED GEOLOGIST, SIR CHARLES LYELL.

' There are many hearts in the United States that will be saddened by the death of this admirable woman. She had twice visited our country in company with her husband, and in every part of it had made warm and lasting friends. And such she could not but make, for she had a rare union of the qualities which ensure confidence with those that win affection. Strength and sweetness were hers, both in no common measure. The daughter of Leonard Horner, and the niece of Francis Horner, in her an excellent understanding had been carefully trained, and she had that general knowledge and those intellectual tastes which we expect to find in an educated Englishwoman : and from her childhood she had breathed the refining air of taste, knowledge, and goodness. Her marriage with an eminent man of science gave a scientific turn to her thoughts and studies, and she became to her husband not merely the truest of friends and the most affectionate and sympathising of companions, but a very efficient helper. She was frank, generous, and true; her moral instincts were high and pure ; she was faithful and firm in friendship; she was fearless in the expression of opinion, without being aggressive; and she had that force of character and quiet energy of temperament that gave her the power to do all that she had resolved to do.

'All that we have said she might have been, and yet not have been winning and sweet; but this she was, and to such an extent that those who saw her only casually, carried away hardly any other impression of her. She had more than a common share of personal beauty; but had she not been beautiful she would have been lovely, such was the charm of her manners, which were the natural expression of warmth and tenderness of heart, of quick sympathies, and of a tact as delicate as a blind man's touch. This woman, so widely informed, so true, so strong, so brave, seemed all compact of softness, sweetness, and gentleness; a very flower that had done no more than drink the sunshine and the dew. In her smile, her greeting, the tones of her voice, there was a charm which cannot be described, but which all who knew her have felt and will recall.

'No tribute to Lady Lyell would be complete that did not speak of her strong attachment to America and American friends. Her feelings in this direction were in harmony with her warmth of heart and her generous spirit. She was a fervid and enthusiastic friend to America. During the war there was not a woman or a man in England that stood by the Union and the Government more ardently and fearlessly than she. Never, at any period, was she silent when America was disparaged. No one has died in England upon whose monument might be more appropriately carved the words, " Here lies the friend of America."

'We have dwelt somewhat upon Lady Lyell's strong attachment to America, not merely because it entitles her to the grateful remembrance of all Americans, but because it was illustrative of her independence of judgment and her generous and catholic nature. For be it remembered that she was an Englishwoman, and never for a moment could have been taken for anything but an Englishwoman; but she came to America with no insular prejudices and no conventional standard. She did not look at this great country with the eye of a lady in waiting. In her view, usages and customs were not wrong simply because they were not English. She saw and appreciated all that was good in us; and if she saw anything that she could not approve, she was willing to forget it.

'Lady Lyell had reached an age when the end of life begins to be a natural event, and cannot be called premature; but such was her warmth of heart, so fresh were her feelings, her faculties so bright, that her death falls upon the hearts that loved her with the shock of unexpectedness, and in parting from her, her friends must feel that to them something of the light of life has passed away from earth.'

APPENDIX C.

Vol. II. p. 463.

[The following Tributes, written after the death of Sir Charles Lyell, are extracts from some of the many notices which appeared in various journals, and are offered as being eminently characteristic of him, and his work.]

'To sketch the life and labours of Sir Charles Lyell would be much the same thing as sketching the development of the modern school of British geology during well-nigh half a century. The task to which he devoted his noblest energies was that of establishing the principles of geology on a sound and philosophical basis. His leading lesson was a belief in the uniformity of the laws of nature : a belief which led him to argue that by studying the changes which are being wrought upon the surface of the earth by the silent action of forces now in operation, we put ourselves in possession of a key to the interpretation of those ancient records which it is the special business of the geologist to decipher. Sir Charles, indeed, developed with singular success the great truths which were first enunciated by Dr. Hutton of Edinburgh, and eloquently illustrated by his friend Professor Playfair. Hutton died in 1797, and it is curious to note that the same year which witnessed his death gave birth to one who was destined to expound his doctrines with such force of argument as to carry them successfully against all opposition, and establish them as fundamental principles of the science.

'The earliest scientific observations of the young geologist appear to have been made on the rocks of his native country, since we find that his first paper, contributed in 1825 to the " Edinburgh Journal of Science," was one on a dike of serpentine in the county of Forfar. The first volume of his celebrated " Principles of Geology " was published in 1830.

So great has been the popularity of this work, that it has passed through no fewer than eleven editions; and during his last illness, the venerable author was engaged upon a twelfth.

'In casting a glance over the life of Sir Charles Lyell, it will be seen that he was characterised by singular steadiness of purpose. The great doctrine of uniformitarianism which he advocated in 1830, he nobly supported to the day of his death, although modified, of course, by the progress of scientific inquiry. He made everything subordinate to his one ruling idea, that of establishing the principles of geology upon a thoroughly logical basis. Nor were his honesty and boldness less marked than his steadiness and concentration. A staunch advocate of perfect freedom of scientific opinion, he fearlessly pushed his principles to their legitimate conclusions. Having first satisfied himself of the soundness of his fundamental postulates, and employing a vigorous logic at each successive step of his reasoning, he cared but little whither his conclusions carried him; whether they chanced to fall in unison with general belief, or cut directly across the grain of popular prejudice. Toleration had been taught him by bitter experience in early life. Like most advanced thinkers he had suffered keenly from the harsh criticisms of the narrow-minded; he had shared the fate which usually falls to

> Teachers whose minds move faster than the age,
> And faster than society's slow flight.

'Perhaps the most striking characteristic of Sir Charles Lyell, was his remarkable mental plasticity—a power which made him ever ready to receive new impressions, and never too proud to correct his old views, or confess to a change in his previous opinions. Not that he craved for novelty merely for novelty's sake. But if he considered that fresh evidence on a given subject justified the alteration of a previously formed opinion, he frankly turned round and renounced his old views. This was nowhere more strikingly seen than in his change of attitude towards the great question of the Origin of Species after the publication of Mr. Darwin's epoch-marking work. Whenever Sir Charles considered that a case had been fairly made out, he was too

noble to shut his eyes against the evidence, but freely
accepted the new conclusion, even to the overthrow of his
previous work. It was the advancement of the philosophy
of geology, not the advancement of self, that he was constantly seeking. To the very last he retained this plasticity
of mind; a characteristic which led him so freely, yet so
cautiously, to bend before new arguments, and to stretch
his old views to meet the requirements of modern research ;
thus strikingly unlike so many men of genius, who having
developed in early life to a certain point, are content to
spend the rest of their life in a state of intellectual crystal-
lisation.' [1]

'The first generation of geologists has passed away.
Buckland, Conybeare, Sedgwick, Murchison, Phillips, all—
with the exception of the first two—have died within the
last few years; and now Sir Charles Lyell, full of years and
honours, has been added to the number.

'Knowledge has widened all round; larger views of nature
and the universe have gained credit everywhere; and geology
has fully shared in the general expansion. In this process
no one has had a greater share than Sir Charles Lyell.
For he was much more than a mere geologist. He had a
well-trained and philosophic mind, which enabled him to
take large views of every subject presented to his intellect,
to see its various bearings and its points of alliance or contrast with other ranges of thought. All his work was done
leisurely, fully, and completely, in large books, and not in
fragments of essays and papers; and every book was absolutely finished up to the point which knowledge had reached
when he put it forth. Sir Charles Lyell, in spite of his
great age, has been singularly open to fresh accessions of
knowledge and fresh generalisations from the increasing
store of facts; and it is no light testimony to the original
soundness of his views, that they have easily admitted an
assimilation of all fresh discoveries and a re-adjustment to
newly-accepted theories. A striking example of this was
given in his late conversion to Mr. Darwin's doctrine of
Natural Selection. Nine editions of the "Principles of

[1] By F. W. Rudler in the *Academy*, February 27, 1875.

Geology " had carried his name and reputation over the
civilised world, and along with it his exposition of the doctrine of Special Creations. The only explanation which
then seemed possible to him of the perpetual change of life
revealed by successive strata was, that when the material
conditions of any district became so changed that the old
inhabitants died out, a new creative *fiat* went forth, by
virtue of which the district was again peopled with fresh
inhabitants especially adapted to its new conditions. When,
however, it was shown that causes were at work which
slowly and gradually modified the characters of plants and
animals, so that they became adjusted by a self-adapting
process to the changing circumstances around them, he
gladly adopted a view which was so much in harmony with
his general principles; and he put forth a tenth edition, in
which the old theory was formally renounced, and the new
one taken up. It was justly characterised by Dr. Hooker,
in his address to the British Association at Norwich, as a
bright example of heroism, that an author could thus
abandon, "late in life, a theory which he had for forty years
regarded as one of the foundation-stones of a work that had
given him the highest position attainable among contemporary scientific writers; " and it was no less justly observed
that the superstructure must be very solid and coherent
which could allow the builder thus " to undermine it and substitute a new foundation," and yet, after all, survey his
edifice, and behold it, " not only more secure, but more har-
monious in its proportions than it was before." ' [2]

' Lyell was not only a keen investigator of natural phenomena, he was also a shrewd observer of human nature,
and his four interesting volumes of travel in America are
full of clever criticism and sagacious forecasts. His mind,
always fresh and open to new impressions, by sympathy
drew towards it and quickened the enthusiasm of all who
studied nature. Had he done nothing himself, he would
have helped science on by the warmth with which he hailed
each new discovery. How many a young geologist has been
braced up for new efforts by the encouraging words he heard

[2] From the *Guardian*, March 3, 1875.

from Sir Charles, and how many a one has felt exaggeration checked, and the faculty of seeing things as they are strengthened, by a conversation with that keen sifter of the true from the false!

'The little wayside flower, and from early associations still more the passing butterfly, for the moment seemed to engross his every thought. But the grandeur of the sea impressed him most; he never tired of wandering along the shore, now speaking of the great problems of earth's history, now of the little weed the wave left at his feet. His mind was like the lens that gathers the great sun into a speck, and also magnifies the little grain we could not see before. He loved all nature, great and small.

'In the companion of his life, sharing his labour, thinking his success her own, Sir Charles had an accomplished linguist who braved with him the dangers and difficulties of travel, no matter how rough; the ever-ready prompter when memory failed, the constant adviser in all cases of difficulty. Had she not been part of him she would herself have been better known to fame. The word of encouragement that he wished to give lost none of its warmth when conveyed by her; the welcome to fellow-workers of foreign lands had a grace added when offered through her. She was taken from him when the long shadows began to cross his path; but it was not then he needed her most. When in the vigour of unimpaired strength he struggled amongst the foremost in the fight for truth, then she stood by and handed him his spear or threw forward his shield. He had not her hand to smooth his pillow at the last, but the loving wife was spared the pain of seeing him die.

'His was a well-balanced, judicial mind which weighed carefully all brought before it. A large type of intellect— too rare not to be missed. But it was well that circumstances did not combine to keep the young laird on his paternal lands among the hills of Forfarshire : it was well for science that he was induced to prefer the quieter study of nature to the subtle bandying of words or the excitement of forensic life. Failing health had for some time removed him from debates. Still, to the last his interest in all that was going on in the scientific world never failed, and nothing pleased him more than an account of the last discussion at the Geological Society, or of any new work done.' [3]

At a meeting of the Edinburgh Geological Society, March 4, 1875, the President, Mr. D. Milne Home, said

' That since their last meeting they had met with a severe loss by the death of the distinguished man who held in their Society the honorary office of patron. Sir Charles Lyell's death he was sure all of them deeply regretted, not merely as geologists but as Scotchmen. He was the oldest of a very distinguished band, who, during the last fifty years, had done more for geology than had been effected in any other country. In fact, the very foundations of the science were, it might be said, laid in the city of Edinburgh by Sir James Hall, Hutton, Playfair, and Jamieson. Many others, animated by their example and stirred by the novelty of their investigations, came forward to help in carrying on this good work.

' But of all the labourers in the geological field who had done good service, he ventured to say that no one of them, nor even all of them together, had done so much to extend the geological edifice and raise it to its present conspicuous height, as Sir Charles Lyell. How could it be otherwise? His whole life after manhood was devoted to geology. He probably collected more facts and drew more important conclusions than had ever been done by any other geologist. His published works formed a treasury of valuable information, and his method of investigation gave important lessons which all labourers in the same field would do well to profit by.

' In working out his speculations, Sir Charles Lyell always endeavoured to obtain a broad basis on which to rest them. He pointed out the absurdity of drawing conclusions applicable to the whole earth from phenomena observable only in one country. To attain that object he himself travelled, hammer in hand, over half the earth's surface. In search of facts he visited more than once almost every country in Europe ; and was twice in North America and Canada.

[3] From *Nature*, March 4, 1875. By Prof. Hughes.

' The great ends he accomplished were due to methodical industry, to delight in making a discovery, to love of truth, and to untiring perseverance. These were the simple weapons with which he fought the battle of life, and by means of which he amassed for the benefit of his race a rich store of truths previously unknown, and laid up for himself a reputation which would long endure. He had mentioned the love of truth as one of the motives by which Lyell was inspired. That feeling was in him so strong and sacred that whenever he discovered he was in error, whether error of fact or of inference, he never was happy till he had an opportunity of publicly avowing and correcting it.

' Sir Charles Lyell visited Scotland last year, and went to his paternal estate of Kinnordy, in Forfarshire, in the neighbourhood of the Sidlaw Hills and the more lofty Grampian range, with which, as a boy, he had been familiar. He there took the opportunity of revisiting some of the haunts of his earlier days, and the spots where the love of his favourite science had first been imbibed, and had the satisfaction of feeling that all through his life he had been faithful to his first love, and under its impulse had done service to the world, and also done credit to the land of his birth.' [4]

[4] From the *Scotsman*, March 5, 1875.

APPENDIX D.

BEQUEST TO THE GEOLOGICAL SOCIETY.

Extract from, and Codicil to, the Will of the late Sir Charles Lyell, Bart.

I give to the Geological Society of London the die executed by Mr. Leonard Wyon of a medal to be cast in bronze and to be given annually and called the Lyell Medal, and to be regarded as a mark of honorary distinction and as an expression on the part of the governing body of the Society, that the medallist (who may be of any country or either sex) has deserved well of the science. I further give to the said Society the sum of two thousand pounds (free of legacy duty) to be paid to the President and Treasurer for the time being, whose receipt shall be a good discharge to my Executors ; and I direct the said sum to be invested in the name of the said Society or of the Trustees thereof in such securities as the Council shall from time to time think proper, and that the annual interest arising therefrom shall be appropriated and applied in the following manner :—Not less than one third of the annual interest to accompany the medal, the remaining interest to be given in one or more portions at the discretion of the Council for the encouragement of geology or of any of the allied sciences by which they shall consider geology to have been most materially advanced, either for travelling expenses or for a memoir or paper published or in progress, and without reference to the sex or nationality of the author or the language in which any such memoir or paper shall be written.

And I declare that the Council of the said Society shall be the sole judges of the merits of the memoirs or papers for which they may vote the Medal and Fund from time to time.

And I direct that the legacy hereinbefore given to the said Society, shall be paid out of such part of my personal estate as may be legally applicable to the payment of such bequests.

As a Codicil to my Will dated January 1874, in which I directed that the Medal should be awarded annually by the Council of the Geological Society, I think it would be preferable that, instead of requiring it and the interest of the 2,000*l.* to be given annually, 1 should leave it to the discretion of the Council to suspend the awarding of the Medal for one year, as it may sometimes be a source of embarrassment when there are several medals to bestow, to be forced to find a fit recipient. In this case the Council would have in the year following a larger sum from the interest of the 2,000*l.*, as well as two medals, to give away—which might be an advantage, because it has sometimes happened that two persons have been jointly engaged in the same exploration in the same country, or perhaps on allied subjects in different countries, and the Council may think that the labours of both of them may deserve to be crowned by a mark of their approbation. In this case a medal may be given to each, with such proportion of the interest as the Council may decide, always not being less to each medal than one third of the annual interest of the 2,000*l.* as directed in my Will.

APPENDIX E.

GEOLOGICAL PAPERS AND WORKS BY SIR CHARLES LYELL.

' On a recent Formation of Freshwater Limestone in Forfarshire, and on some recent Deposits of Freshwater Marl,' *Geological Society's Transactions*, 1825.

' On Serpentine Dyke in Forfarshire,' *Edinburgh Journal of Science*, 1825.

' On various Scientific Institutions in England,' *Quarterly Review*, 1825.

' On Fossil Bones of the Elephant and other Animals found near Salisbury,' *Geological Society's Proceedings*, 1826.

' On the Strata of the Plastic Clay Formation between Christchurch, Hants, and Studland Bay, Dorset,' *Geological Society's Transactions*, 1826.

' On the Freshwater Strata of Hordwell Cliff, Beacon Cliff, and Barton Cliff, Hampshire,' *Geological Society's Transactions*, 1826.

Review of Scrope's ' Geology of Central France,' *Quarterly Review*, 1827.

' Lyell and Murchison on Excavation of Valleys,' *Edinburgh New Philosophical Journal*, 1829.

' Lyell and Murchison on Lacustrine Deposits of Cantal,' *Annales des Sciences Naturelles*, 1829.

' Murchison and Lyell on Freshwater Formation of Aix in Provence,' *Edinburgh Philosophical Journal*, 1829.

' PRINCIPLES OF GEOLOGY,' vol. i. 8vo.   January, 1830.

   ,,   ,,   ,,   ,, ii. ,,   ,,   1832.

   ,,   ,,   ,,   ,, i. 2nd edition, 8vo.   ,,   1832.

   ,,   ,,   ,,   ,, ii.   ,,   ,,   1833.

   ,,   ,,   ,,   ,, iii. 1st edition, 8vo.   May 1833.

   ,,   ,,   ,,   New edition (called the third) of the whole work, in four vols. 12mo.   May 1834.

' On Freshwater Formation of Cerdagne in the Pyrenees,' *Magazine of Natural History*, 1834.

' On the Proofs of a gradual Rising of the Land in certain Parts of Sweden, The Bakerian Lecture,' *Philosophical Transactions*, 1834.

' On the Change of Level of the Land and Sea in Scandinavia,' *British Association Report*, 1834.

' PRINCIPLES OF GEOLOGY,' 4th edition, four vols. 12mo.   June 1835.

' On Relative Ages of Crag in Norfolk and Suffolk,' *Magazine of Natural History*, 1835.

'On the Cretaceous and Tertiary Strata of the Danish Islands of Seeland and Möen,' 1835.

'On the Occurrence of Fossil Vertebræ of Fish of the Shark Family in Loess of the Rhine,' *Geological Proceedings,* 1835.

'Address as President of the Geological Society,' 1836.

   "   "   "   "   1837.

'PRINCIPLES OF GEOLOGY,' 5th edition, in four vols. 12mo., March 1837.

'On Phenomena connected with the Junction of Granitic and Transition Rocks near Christiania in Norway,' *British Association Report,* 1837.

'On Vertical Lines of Flint, traversing Horizontal Strata of Chalk near Norwich,' *British Association Report,* 1838.

'ELEMENTS OF GEOLOGY,' 1st edition, in one vol., July 1838.

'On the Occurrence of Graptolites in the Slate of Galloway,' *Geological Proceedings,* 1838.

'Remarks on Captain Bayfield's Canada Shells,' *Geological Transactions,* 1839.

'On Remains of Mammalia in the Crag and London Clay of Suffolk,' *British Association Report,* 1839.

'On Sandpipes in Chalk near Norwich,' *Philosophical Magazine,* 1839.

'On Fossil Teeth of Leopard, Bear, &c. at Newbourn, Suffolk,' *Annals of Natural History,* 1839.

'On Fossil Quadrumana, Marsupials, &c., in London Clay, near Woodbridge, Suffolk,' *Annals of Natural History IV.,* 1839.

'PRINCIPLES OF GEOLOGY,' 6th edition, three vols. 12mo., June 1840.

'On Ancient Sea-cliffs in the Valley of the Seine in Normandy,' *British Association Report,* 1840.

'On the Boulder Formation and Mud Cliffs of Eastern Norfolk,' *Geological Magazine,* 1840.

'On the Geological Evidence of the former Existence of Glaciers in Forfarshire,' *Geological Proceedings,* 1840.

'On the Genus Conus in the Lias of Normandy,' *Annals of Natural History VI.,* 1840.

'On the Faluns of the Loire,' *Geological Society's Proceedings,* 1841.

'On the Freshwater Fossil Fishes of Mundesley as determined by Agassiz,' *Geological Society's Proceedings,* 1841.

'Remarks on the Silurian Strata between Aymestry and Wenlock,' *Geological Society's Proceedings,* 1841.

'Notes on the Silurian Strata near Christiania in Norway,' *Geological Proceedings,* 1841.

'ELEMENTS OF GEOLOGY,' 2nd edition, in two vols. 12mo.. June 1841.

'On the Carboniferous and Older Rocks of Pennsylvania,' 1841.

'On the Recession of the Falls of Niagara,' *Geological Society's Proceedings,* 1842.

'On the Elevated Beaches and Boulder Formations of the Canadian Lakes and Valley of St. Lawrence,' *Geological Society's Proceedings,* 1842.

'On Fossil Foot-prints of Birds, Connecticut,' *Geological Society's Proceedings,* 1842.

'On the Tertiary Formations in Virginia,' *Geological Society's Proceedings*, 1842.

'On Tertiary Strata of Martha's Vineyard,' *Geological Society's Proceedings*, 1843.

'On Mastodon at Big-bone-Lick, Kentucky,' *Geological Society's Proceedings*, 1843.

'On Coal and Gypsum of Nova Scotia,' *Geological Society's Proceedings*, 1843.

'On Loess of the Rhine,' *Edinburgh Philosophical Journal*, 1843.

'On Chalk of New Jersey,' *Geological Journal*, 1844.

'On Age of Plumbago and Anthracite at Worcester, Massachusetts,' 1844.

'Report on Haswell Colliery, Lyell and Faraday,' *Geological Journal*, 1844.

'On Miocene Strata of Maryland, Virginia,' &c., *Geological Journal*, 1845.

'On White Limestone, and Eocene Formations of Virginia, Carolina, &c.,' *Geological Journal*, 1845.

'On Lava-currents, Auvergne,' *Geological Journal*, 1845.

'TRAVELS IN NORTH AMERICA,' with Geological Observations, two vols., 1845.

'On Coal-Field of Tuscaloosa, Alabama,' *Silliman's Journal*, 1846.

'On Alabama Coal-fields,' *Geological Journal*, 1846.

'On Newer Deposits of Southern States, Claiborne,' *Geological Journal*, 1846.

'On Fossil Foot-prints, allied to Cheirotherium, in Pennsylvania,' *Geological Journal*, 1846.

'On Delta of Mississippi,' *Lecture to the British Association*, 1846.

'PRINCIPLES OF GEOLOGY,' 7th edition, in one vol. 8vo., February 1847.

'Age of Volcanos in Auvergne, as determined by Fossil Mammalia,' *Lecture at Royal Institution*, 1847.

'On Structure and Probable Age of Coal-field of James River, Virginia,' *Geological Society*, 1847.

'On Craters of Denudation, with Observations on the Structure and Growth of Volcanic Cones,' *Geological Society's Proceedings*, 1849.

'On Recent Foot-prints on Red Mud in Nova Scotia,' *Geological Journal*, 1849.

'A SECOND VISIT TO THE UNITED STATES OF NORTH AMERICA,' two vols., 1849.

'Lecture on Delta of Mississippi at the Royal Institution,' 1849.

'PRINCIPLES OF GEOLOGY,' 8th edition, in one vol. 8vo., May 1850.

'On Forests of Erect Fossil Trees in Coal Strata of North America,' *Lecture at Royal Institution*, 1850.

'President's Address to Geological Society of London,' 1850.

" " " " " 1851.

'ELEMENTS OF GEOLOGY,' 3rd edition (or Manual of Elementary Geology), in one vol. 8vo., January 1851.

'On Impressions of Rain-drops in Ancient and Modern Strata,' *Lecture at Royal Institution*, 1851.

'On Fossil Rain-marks of the Recent Triassic and Carboniferous Periods,' *Geological Quarterly Journal,* 1851.

'On Blackheath Pebble-bed, and on Certain Phenomena in the Geology of the Neighbourhood of London,' *Royal Institution's Proceedings,* 1851.

'ELEMENTS OF GEOLOGY,' 4th edition (or Manual), in one vol. 8vo., January 1852.

'On Tertiary Strata of Belgium and French Flanders,' *Geological Journal,* 1852.

'PRINCIPLES OF GEOLOGY,' 9th edition, in one vol. 8vo., June 1853.

'On Remains of Dendrerpeton and Land Shells in Nova Scotia, by Sir C. Lyell and J. W. Dawson, with Notes by Wyman,' *Quarterly Geological Journal,* 1853.

'On Geology of Madeira,' *Quarterly Geological Journal,* 1853.

'ELEMENTS OF GEOLOGY,' one vol. 8vo., February 1855.

'On Erratic Blocks West of Massachusetts,' *Royal Institution Lecture,* 1855.

'On Successive Changes in Temple of Serapis,' *Royal Institution Lecture,* 1856.

'On Stony Lava on Steep Slopes on Etna,' *Royal Society's Proceedings,* 1858.

'On Consolidation of Lava and on Volcanos,' *Royal Institution Lecture,* 1859.

'ANTIQUITY OF MAN,' in one vol. 8vo., 1st edition, February 1863.

  ,,   ,,   ,,   2nd   ,,   April 1863.

  ,,   ,,   ,,   3rd   ,,   November 1863.

'ELEMENTS OF GEOLOGY,' 6th edition, in one vol. 8vo., January 1865.

'PRINCIPLES OF GEOLOGY,' 10th edition, in two vols. 8vo., the first in November 1866, the second in 1868.

'STUDENT'S ELEMENTS OF GEOLOGY,' one vol. 12mo., 1871.

'PRINCIPLES OF GEOLOGY,' 11th edition, in two vols. 8vo., January 1872.

'ANTIQUITY OF MAN,' 4th edition, one vol. 8vo., January 1873.

INDEX.

THE END.